Exploring Statistics

with

Key Curriculum Press
Key College Publishing

Writer:	Cindy Clements
Contributors:	Tim Erickson, William Finzer
Project Editor:	Heather Dever
Editorial Assistants:	Aneesa Davenport, Christa Edwards
Reviewers:	Beth Fox-McManus, Jennifer North Morris, Murray Siegel
Accuracy Checker:	Christopher David
Production Director:	McKinley Williams
Production Editor:	Angela Chen
Copyeditor:	Margaret Moore
Production Coordinators:	Charice Silverman, Thomas Brierly
Compositor:	Graphic World
Art Editor:	Jason Luz
Text Designer:	Marilyn Perry
Cover Illustration:	Seventeenth Street Studios
Printer:	Data Reproductions
Executive Editor:	Casey FitzSimons
Publisher:	Steven Rasmussen

®Key Curriculum Press is a registered trademark of Key Curriculum Press. ™Fathom Dynamic Data and the Fathom logo are trademarks of KCP Technologies. All other brand names and product names are trademarks or registered trademarks of their respective holders.

Some of the activities in this book were adapted from the following sources: *Statistics in Action with Fathom* by William Finzer (Key Curriculum Press, Emeryville, CA, 2004), a companion volume to *Statistics in Action: Understanding a World of Data* by Ann E. Watkins, Richard L. Scheaffer, and George W. Cobb (Key Curriculum Press, Emeryville, CA, 2004); *Data in Depth: Exploring Mathematics with Fathom* by Tim Erickson (Key Curriculum Press, Emeryville, CA, 2001); and *Teaching Mathematics with Fathom* (Key Curriculum Press, Emeryville, CA, 2005).

Limited Reproduction Permission

© 2007 Key Curriculum Press. All rights reserved. Key Curriculum Press grants the teacher who purchases *Exploring Statistics with Fathom* the right to reproduce activities and Fathom documents for use with his or her own students. Unauthorized copying of *Exploring Statistics with Fathom* or of the *Exploring Statistics* Fathom documents is a violation of federal law.

***Exploring Statistics with Fathom* CD-ROM**

Key Curriculum Press guarantees that the *Exploring Statistics with Fathom* CD-ROM that accompanies this book is free of defects in materials and workmanship. A defective CD-ROM will be replaced free of charge if returned within 90 days of the purchase date. After 90 days, there is a $10.00 replacement fee.

Key Curriculum Press
1150 65th Street
Emeryville, CA 94608
510-595-7000
editorial@keypress.com
www.keypress.com

10 9 8 7 6 5 4 3 2 1   10 09 08 07 06

ISBN 1-55953-851-1

Acknowledgments

Fathom Dynamic Data™ software has been helping statistics students visualize data and statistical models for nearly ten years. During this time, many authors have written excellent activities using Fathom's unique features. However, no activity book has been specifically devoted to covering the range of introductory statistics using Fathom. Key Curriculum Press thanks the authors of these activities, and acknowledges their previous publications, without which this book would not be possible.

This book includes (in revised form) a number of activities that originally appeared in *Statistics in Action with Fathom* by William Finzer (Key Curriculum Press, Emeryville, CA, 2004), a companion volume to *Statistics in Action: Understanding a World of Data* by Ann E. Watkins, Richard L. Scheaffer, and George W. Cobb (Key Curriculum Press, Emeryville, CA, 2004). The current volume expands those activities into more complete lessons in statistics, and includes other activities: some revised from *Data in Depth: Exploring Mathematics with Fathom* by Tim Erickson (Key Curriculum Press, Emeryville, CA, 2001) and *Teaching Mathematics with Fathom* (Key Curriculum Press, Emeryville, CA, 2005), as well as original activities written for this book by Cindy Clements.

Key also thanks Ann E. Watkins, Richard L. Scheaffer, and George W. Cobb for their gracious permission to use their ideas in these Fathom activities.

Contents

CD-ROM Contents	vi
Using These Activities	vii
Chart of Activities	xii

1: Distributions of Data

Variation in Measurement—Tennis Balls	3
Variation in Estimates—Meters vs. Feet	7
Exploring Graphs—CensusAtSchool	11
Summary Measures—CensusAtSchool	17
Box Plots and Transformations—CensusAtSchool	25
Deviation from the Mean—Hand Spans	33
Exploring the Normal Distribution	37

2: Relationships Between Data

Least Squares Lines—Pinching Pages	45
Least-Squares Lines and Correlation—Was Leonardo Correct?	51
Exploring Correlation	55
Creating Error	61
Nonlinear Transformations—Equilateral Triangles	69
Modeling Data—Copper Flippers	75

3: Collecting Data

Importing Data—U.S. Census	83
Sampling Bias—Time in the Hospital	89
Choosing a Representative Sample—Random Rectangles	95
Randomization in Experiment Design	99
Designing an Experiment, Part 1—Bears in Space	105
Designing an Experiment, Part 2—Block Those Bears	113
Comparing Experiment Designs—Sit or Stand	117

4: Sampling Distributions

Introduction to Sampling Distributions—Random Rectangles	127
Sampling Distributions of the Sample Mean—Pocket Pennies	135
Sampling Distributions of the Sample Proportion—Seat Belts	141
Sampling Distributions of the Sample Sum and Difference—Dice Rolls	147
Using Sampling Distributions—By Chance or By Design?	153

5: Probability

Constructing a Probability Model—Spinning Pennies	161
The Law of Large Numbers	165
Addition Rule—Scottish Children	173
Establishing Independence with Data	179
Exploring Sampling—With or Without Replacement	185
Binomial Distributions—Can People Identify the Tap Water?	193
Geometric Distributions—Waiting-Time Problems	199
Geometric Distributions: Expected Value	205
Transformations and Simulation Explorations	209

6: Inference for Means and Proportions

Reasonably Likely Values and Confidence Intervals	217
Confidence Interval for a Proportion—Capture Rate	225
Significance Test for a Mean—Body Temperature	231
Significance Test for a Mean: Types of Errors	239
Confidence Interval for a Mean: σ Unknown—SAT Scores	245
Confidence Interval for a Mean: Not Normal—Brain Weights	251
Significance Test for a Proportion—Spinning and Flipping Pennies	259
Inference Between Two Proportions—Plain and Peanut M&M's	267
Inference for the Difference of Two Means: Unpaired—Orbital Express	275
Inference for the Difference of Two Means: Paired—Hand Spans	283

7: Chi-Square Tests

Measuring Fairness—Constructivist Dice	291
Chi-Square Goodness-of-Fit Test—Plain M&M's	297
Chi-Square Test of Independence	305

8: Inference for Regression

Sampling Distribution of the Slope—How Fast Do Kids Grow?	315
Variation in the Slope	321
Inference for a Slope—How Tall Are You Kneeling?	327

CD-ROM Contents

Activity Documents

The Fathom documents that support the activities are included in this folder. For instance, you'll find the documents for the activity Variation in Measurement—Tennis Balls in the folder **Activity Documents | 1 Distributions of Data.** Activity documents can include simulation setups, sample data, or presentation documents. You'll find details on how to use the documents in the Activity Notes. The documents are locked as read-only on the CD to prevent students from saving changes in the original, provided files. These locks will be preserved when you copy the documents to a Macintosh computer. However, if you are using a Windows computer or a server, the locks will not be preserved. You may want to lock the documents on the computer yourself, or put them in a read-only space on the server.

Activity PDFs

This folder contains PDF files of all the activities, which you can use to print an activity. (To read PDF files, use Adobe® Reader®, a free download from www.adobe.com.)

Chart of Activities.pdf

This PDF document contains the Chart of Activities (see pages xii–xv).

Textbook Correlations

The activities in this book have been correlated with several statistics texts, as well as with the AP Course Topic Outline. Check this folder to determine whether a correlation for your text is included.

Using These Activities

Using the activities in this book, students in any introductory statistics course (general high school, Advanced Placement, or college) can directly collect, organize, and interpret numerical data without the endless hours of calculation required when using pencil and paper. Students understand and retain statistics concepts through active engagement with visual models.

These activities allow students to

- drag data to see the effect one value has on the whole data set
- build simulations and watch sampling happen
- run experiments hundreds of times in an instant

Through these and similar experiences, students encounter the excitement of statistics with unprecedented clarity and concreteness.

Your students will benefit most from using Fathom when they study topics involving simulation. These activities show you how to use Fathom to design simulations, summarize data using graphs and tables, and answer the important "What if?" questions.

When to Use an Activity

These activities are designed to be independent, so that you only need to use the ones that fit your particular course of study. You don't have to present them in sequence, even though we've organized them into units. You should use them in whatever order best matches your own curriculum and statistics text.

You don't need to find extra time to fit these activities into your course of study because dynamic exploration actually reduces the amount of time that students need to master a particular topic.

The Chart of Activities on pages xii–xv gives a brief description of each activity, including a correlation to the AP Course Topic Outline, the time required, and the objective. Refer to this chart to identify appropriate activities quickly.

The CD-ROM in the back of the book contains correlations of these activities with a number of popular statistics texts. You can use the correlations to pick activities appropriate for a particular lesson in your text.

Each activity includes easy-to-follow, step-by-step procedures. As students progress through the steps, they will be asked to answer questions, enter data, make graphs, summarize concepts, and make inferences. Students can record information on their own paper, or write their answers in the Fathom document and print out the document to hand in or submit the document electronically.

Many activities are supplemented with Explore More sections, and with additional Extensions in the Activity Notes. These sections are optional, and let students further explore or generalize concepts developed in the main activity. Often the Explore More questions ask students to analyze real-world situations, if that wasn't done in the activity. You can also use these questions as class presentations or to wrap up the activity. These questions are challenging, and some require a moderate level of proficiency with Fathom. Procedures for many Explore More questions are provided in the Activity Notes.

Components of an Activity

- **Student Worksheets:** These blackline masters, suitable for duplicating and passing out to students, contain steps for students to follow and questions for them to answer. Duplicate them either by copying the printed sheet in this book or by printing the appropriate pages from the PDF file included on the CD-ROM. Occasionally, additional worksheets are included. You can pass these out to students with the activity or use them as transparency masters.

- **Activity Documents:** All Fathom documents needed to complete the activities are included on the CD-ROM in their unit folder. These documents may contain data or simulation setups. Additionally, sample data is often provided for student-collected data. Some activities also include presentation documents with simulations or graphs already made. The Settings section in the Activity Notes (see Activity Settings below) lists the documents available for an activity.

- **Activity Notes:** Each activity includes notes describing the activity, providing suggestions for using it, and giving detailed answers to the questions on the student worksheets. These notes specify the statistical objectives of the activity, the expected time required, any materials needed, statistical prerequisites and skills covered, a correlation to the AP Course Topic Outline, and the classroom settings in which the activity is useful. The Procedure section gives information on how to implement the activity in different ways. Finally, Discussion Questions and Extensions are provided as follow-up for most activities.

Activity Settings

The activities in this book are flexible enough that they can be used in a variety of instructional settings. The Activity Notes give details on which settings are appropriate for an activity.

- **Paired/Individual Activity:** Your entire class works in pairs or individually at a number of computers, possibly collecting data first. This setting requires either a computer lab or a classroom with many computers available, and gives students the most direct experience with the statistics they are exploring. It's best to group two, or at most three, students at each computer to promote interaction, discussion, and cooperative problem-solving. (Some students may be "keyboard hogs." Halfway through the activity you should direct students to switch roles so that a different student handles the keyboard and mouse.)

  To use an activity in this way, make copies of the student activity sheets and pass them out. It can be useful periodically to call the whole class together to discuss what they've done so far and to deal with any questions or difficulties they've encountered.

If you have a projection system available, you may want to demonstrate briefly certain techniques or steps, or to answer common student questions. Avoid the temptation of having students imitate your work on their computers, and don't spend too much time showing them what to do. Students benefit most from what they observe or figure out for themselves. When you demonstrate using a presentation system, consider having a student actually operate the computer.

Reserve time at the end of the activity for students to discuss and summarize as a whole class what they have done and observed. See the Discussion Questions in the Activity Notes for some ideas.

- **Small-Group Activity:** In this setting, three to six students work together to collect data. They can then analyze it together at a single computer or in pairs as described above.

- **Whole-Class Presentation:** In this setting, you use the Fathom-based activity and a single computer with a projection device to present a lesson and elicit student discussion. For the most part, you can simply work through the activity as a class, though sometimes you may wish to have students collect data themselves before the presentation. Some activities have a separate document for presentations. Usually in these documents everything is made for you and you can go through the activity questions, omitting the steps. The Procedure section of the Activity Notes gives details. In this setting, you may want to have Fathom display text in larger fonts, to help students see what is going on. Choose **Preferences** from the **Edit** menu (Win) **Fathom** menu (Mac), and then choose **Larger** or **Largest** from the **Font Size** menu.

  A whole-class presentation is not simply a Fathom-based demonstration that students watch. Student questions and discussion form an important part of the presentation, and you should look for ways to encourage participation. Consider having a student operate the computer during the presentation, allowing you to interact more effectively with the class and to encourage student questions and discussion. A rhythm and flow between class discussion and computer-based manipulation of an statistical model can be wonderfully effective for promoting student involvement and understanding.

Even if you have a presentation system in your classroom and consistent access to a computer lab, we suggest you vary the ways you use the activities. Students have different learning styles and will benefit from using the materials in a variety of ways.

No matter how you use a particular activity, involve students as actively as you can in the statistics. Ask them to describe what they've done and what they've observed, and ask "what-if?" questions to encourage them to speculate and make conjectures.

Fathom makes it very easy to generate graphs. It's important, though, to ask students to predict the graph before making it. At the same time, you should encourage your students to study the graphs. Fathom dynamically links the data in the case tables with the graph. Therefore, students should move points on the graph or change values in the case table and explore what happens when the values are changed. Finally, students can quickly change their graph type and compare the advantages of a box plot to a histogram, or a dot plot to an ntigram.

It helps to establish an unobtrusive yet obvious mechanism for students to indicate that they have questions. In workshops presented by Key Curriculum Press, we often give participants red plastic cups that they can display prominently when they have a question. Then you can see at a glance where to direct your attention next, without the disruption of having students call out.

Collecting Data

Many activities have students collect data. Any materials needed are listed on the student worksheet and in the Activity Notes. Additionally, the Procedure section of the Activity Notes gives suggestions on the most efficient way to collect the data. Sample data are usually provided, as well.

Some activities ask students to gather data as a group and then combine their data with the entire class. Some of these activities provide a template for entering data, so students will have the same attribute names. If the data sets are small and you have access to a projector, have students come up to a central computer and enter their data for all to see. Otherwise, you can have a student combine the data and distribute the Fathom document during the next class session. Also, if you have the Fathom Surveys extension (available online Spring 2006), you can upload a survey with the attributes and have each student fill in their data on their computer. They can then download the complete data set by dragging the URL into Fathom.

Some student-collected data, such as hand and eye dominance, are used in several activities, as noted in the Chart of Activities. If you plan to do more than one of these activities, you'll probably want students to save their files, to avoid having to collect the data twice.

Splitting the Class Randomly

There will be several occasions when you must split your class randomly. You can do this in low-tech ways or use Fathom. Some low-tech ways to split the class:

- Have a slip of paper for each student in the class marked with the different options. Mix them up and have each student draw a slip.

- Have students count off, and then randomly assign the evens to one group and the odds to another (or use the multiples of however many groups you need). If groups are to do different treatments, assign the treatments by flipping a coin or rolling a die. While this method is quick, it does not ensure a random division, as there may be some pattern in how students are seated.

- If only two groups are needed (or a multiple of two), have each student flip a coin and those with heads do one treatment, while those with tails do the other. This is random, but you may not end up with groups approximately equal in size.

You can also use Fathom to split the class.

1. Open the document **RandomlySelect.ftm** in the folder **1 Distributions of Data.**

2. Enter student names in the case table.

3. On the **Sample** panel of the inspector, change the number of cases to the group size you want. Click **Sample More Cases** to choose a group. Click again to choose another group.

4. If you want to change the groups, check Replace existing cases and click **Sample More Cases.**

5. Save this Fathom document to help you randomly divide the class in the future.

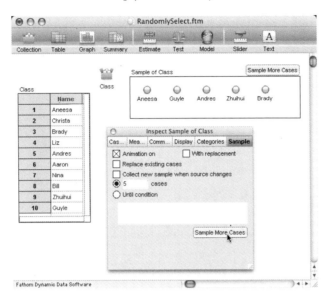

Using Fathom

Although you should be familiar with Fathom, you do not need to be an expert user. If you are unfamiliar with Fathom and would like to evaluate it, you can download an Instructor's Evaluation Edition at the Fathom Resource Center web site: www.keypress.com/fathom/.

For more information, consult Fathom Help and the *Fathom Learning Guide.*

All of the activities will be easier for students if they cooperate, so encourage students to help each other as they analyze and explore. In addition, they can often answer their own technical questions by using Fathom Help.

For additional ways to use Fathom in your class, see *Data Are Everywhere: Project Ideas for Fathom.* This book, which comes with Fathom, contains many projects that students can complete on their own. Additionally, Fathom comes with over 300 sample documents. Some of these, located in the **Statistics** and **Mathematics** folders, are simulations or demonstrations that can aid students' understanding of abstract concepts. Most of the rest of the documents contain data taken from all areas of study, including science, social science, language and the arts, and sports. To open a sample document, choose **File | Open Sample Document.** To browse descriptions of these documents, choose **Help | Sample Documents.**

Chart of Activities

Use this chart to quickly find an activity matching your needs. *Note:* Some student-collected data is used in more than one activity. To save time, you may wish to have students save their data between activities. See the footnotes for details.

Activity	AP Course Topic Outline				Activity Time	Description
	I	II	III	IV		
1: Distributions of Data						
Variation in Measurement—Tennis Balls	A, C				40–50	Measure tennis balls and learn about measurement error.
Variation in Estimates—Meters vs. Feet	A, C				30–40	Compare estimates in meters and feet and learn about bias in estimates.
Exploring Graphs—CensusAtSchool	A, C, E				40–50	Compare distributions using different types of graphs.
Summary Measures—CensusAtSchool	A, B (1–2), C, E				40–50	See how summary measures vary dynamically as data changes.
Box Plots and Transformations—CensusAtSchool	A, B (2–5), C, E				40–50	Compare distributions using box plots and see how summary measures change with transformations.
Deviation from the Mean—Hand Spans[1]	B (1, 2, 5)				40–50	Create a statistic to measure deviation from the mean.
Exploring the Normal Distribution	B (3, 5)		C		40–50	Use a dynamic model to explore features of the normal distribution.
2: Relationships Between Data						
Least-Squares Lines—Pinching Pages	D (1–4)				40–50	Manually fit a line to measurement data, and look at residual squares.
Least-Squares Lines and Correlation—Was Leonardo Correct?[2]	D (1–4)				40–50 or 80	Determine the strength of relationships in classic ratios involving body proportion.
Exploring Correlation	D				20–30	Manipulate data to explore features of correlation.
Creating Error	D				60	See how data consists of a signal plus error, and how the error can obscure the shape of relationships.
Nonlinear Transformations—Equilateral Triangles	D				40–50	Transform data and fit a model to determine a formula for the area of an equilateral triangle.
Modeling Data—Copper Flippers	D				40–50	Work with exponential models and logarithmic transformations.

[1] Hand spans (both left and right) are also used in Inference for the Difference of Two Means: Paired—Hand Spans (unit 6).
[2] Height and kneeling height are also used in Inference for a Slope: How Tall Are You Kneeling? (unit 8).

Activity	AP Course Topic Outline				Activity Time	Description
	I	II	III	IV		
3: Collecting Data						
Importing Data—U.S. Census	all	A, D			30–50	Work with distributions using data about individual U. S. citizens.
Sampling Bias—Time in the Hospital	B (1, 2), C	B			40–50	Explore sampling bias using a pre-made simulation.
Choosing a Representative Sample—Random Rectangles	A–C	B–C			40–50	Look at different types of sampling and their distributions.
Randomization in Experiment Design	E	A–C			50–60	Compare three ways to assign treatments to subjects.
Designing an Experiment, Part 1—Bears in Space	C	A, C			100–120	Compare within-treatment variability to between-treatment variability.
Designing an Experiment, Part 2—Block Those Bears	C	C			50	Use blocking to control variability.
Comparing Experiment Designs—Sit or Stand	all	C, D			50–70, or 30–40	Compare the merits of different experiment designs.
4: Sampling Distributions						
Introduction to Sampling Distributions—Random Rectangles	all	B (4)	D (1, 2, 6)		40–50	Explore features of the sampling distributions of the sample mean, median, and maximum area of rectangles.
Sampling Distributions of the Sample Mean—Pocket Pennies		B (4)	C, D (2, 3, 6)		30–50	Look at how the sampling distribution changes with sample size.
Sampling Distributions of the Sample Proportion—Seat Belts			C, D (1, 3, 6)		20–40	Build a simulation to look at the sampling distribution.
Sampling Distributions of the Sample Sum and Difference—Dice Rolls			B, C, D (4, 5, 6)		20–40	Compare these sampling distributions.
Using Sampling Distributions—By Chance or By Design?			A, D		40–50	Use sampling distributions to decide whether actual events can be reasonably attributed to chance.
5: Probability						
Constructing a Probability Model—Spinning Pennies[3]			A, C, D		40–50	Make a probability model and compare it to collected data.
The Law of Large Numbers			A (1, 2)		40–80 or 30–40	Look at variation in sampling in small and large samples.
Addition Rule—Scottish Children			A		30–50	Work with the addition rule for disjoint and non-disjoint categories.
Establishing Independence with Data[4]			A (1, 3, 5), B (1)		30–50 or 25	Work with the definition of independence and real-world data.
Exploring Sampling—With or Without Replacement			A (1–3, 5), B (1), D (6)		30–45	Compare probabilities when sampling with and without replacement.

[3] Penny spins are also used in Significance Test for a Proportion—Spinning and Flipping Pennies (unit 6).
[4] Hand and eye dominances are also used in Chi-Square Test of Independence (unit 7).

Activity	AP Course Topic Outline				Activity Time	Description
	I	II	III	IV		
5: Probability continued						
Binomial Distributions—Can People Identify the Tap Water?		A (3), C	A (1, 4, 5), C (3)		35–50	Design a study with a binomial sampling distribution.
Geometric Distributions—Waiting-Time Problems		A (4), B	A, B (1), D (6)		35–50	Simulate waiting-time problems and explore their distributions.
Geometric Distributions: Expected Value			A (4, 6)		20–35	Discover the formula for the expected value of a geometric distribution.
Transformations and Simulation Explorations			A, B, D (5, 6)		30–50	Build probability simulations to solve problems with real-world data.
6: Inference for Means and Proportions						
Reasonably Likely Values and Confidence Intervals			D (1, 6)	A (1–4)	40–90	Construct a chart of reasonably likely outcomes by simulation and by theoretical values.
Confidence Interval for a Proportion—Capture Rate			A (5), D (1)	A (1–5)	30	Test whether 95% of 95% confidence intervals capture the population proportion.
Significance Test for a Mean—Body Temperature			D (2, 6, 7)	A (1–3, 6), B (1, 4)	40–50	Explore hypothesis tests and confidence intervals, and the relationship between them.
Significance Test for a Mean: Types of Errors			D (2, 6, 7)	B (1, 4)	30–40	Deepen understanding of Type I and Type II errors and how to reduce their probability.
Confidence Interval for a Mean: σ Unknown—SAT Scores			C, D (2, 6, 7)	A (1–3, 6)	30	See how estimating the standard deviation affects the confidence interval.
Confidence Interval for a Mean: Not Normal—Brain Weights			C, D (2, 6, 7)	A (1–3, 6)	40–50	Explore the effect of skewness on confidence intervals.
Significance Test for a Proportion—Spinning and Flipping Pennies[5]			A (1, 4, 5), C, D (1, 6)	A (1–4), B (1, 2)	40–50	Perform informal and formal tests of significance.
Inference Between Two Proportions—Plain and Peanut M&M's			A (1, 4–6), B, C, D (4, 6)	A (1–3, 5), B (1, 3)	40–50	Build a simulation to explore the difference of two proportions.
Inference for the Difference of Two Means: Unpaired—Orbital Express	A–C	A (3), C	A (5, 6), B, C, D (5, 6)	Optional: A (1, 6, 7), B (1, 4 or 5)	60–120	Develop a measure to test the difference between two treatments.
Inference for the Difference of Two Means: Paired—Hand Spans[6]	A–C, D (1–3)		B, C, D (5, 6)	A (1, 6, 7), B (1, 4, 5)	50–90	See how pairing changes the standard errors between means when using paired differences, unpaired differences, and independent samples.

[5]Penny spins are also used in Constructing a Probability Model—Spinning Pennies (unit 5).
[6]Hand spans (either hand) are also used in Deviation from the Mean—Hand Spans (unit 1).

Activity	AP Course Topic Outline				Activity Time	Description
	I	II	III	IV		
7: Chi-Square Tests						
Measuring Fairness—Constructivist Dice	E (1, 4)		A (1, 4, 5), D (6, 8)	A (1, 2), B (1, 6)	40–100	Make and roll dice and create a measure to assess fairness.
Chi-Square Goodness-of-Fit Test—Plain M&M's	E (1, 4)		A (1, 5), D (8)	B (1, 6)	40–50	Explore essential concepts in using a goodness-of-fit test with non-equally likely categories.
Chi-Square Test of Independence[7]	E		A (1, 5), D (8)	B (1, 6)	30–50	Determine when data is probably independent even if it fails the definition of independence.
8: Inference for Regression						
Sampling Distribution of the Slope—How Fast Do Kids Grow?	D (1–3)		C	A (1)	30–45	Look at the assumptions of the linear regression model and conditional distributions.
Variation in the Slope	D (1–3)		C	A (1)	30–45	Discover what causes variation in the slope of the least-squares regression line.
Inference for a Slope—How Tall Are You Kneeling?[8]	D		D (7)	A (1, 3, 8), B (1, 7)	40–50	Check conditions, perform a test of significance, and construct a confidence interval for a slope.

[7] Hand and eye dominances are also used in Establishing Independence with Data (unit 5).
[8] Height and kneeling height are also used in Least-Squares Lines and Correlation—Was Leonardo Correct? (unit 2).

Distributions of Data

Variation in Measurement—Tennis Balls

You will need
- tennis ball
- centimeter ruler
- **TennisBalls.ftm**

In this activity you'll see how accurately you and your classmates can measure the diameter of a tennis ball.

COLLECT DATA

1. With your partner, plan a method for measuring the diameter of the tennis ball with the centimeter ruler.

2. Using your method, make two measurements of the diameter of your tennis ball to the nearest millimeter.

3. Combine your data with that of the rest of the class.

Q1 You're going to make a dot plot of these measurements. Speculate first, though, about the shape you expect for the distribution. Describe or sketch the shape you expect.

Now you'll use Fathom to enter, plot, and analyze the data.

4. Open the Fathom document **TennisBalls.ftm.** In this document you will find a collection of randomly generated measurements that you will use for comparison. But first, you'll want to enter your class's measurements into a new collection.

Make sure the Random Tennis Balls collection is not selected when you drag the new case table.

5. Drag a new case table from the shelf and drop it in the document.

6. Click on the column label <new> and type Diameter.

In Fathom, data are stored in a collection. The variables for each case are called attributes. A case table allows you to see the attributes for all cases in the collection.

7. Starting in the first empty cell under *Diameter*, enter your class's data. To enter the units, choose **Show Units** from the **Table** menu. Type millimeters in the units row of the *Diameter* column. The units will be assigned to each value you enter.

8. Double-click the collection name Collection 1 and rename the collection Tennis Balls.

Variation in Measurement—Tennis Balls
continued

INVESTIGATE

Shape, Center, and Spread

9. Drag a new graph from the shelf.

10. In the case table, click and hold the attribute name *Diameter*. The cursor icon becomes a hand. Drag *Diameter* to the horizontal axis of the graph and release.

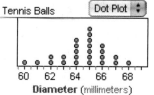

Q2 What is the approximate shape of the plot? Are there clusters and gaps or unusual data values (outliers) in the data?

Q3 Choose two numbers that seem reasonable for completing the sentence "Our typical diameter measurement is about _____, give or take about _____." (There is more than one reasonable set of choices.)

Q4 Discuss some possible reasons for the variability in the measurements. How could the variability be reduced? Can the variability be eliminated entirely?

A Randomly Generated Distribution

Now you will compare your class's collection to a randomly generated collection.

11. Drag a new graph from the shelf.

12. Double-click the Random Tennis Balls collection to show the collection's inspector. Drag the attribute name *Diameter* to the horizontal axis of the graph.

Q5 What is the approximate shape of the Random Tennis Balls plot? Are there clusters and gaps or unusual data values (outliers) in the data? How does the shape of this randomly generated distribution compare to the shape of your class's distribution?

Q6 Choose two numbers that seem reasonable for completing the sentence "The typical diameter measurement is about _____, give or take about _____." (There is more than one reasonable set of choices.) How do the center and spread of this randomly generated distribution compare to the center and spread of your class's distribution?

Q7 Give some reasons for the differences between your class's distribution and the randomly generated distribution.

Q8 Click the **Rerandomize** button on the Random Tennis Balls collection several times (slowly). How does the shape change for the different distributions? Is the shape always mound-shaped? How does the center change? Is the center always in the same location? How does the spread change? What is the smallest diameter you get if you rerandomize ten times? The largest?

Variation in Measurement—Tennis Balls

Activity Notes

Objectives

- Understanding the concept of measurement error, including equating the term *error* with "variation" rather than "mistake"
- Understanding that variation always occurs in the process of gathering data and may not represent an error on the part of anyone associated with the process
- Recognizing that the distribution that results from variation is typically mound-shaped

Activity Time: 40–50 minutes

Setting: Paired Activity (use **TennisBalls.ftm** and collect data) or Whole-Class Presentation (use **TennisBallsPresent.ftm**)

Materials

- One tennis ball (a dead one is fine) for each pair of students
- One centimeter ruler for each pair of students

Statistics Prerequisites

- Dot plots

Statistics Skills

- Center and spread
- Shape, outliers, clusters, and gaps
- Variation and error
- Comparing graphs in terms of shape, center, and spread

AP Course Topic Outline: Part I A, C

Fathom Prerequisites: none

Fathom Skills: Students make graphs and case tables, enter data, and assign units to attributes.

General Notes: This activity is a great way to introduce students to Fathom. No prior experience with Fathom is required: The data are gathered and entered by hand, and the student worksheet leads students through the process step by step. The main uses of Fathom in this activity are to aid comparison of students' distributions with many randomly generated distributions and to simplify graph making. Students should work in pairs for this activity. In general, measurement distributions should be somewhat symmetrical and mound-shaped. Although tempting, it is premature to show students how to calculate specific statistics, such as the mean or standard deviation. Introducing these concepts at this point would defeat the open-ended nature of this activity.

Procedure: Each pair of students collects data, then the class combines data and analyzes the shape, center, and spread of the distribution. One way you can combine data is to have Fathom running on a computer connected to a projector. As each pair finishes, they can come up and enter their names and measurements. Each pair of students also should enter all the measurements into their own computer so that they can explore the data themselves.

Tennis Balls

	Pair	Diameter
units		millimeters
1	Glen/Vin	65 mm
2	Glen/Vin	66 mm
3	Jill/Buzz	67 mm
4	Jill/Buzz	68 mm
5	Matteo/Liz	64 mm
6	Matteo/Liz	64 mm
7	Maggie/Joan	65 mm

If you don't wish to have students collect data, you can have a whole-class discussion with the document **TennisBallsPresent.ftm**, which has sample data and all the graphs made in the activity.

COLLECT DATA

Q1 Answers will vary.

INVESTIGATE

Q2 Graphs may peak at a central value.

Q3 Students will probably choose a central value as the "typical diameter measurement" and half of the range as a measure of spread.

Q4 Reasons for variability include differences between tennis balls, differences in method of measurement, and error in reading the measurement. Standardizing the measuring process or measuring only one brand

Variation in Measurement—Tennis Balls
continued

Activity Notes

of tennis balls could reduce the variability. However, you cannot eliminate variability entirely.

Q5–Q7 The distributions should be somewhat symmetrical and mound-shaped; the central value should be close to 65.1 mm and half of the range as a measure of spread (around 1.6 mm). Students should recognize that sources of variability and differences between the class's measurements and the randomly generated ones include differences between tennis balls, differences in method of measurement, and error in reading the measurement. The random process is actually modeling the variation in the manufacturing process of the tennis ball, while the class's distribution models that process as well as the error in measuring process. So the class's distribution should have a little wider range and spread.

Q8 The shape is not always mound-shaped. The central value (around 65) varies slightly. The spread of the values does change. Students should recognize this as variation in the manufacturing process.

DISCUSSION QUESTIONS

You should project either class data or **TennisBalls Present.ftm** while you discuss these questions.

- Change the graph of the class's distribution from a dot plot to a histogram. Do the shape, center, and spread of the histogram look to be the same as what you got in Q1 and Q2 with your dot plot? What information does the dot plot show that the histogram does not? What information does the histogram show that the dot plot does not? [The dot plot shows the individual measurement values for the whole class, whereas the histogram shows only the range of values collected and how many values were collected in certain intervals. Some students may notice that if they make the bin width of the histogram one, they do not lose any information. The histogram can show the general shape, center, and spread more clearly than a dot plot. Patterns are sometimes more easily seen.]

- The United States Tennis Association (USTA) uses only tennis balls with diameters measuring 63.5 mm to 66.7 mm. [Source: www.tennislovers.com] What proportion of the class's measurements fall within this range? [One way to compare the class measurements with the United States Tennis Association (USTA) standards is to plot the USTA diameter requirements, then calculate the proportion by hand. To plot a value, select the graph and choose **Graph | Plot Value**. In the formula editor, type the value and click **OK**. Now count the dots between the vertical lines that represent the range of USTA values and divide by the total number of cases.]

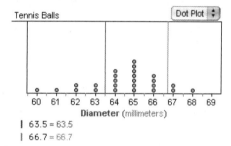

EXTENSION

Have students repeat the activity but measure the length (in centimeters) of their statistics text. Compare the distribution of tennis ball diameters with the distribution of book lengths. Why might the spread of tennis ball diameters be larger than that of book lengths? [The tennis ball diameters will likely have more spread than the book lengths because measuring a curved surface is more susceptible to error.]

Variation in Estimates—Meters vs. Feet

You will need
- Estimates.ftm

In this activity you will see if you and your class estimate lengths more accurately in feet or in meters.

COLLECT DATA

1. Your instructor will randomly split the class into two groups.

2. If you are in the first group (G1), estimate the length of your classroom in feet. If you are in the second group (G2), estimate the length of the room in meters. Do this by looking at the length of the room; no pacing the length of the room is allowed.

3. Open the Fathom document **Estimates.ftm.** In this document you will find a collection of estimates from an Australian class that you will use for comparison. But first, you'll want to enter your class's estimates into a new collection.

Make sure the Australia collection is *not* selected when you drag the new case table.

4. Make a new, empty case table and define two new attributes: *Group* and *Estimate*. For each person in your class, enter their group (G1 or G2) and their estimate of the length of the room. Each case (person) represents a single estimate, identified by the person's group.

Estimates

	Group	Estimate
10	G1	17.5
11	G2	4.4
12	G2	5.8
13	G2	6.4
14	G1	19.6

INVESTIGATE

Converting the estimates to a common unit is necessary to make an appropriate and meaningful comparison of the two distributions. Fathom can be used to do the unit conversions.

5. Double-click the collection to show its inspector. On the **Cases** panel, click <*new*> and create an attribute called *Estimate_in_meters*.

The "if" statement checks whether the case is in G1. If so, it converts the measurement in feet to a measurement in meters. If the case is not in G1, it keeps the original measurement.

6. Double-click in the formula cell to show the formula editor. Enter the formula with an "if" statement that will convert the estimates only from the group using feet. (See the example here.) Click **OK**.

Estimate_in_meters	5.45592	$\text{if}(Group = \text{"G1"})\begin{cases} \dfrac{(Estimate)(12)(2.54)}{100} \\ Estimate \end{cases}$

Variation in Estimates—Meters vs. Feet
continued

7. Make a new graph and drag *Estimate_in_meters* to the horizontal axis. Then drag *Group* to the vertical axis to split the graph into two graphs with the same horizontal axis. Choose an appropriate and meaningful plot to display the data by using the pop-up menu in the top right of your graph.

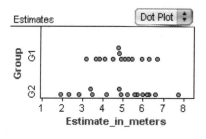
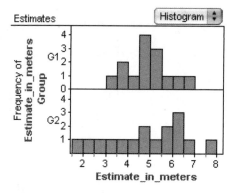

Q1 Describe the two distributions from your class in terms of shape, center, and spread. How are the two distributions similar and how are they different?

Q2 Are there clusters and gaps or unusual data values (outliers) in either of the distributions? Does one group's distribution have more clusters or gaps than the other group's distribution?

Q3 Do you think the students in your class tend to estimate more accurately in feet or in meters? What is the basis for your decision?

Q4 Why should you split the class randomly into two groups instead of simply letting the left half of the room estimate in feet and the right half in meters?

A similar experiment was done by a group of students in Australia. You have their data in the Australia collection.

8. As you did in steps 5–7, find an appropriate and meaningful way to plot the Australian data so that you can compare the estimates in feet and the estimates in meters. This time, however, convert the estimates of group 2 from meters to feet.

Q5 Answer Q1–Q3 for the Australia collection.

Q6 Compare the results from your class's experiment to the results from the Australian class. What similarities are there between the two classes? What differences are there?

Variation in Estimates—Meters vs. Feet   Activity Notes

Objectives
- Choosing an appropriate and meaningful way to display data
- Discussing the importance of random assignment in any experiment
- Introducing the concept of bias in measurement: Some methods of measurement are more likely to result in estimates that are too high, and the measurement error is tied to the units of the measurement system.

Activity Time: 30–40 minutes

Setting: Paired/Individual Activity (use **Estimates.ftm** and collect data) or Whole-Class Presentation (use **EstimatesPresent.ftm**)

Materials
- *Optional:* Yardsticks and metersticks

Statistics Prerequisites
- Making dot plots
- Familiarity with describing shape, center, and spread in graphical displays (not with formulas)

Statistics Skills
- Comparing distributions in terms of shape, center, and spread
- Comparing clusters and gaps, outliers and influential points
- Organizing data
- Making stacked dot plots, histograms
- Introduction to the concept of bias in measurement

AP Course Topic Outline: Part I A, C

Fathom Prerequisites: Students should be able to make graphs and case tables, and enter data.

Fathom Skills: Students organize case tables in different ways, make different types of graphs, including split graphs, and use formulas to do unit conversions.

General Notes: This activity is a great way to introduce students to different ways of organizing data in case tables and graphs within Fathom. The dynamic nature of Fathom helps students see the relationships between the different types of graphs.

Procedure: If you don't wish to have students collect data, you can have a whole-class discussion with the document **EstimatesPresent.ftm**, which has sample data and all the graphs made in the activity.

COLLECT DATA

2. Watch to be sure that students who measure in meters do not estimate in feet and then convert to meters.

4. There are two basic methods for organizing the estimates in Fathom. First, students can define two attributes—*G1* and *G2*, one for each group—and enter the feet and meter estimates in separate columns. But this can be problematic because students will end up making two separate graphs for G1 and G2, each with its own axis. The method given in step 4 of the activity is better because it makes it easy to create a split graph, which makes comparing the data easier. A split graph gives you two graphs with the same horizontal axis.

Note: Units are not defined in Fathom for this activity because the *Estimate* attribute uses both feet and meters. Fathom cannot define two units within one attribute.

INVESTIGATE

Q1–Q4 Be sure to discuss student answers to these questions. It's possible that there is some difference between students on the two sides of the room. For example, students on the right side of the room might have a better view of the length than students on the left side. Or perhaps students on one side of the room have to walk farther from the door to their desks, so they have a better sense of the size of the room. When there is any possibility of bias like this—and there almost always is—it's best to randomize.

Foremost, students should notice that there is more variation in the measurements for the less familiar system (the metric system for American students).

Q5 Students compare their results to those of a similar experiment done in Australia. [Source: Hand et al., *Small Data Sets*, Chapman and Hall, 1994, p. 2] For those students, the meter was the familiar unit.

Variation in Estimates—Meters vs. Feet
continued

Activity Notes

Histograms of the two groups are shown below. The length of their room was 13.1 m, or 43 ft. Both distributions have centers remarkably close to the true value, and both are similarly skewed. Using 1 m = 3.28 ft, the third histogram shows the meter data rescaled to feet. Now you can see that estimating in meters produced much more variation than estimating in feet.

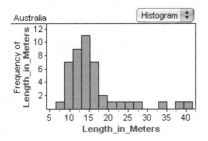

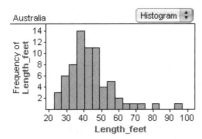

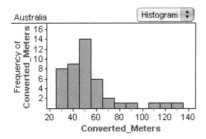

DISCUSSION QUESTIONS

- Measure the length of the room with yardsticks or metersticks. Does knowing the actual length change your answers to Q3 and Q4? [This is a nice way to close the activity. After one or more students measure the room's actual length, show students how to plot this value on the graph: choose **Graph | Plot Value**, type the value in the formula editor, and click **OK**. Discuss whether this true value changes students' opinions about which group estimated more accurately.]

- Try using different types of plots to look at the data in this activity, particularly a box plot. What are the advantages and disadvantages of each type of plot? Which plots give the most information about the shape of the distributions? About the spread? About the center? About gaps and clusters? [Students should find that different plots emphasize different features of the distributions. The box plot emphasizes outliers, clusters, and gaps.]

Exploring Graphs—CensusAtSchool

You will need
- CensusAtSchool.ftm

CensusAtSchool is an international project that collects and disseminates data about students from the United Kingdom, South Africa, Australia, and New Zealand. (For more information, visit its web site at www.censusatschool.ntu.ac.uk.) In this activity you'll use Fathom to learn about a random sample of students from this project.

EXAMINE DATA

1. Open the Fathom document **CensusAtSchool.ftm.** In this document you will find a collection of 500 students from the UK, South Africa, and Queensland, Australia, who were randomly selected from the CensusAtSchool project.

You can resize the column widths by dragging a boundary between two column headings.

2. Double-click the collection to show the inspector. The **Cases** panel shows all of the attributes and the values for one case. Browse the cases by clicking the arrows at the bottom of the inspector.

INVESTIGATE

Plotting Categorical Variables

There are many different relationships to explore. First look at where these students are from.

3. Drag a new graph from the object shelf and drag *Place* to the horizontal axis.

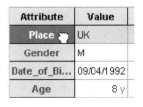 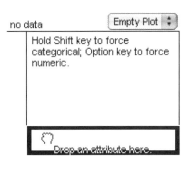 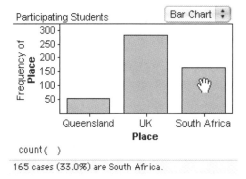

To see exactly how many students are from South Africa, for example, hold the cursor over the bar representing the South African students, and read the status bar in the lower-left corner of the Fathom window.

4. Use the pop-up menu at the top right of the graph to change the graph to a ribbon chart. The data appear as a single big block, representing all of the values, divided into sections; the axis is labeled with percentages.

Exploring Graphs—CensusAtSchool
continued

5. Drag the attribute *Gender* to the middle of the graph to split it. You can use the tick marks on the vertical axis to estimate the proportion of students in South Africa who are female: a bit more than 50%. For a more exact proportion, move the mouse on top of a region in the graph and read the statement that appears in the status bar.

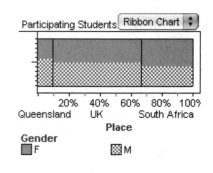

Q1 Using your split ribbon chart, compare the distributions for the different countries with respect to gender. What percentage of students are female? What percentage of South African students are female?

Q2 Drag the attribute *Computer* to the middle of your ribbon chart and compare the distributions for the different countries with respect to the percentage of students who have computers.

Q3 Change the graph back to a bar chart. What does the ribbon chart tell you that the bar chart does not? What does the bar chart tell you that the ribbon chart does not?

Q4 Pick another categorical variable to explore with a ribbon chart and a bar chart. Describe any differences or similarities between the countries that you find for that categorical variable.

Q5 Experiment with Fathom to try to get this plot, which shows stacked bar charts, sorted by computer usage and travel method. How can you make this kind of bar chart? What does it look like as a ribbon chart?

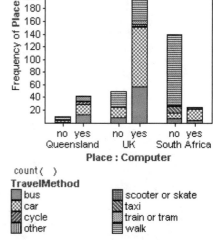

Plotting Quantitative Variables

Next you'll look at noncategorical variables and their plots.

6. Make a new graph and drag *Age* to the horizontal axis.

Notice that when you select a data point, the inspector shows that case. You can read all of that student's values.

Q6 Describe the distribution of *Age* in terms of shape, center, and spread. Are there clusters and gaps or unusual data values (outliers) in the distribution?

Exploring Graphs—CensusAtSchool
continued

Q7 Select the oldest student (outlier) by clicking his or her dot in the dot plot. Where is that student from? Give a few facts about that person.

Is there any relationship between where students are from and their ages? One way to look for relationships is to make use of the fact that selection in one graph is reflected in another.

7. Make a second graph next to your dot plot and drag *Place* to its horizontal axis.

8. Select the bar representing Queensland students by clicking it. The bar turns red. Notice that some of the dots in the dot plot have also turned red and that the students from Queensland are highlighted in the collection.

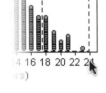

9. Select the students who are 18 or older by drawing a selection rectangle around them.

Q8 The first graph in step 8 showed that the Queensland students are all in the middle; none are the youngest or oldest. What did the graph in step 9 show?

Splitting Graphs

To get a better sense of the range in age of the students, try looking at the three distributions in one graph.

You can drag attribute names from anywhere you see them.

10. Drag *Place* to the vertical axis of the *Age* graph. Putting a categorical attribute on a numeric graph *splits* the graph, allowing you to compare the age distributions of the three groups directly.

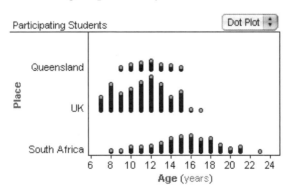

Q9 Compare the three distributions in terms of shape, center, and spread. In what ways are the three distributions similar and in what ways are they different?

Q10 Are there clusters and gaps or unusual data values (outliers) in any of the distributions? Does one group's distribution have more clusters or gaps than the other group's distribution?

Exploring Graphs—CensusAtSchool
continued

Histograms and Bin Width

11. Select the split dot plot, then choose **Graph | Remove Y Attribute: Place.**

12. Change the graph to a histogram. Select one of the histogram bars. Put the cursor over the selected bar to read about it in the status area.

Sometimes the choice of bin width affects the overall look of the data, for example, whether there's one tallest bin or more than one.

> If you drag the left edge of the first bin, you can choose where your first bin begins.

13. In the histogram, drag a bin edge to make the bins wider. Each bin now represents more cases, so some of the bins are too tall for you to see their tops. Drag the 70 on the vertical axis down until you can see the tops of all the bins. Notice how some of the bumpiness in the histogram has been washed out with the wider bins. Selecting a bin now selects a larger interval of cases.

> You can also make multiple graphs to compare the results when you change the bin width. To do this, choose **Object | Duplicate Graph.**

Q11 Drag the edge of a bin until your histogram has a bin width of one. Then change the bin width to two. Describe the shape, center, and spread of *Age* based first on the histogram with a bin width of one, then based on the histogram with a bin width of two. How much difference does the bin width make for this data set?

Q12 Does a histogram with narrow bars give you more or different information than a histogram with wide bars? Explain.

Q13 Make a histogram with a bin width of four. Describe the shape, center, and spread of *Age* based on this histogram. Which histogram gives you the best view of the data? Explain.

Exploring Graphs—CensusAtSchool   Activity Notes

Objectives

- Choosing an appropriate and meaningful way to display data
- Learning how to compare distributions, either with proportions or with shape, center, and spread, based on their plots
- Discovering some of the differences between categorical and quantitative data and how to work with the differences

Activity Time: 40–50 minutes

Setting: Paired/Individual Activity or Whole-Class Presentation (use **CensusAtSchool.ftm** for either)

Statistics Prerequisites

- Familiarity with estimating percentages or proportions
- Some familiarity with one-variable graphs
- Familiarity with shape, center, and spread in graphical displays (not with formulas)

Statistics Skills

- Comparing distributions of categorical variables using proportions
- Comparing and contrasting different types of plots
- Comparing quantitative distributions in terms of shape, center, and spread
- Comparing clusters and gaps, outliers and influential points
- Organizing data
- Making stacked dot plots, histograms
- Seeing the effect of bin width on the shape of the distribution

AP Course Topic Outline: Part I A, C, E

Fathom Prerequisites: none

Fathom Skills: Students inspect a collection; create and split a variety of graphs including bar charts, ribbon charts, stacked bar charts, dot plots, and histograms; use the cursor to find summary statistics or values on plots; use selection (rectangles), duplicate graphs, and dragging to explore relationships or distributions; and rescale the bin width of histograms.

General Notes: This activity uses the CensusAtSchool data to introduce students to Fathom. It is based on Tour 1 in the Fathom *Learning Guide* and requires little to no knowledge of Fathom or statistics. This activity focuses on how to plot various types of data, use the plot to find values, and how to choose an appropriate plot for data. Fathom makes it easy for students to work with a large data set and quickly compare distributions. Additionally, by dynamically dragging bin edges, students get a feel for how changing the bin width affects the look of a histogram.

Procedure: Try working with the document **CensusAtSchool.ftm** before giving it to your students. This file has 500 cases. If it's too slow on your computers, select about half the cases and delete them. To select many cases, click on the first row number in the case table. Then scroll halfway down and Shift-click on another row number. All the cases in between will be selected. Choose **Edit | Delete Cases.**

Moving, Resizing, and Deleting Objects: At some point in the activity, you may need to help students clean up their documents. Here are some tips:

- Now is a good time to tell your students that anytime they make a mistake in Fathom, they can choose **Edit | Undo [action].** They can also redo anything they've undone by choosing **Edit | Redo [action].**
- You can move objects in the Fathom window by dragging the top of its frame.
- To clear space, make the collection (or any object in Fathom) a small icon that won't take up much room by dragging the lower-right corner of the collection up and toward the opposite corner until the object becomes *iconified* (that is, a little box of gold balls).

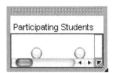 

- To actually delete an object, select the object and choose **Object | Delete [object].** *Note:* Deleting the collection would delete all the data. However, Fathom has unlimited undo, so the students can always get the

Exploring Graphs—CensusAtSchool
continued

Activity Notes

collection back. You can have your students test this by deleting the collection. All the graphs will become empty. To get the collection back, choose **Edit | Undo Delete Collection,** then double-click the collection to get the inspector back.

INVESTIGATE

Q1 The distributions for the different countries with respect to gender are all close to being half female and half male, although Queensland has the most males and South Africa has the most females in the study. Specifically, for South Africa, 40.9% are male and 59.1% are female; for the UK, 48.4% are male and 51.6% are female; and for Queensland, 56.9% are male and 43.1% are female. The total percentage of students who are female is 53.2%.

Q2 The distributions for *Computer* for Queensland and the UK are roughly the same—over 80% of the students have computers. However, in South Africa only 15.2% of the students have computers.

Q3 The bar chart shows, in a more obvious sense, how many more students are in the study from the UK and South Africa and how many of those students have computers, in terms of counts. The proportions and relative proportions are easier to see in the ribbon chart than in the bar chart.

Q5 Beginning with a bar chart of *Place*, drop *Computer* on the plus that appears on the horizontal axis and then drop *TravelMethod* in the middle of the plot.

Q6 The distribution of *Age* is skewed right, with the center at around 12 yr, the first quartile at about 10, and the third quartile at about 15. There is one outlier and no clusters or gaps.

Q7 The outlier is from South Africa. She is a female, 23 years old, who likes English best.

Q8 This graph shows that South Africa has all the oldest students.

Q9–Q10 The split graph shows that the narrowest range of ages is for the Queensland students; the South African students span the greatest range and have all the oldest students; and the youngest students are from the United Kingdom. All three distributions are fairly mound-shaped, with South Africa and the UK having roughly the same spread, about 2 yr (or an *IQR* of about 4). Queensland is centered at about 12 with a spread of 1. The UK is centered at about 11 and South Africa is centered at about 16. The UK has the most students less than 16 yr, and it looks like South Africa may still have an outlier at 23. There do not appear to be any other clusters or gaps in any of the distributions.

Q11–Q13 The histograms with bin widths of one and two both show that the distribution of *Age* is skewed right. The center and spread are a little easier to see with a bin width of two. The outlier is more obvious with a bin width of one. It is hard to tell on the histogram with a bin width of two if there is an outlier or a gap. With a bin width of four, the outlier and gap are lost. You can still see that the distribution is skewed right but the center is harder to identify. It looks like a bin width of two is better than four. Answers could vary as to whether a bin width of one or two is better, depending on what the student wishes to emphasize.

EXTENSIONS

Have students use the techniques they've learned to answer these questions or any questions they may have about these students.

1. Is math more popular with males or females?
2. Which country has the highest proportion of students with Internet access?
3. Which country's students have to travel farthest to get to school?
4. Which country has the highest proportion of students who walk to school? Are these the students who have the greatest travel time?

Summary Measures—CensusAtSchool

You will need
- CensusAt School.ftm

CensusAtSchool is an international project that collects and disseminates data about students from the United Kingdom, South Africa, Australia, and New Zealand. (For more information, visit its web site at www.censusatschool.ntu.ac.uk.) In this activity you'll use Fathom to learn about a random sample of students from this project.

EXAMINE DATA

1. Open the Fathom document **CensusAtSchool.ftm.** In this document you will find a collection of 500 students from the UK, South Africa, and Queensland, Australia, who were randomly selected from the CensusAtSchool project.

2. Double-click the collection to show the inspector. Drag a new graph from the shelf and drag *Age* to the horizontal axis. Change the graph to a histogram using the pop-up menu.

INVESTIGATE

Plotting Values

Suppose you want to compare the mean and median of the students' ages. You can do that right on the graph.

3. With the graph selected, choose **Graph | Plot Value.** Type the expression whose value you want plotted and then click **OK.**

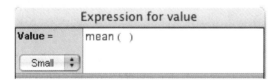
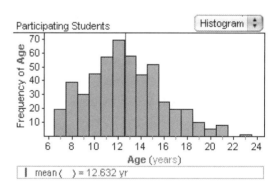

4. Repeat step 3 to plot the median.

Notice that the mouse pointer becomes a hand when you are over a bin (as opposed to a bar boundary), indicating that you can drag the whole bin.

5. Drag one of the bins to the right of the mean in the histogram. As you drag, watch the two plotted values. Drag the bin all the way to the left, then all the way to the right.

Q1 When does the value for the mean change? When does the value for the median change? How can you demonstrate that the median is insensitive to outliers, while the mean is not?

Summary Measures—CensusAtSchool
continued

To restore your data to their original values, choose **File | Revert Collection**.

Q2 Restore your data to their original values. What are the median and mean ages? How are they related in terms of position, and what does that imply about the distribution of *Age*?

6. Split the graph by dragging the attribute *Place* to the vertical axis. Resize your graph so that you can see all three distributions.

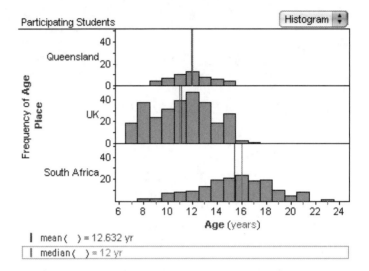

Q3 Compare the three distributions in terms of shape and center based on the mean and the median. Which distribution is the most mound-shaped? Which is the most skewed? How can you tell?

7. Double-click the formula for the median to show the formula editor. Replace median() with mean()−stdDev(). Then plot the value mean()+stdDev(). These values are the mean minus and plus one sample standard deviation.

Q4 Compare the three distributions in terms of shape and spread based on your new plotted values. Which distribution is the most mound-shaped? Which is the most skewed? How can you tell?

You know that about 68% of the data lies within your two new plotted values, or within one standard deviation of the mean.

Q5 For which distributions would you be comfortable using this rule? (*Hint*: If you get stuck, choose **Graph | Scale | Relative Frequency** or **Graph | Scale | Relative Percentage**.)

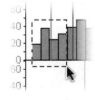

8. Select the group of UK students who are 10 or younger by drawing a selection rectangle around them. They will highlight. Now drag these bins to the right.

18 | 1: Distributions of Data

Summary Measures—CensusAtSchool
continued

Q6 Describe what happens to the values you plotted when you drag these bins close to the center of the distribution. What happens when you drag the bins all the way to the right side of the distribution?

Q7 Based on your experiments with dragging bins, do you think the standard deviation is sensitive or insensitive to outliers? Explain.

When you are done experimenting, restore the data to the original values.

Summary Tables

A summary table is another important tool for exploring data. You can use a summary table to get the numerical values for statistics such as the mean and standard deviation. But you can get any quantity you like with a summary table, as long as you can write a formula to express it.

9. Drag a new summary table from the shelf. Then open the collection's inspector and drag *Age* to the right arrow at the top of the table.

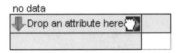
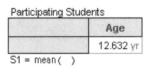

Enter median().

10. Add the median by choosing **Summary | Add Formula.**

Q8 What are the mean and median ages for students in your collection? How do these values compare with your estimates in Q2?

Q9 Drag the attribute *Gender* to the down arrow of the table. The table will split. On average, who is older, males or females? By how much?

Q10 Drag *Place* to the down arrow. You should see *Age* split by *Gender* and *Age* split by *Place*. Which country has the youngest students on average? The oldest?

11. Other functions you can use in a summary table include min(), max(), stdDev(), and iqr() (interquartile range). You can look them up in Fathom Help. Try them out. Use max() to find out how old the oldest male and oldest female are.

Q11 Use your summary table to find out which gender and which country have the greatest average travel times to school. What did you find? Which gender and which country have the smallest foot sizes?

Q12 What happens to your summary table if you drag a categorical attribute like *Computer* and drop it on *Age*? Why do you think that happens? What can you do to fix your table?

Summary Measures—CensusAtSchool
continued

Filters

Suppose you were interested in comparing how students under 15 years old traveled to school in the different countries. One thing you could do is look at a summary table.

12. Make a new summary table with *Place* on the down arrow and *TravelMethod* on the right arrow.

This summary table is for all ages. To examine just the students 14 years or younger, you can *filter* out the older students.

13. With the summary table selected, choose **Object | Add Filter.** Type Age<15yr and click **OK.** The summary table now includes only the cases where the student's age is less than 15 years.

Q13 How do students less than 15 years old travel to school? Is there a difference in travel method for the students in different countries?

It is hard to compare the countries using counts, especially because the UK has so many more students than the other two countries. What you need are proportions.

14. Double-click the formula count() below the table to show the formula editor. Type rowProportion and click **OK.**

The number 0.27083333 in the cell for Queensland for bus means that about 27% of Queensland students who are less than 15 years old take the bus.

rowProportion is a special keyword that applies only to formulas for a summary table. If you oriented your table the other direction, you would want columnProportion.

Participating Students

		bus	car
Place	Queensland	0.27083333	0.35416667
	UK	0.23715415	0.40316206
	South Africa	0.1	0.21666667
Column Summary		0.21883657	0.36565097

S1 = rowproportion

Q14 Using these proportions, how do students less than 15 years old travel to school? Is there a difference in travel method for the students between the three countries? Is your answer here different than your answer to Q13?

Summary Measures—CensusAtSchool | Activity Notes

Objectives

- Choosing an appropriate measure of center and spread based on the shape of the distribution and how the positions of the median and mean change with different shapes
- Learning how to compare distributions, either with proportions or with summary values
- Discovering some of the important tools that can be used to explore data
- Discovering how some patterns in the data are hidden and can be found by looking at different groups within attributes

Activity Time: 40–50 minutes

Setting: Paired/Individual Activity or Whole-Class Presentation (use **CensusAtSchool.ftm** for either)

Statistics Prerequisites

- Familiarity with proportions
- Familiarity with one-variable graphs
- Familiarity with the mean, median, and spread (standard deviation) in a general sense

Statistics Skills

- Comparing distributions of categorical variables using proportions in a two-way table
- Properties of the mean, median, and standard deviation and their sensitivity to outliers
- Comparing distributions in terms of shape, center, and spread using summary values
- Position of the mean and median in relation to skewedness
- Using counts versus proportions in a two-way table for comparing distributions
- Making stacked dot plots, histograms
- The effect of bin width on the shape of the distribution

AP Course Topic Outline: Part I A, B (1–2), C, E

Fathom Prerequisites: Students should be able to make graphs.

Fathom Skills: Students inspect the collection, plot values on graphs, make summary tables, calculate summary statistics on plots or in summary tables, create formulas and filters, and use row or column proportions in summary tables.

General Notes: This activity uses the CensusAtSchool data to introduce students to Fathom. It requires little to no knowledge of Fathom or statistics. This activity focuses on how to calculate summary statistics either in a plot or in a summary table, and how to create filters to examine particular groups. It is a great way to introduce students to different ways of using Fathom to calculate various summary statistics. Fathom makes it easy for students to work with a large data set and quickly compare distributions. Additionally, students can dynamically drag histogram bins and watch how summary statistics are affected by the changing data.

Procedure: Try working with the document **CensusAtSchool.ftm** before giving it to your students. This file has 500 cases. If it's too slow on your computers, select about half the cases and delete them. To select many cases, click on the first row number in the case table. Then scroll halfway down and Shift-click on another row number. All the cases in between will be selected. Choose **Edit | Delete Cases**.

INVESTIGATE

Students plot various summary statistics on a histogram and experiment with changing the distribution by dragging bins to see how that affects the summary statistics.

Q1 The mean will change with the dragged bin. For example, when students drag the bin to the left, the mean will shift to the left. Due to the number of cases in the collection, the median will not change unless the student uses a selection rectangle to select more than one bin to the right of the mean and drags those bins to the left as shown.

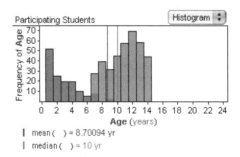

Summary Measures—CensusAtSchool
continued

You can show the mean's sensitivity to outliers by moving any bin all the way to the right, outside the range of the original data. The mean will shift to the right and the median will stay the same.

Q2 Median: 12 yr; mean: 12.632 yr. The mean's position to the right of the median indicates that the distribution is skewed toward the larger values.

Q3 The split graph shows that the narrowest range of ages is for the Queensland students; the South African students span the greatest range and have all the oldest students; and the youngest students are from the United Kingdom. The distribution for Queensland is mound-shaped, with its mean and median almost on top of each other. The distribution for South Africa is skewed left a little, shown by the position of the mean at around 15.4 and the median at 16. The distribution for the UK is the least mound-shaped, although its mean and median are roughly in the same location.

Q4 Queensland is the most mound-shaped, and that is easy to tell with the two plotted values *mean + standard deviation* and *mean − standard deviation*. The mean is in the center of the distribution with roughly the same number of bars in each section. The South African distribution has three bars between the mean and the new plotted values, but on the right side, there are four bars outside and a gap, and on the left side, there are five bars outside the plotted values. The UK distribution has roughly two bars between the mean and the new plotted values, but on the right side, there are four bars outside, and on the left side, there are only two bars outside the plotted values.

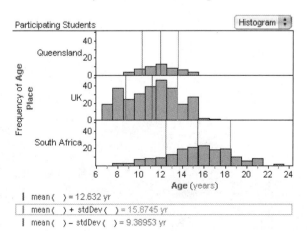

Q5 By choosing **Graph | Scale | Relative Frequency,** it is easier to see that all three distributions can use this rule by counting the percentage of cases either outside or inside the two new plotted values.

Q6–Q7 The standard deviation is sensitive to outliers because outliers will certainly affect the deviations from the mean. As the bins are dragged toward the center of the distribution, the plotted values move toward the mean and the spread gets smaller. As the bins are dragged toward the right side of the distribution, the mean shifts and the spread gets larger as well.

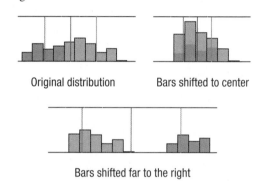

Q8 Median: 12 yr; mean: 12.632 yr. These values are exactly the same as in Q2.

Q9 On average, the females are 0.5 to 1 yr older with mean age 12.85 yr and median 13 yr. The mean for the males is 12.39 yr and the median is 12 yr.

Q10 The UK has the youngest students (mean = 11.13, median = 11), and South Africa has the oldest (mean = 15.44, median = 16).

Q11 Females have the greatest average travel time if you use the mean (18.24 vs. 17.10), but females and males have equal travel times if you use the median (15). South Africa has the greatest average travel time (mean = 23.35 and median = 15), and Queensland and the UK have roughly the same mean and median (mean = 14.9 and median = 10).

Females have the smaller foot size with mean 22.6 cm and median 23 cm. The UK has the smallest foot size with mean 22.7 cm and median 22 cm; This makes sense because the UK also has the youngest students.

Q12 For the mean or median, you'll get an #Argument count error# message. To fix this, students can

Summary Measures—CensusAtSchool
continued

either undo their last move or click on the summary measure with this error message and choose **Edit | Cut Formula.**

13. Students can also add filters to other objects in their document: graphs, case tables, tests, and so on. If they want the filter to affect every object in their document, they should add the filter to the collection.

Q13 The majority of students less than 15 years old travel to school by bus, car, or walking. Just looking at totals, 132 go by car, 131 walk, and 79 take the bus. It is hard to tell with just the counts, but yes, it does look like there is a difference between the countries. There are only 60 South Africans in the sample and 38 of them walk, whereas for the UK there are 253 students in the sample but only 82 walk. Queensland has the fewest walkers, 11 out of 48 students. The car totals are similar, except the UK holds the lead with 102 students out of 253 using a car to get to school, whereas only 13 South Africans use a car to get to school.

Q14 The answers are basically the same as for Q13. The majority of students less than 15 years old travel to school by bus, car, or walking: 36.6% go by car, 36.3% walk, and 21.9% take the bus. For the South Africans, 63.3% walk, whereas for the UK, 32.4% walk. Queensland has the fewest walkers, only 23%. The car totals are similar, except the UK holds the lead with 40.3% using a car to get to school, whereas only 22% of the South Africans use a car.

EXTENSIONS

Have students use the techniques they've learned to answer these questions or others that interest them. (*Note:* If an attribute doesn't have numeric values, you can still write filters, but you have to put the values in quotes. So, for example, (age>15yr) and (gender="F") finds all the females over the age of 15 years.)

1. Is walking more popular with males or females?
2. Which country has the highest proportion of students with Internet access?
3. Which country's students have to travel farthest to get to school?
4. Which country has the highest proportion of students who walk to school? Are these the students who have the greatest travel time?
5. What is the preferred mode of travel for students who are 15 years or older? Is there a difference in preference for these students between the three countries?

Box Plots and Transformations—CensusAtSchool

You will need
- CensusAt School.ftm

CensusAtSchool is an international project that collects and disseminates data about students from the United Kingdom, South Africa, Australia, and New Zealand. In this activity you'll use Fathom to learn about a random sample of students from this project.

EXAMINE DATA

1. Open the Fathom document **CensusAtSchool.ftm.** In this document you will find a collection of 500 students from the UK, South Africa, and Queensland, Australia, who were randomly selected from the CensusAtSchool project.

2. Make a graph with *Age* on the horizontal axis. Choose **Percentile Plot** from the pop-up menu. Your plot should look like a dot plot, but with the dot positions stretched out vertically. The cases are sorted and then plotted in order. The vertical axis shows the *percentile* of the points. That is, if a point is at 40.7, then 40.7% of the cases have an age less than or equal to this case.

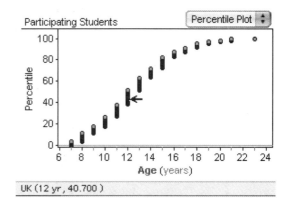

3. Place the cursor on a point. Don't click. When the arrow points straight left, you'll see the coordinates of the point in the status bar at the bottom of the Fathom window. Find the point as close to 50% as you can.

Q1 What is the age at the 50th percentile? What is another name for the 50th percentile?

Q2 What are the ages at the 25th and 75th percentiles? What is another name for these percentiles?

Q3 How old does a student have to be to be in the oldest 10% of the population? Describe how you used the percentile plot to find out. How young does a student have to be to be in the youngest 10%?

Q4 The percentile plot is steep at the beginning, a little less steep in the middle, and shallow at the end. What does that mean for the distribution of students in your sample?

4. Split the graph by dragging *Place* to the vertical axis.

Q5 Use the percentile plots for the three countries to compare their *Age* distributions.

Box Plots and Transformations—CensusAtSchool
continued

Q6 Choose another numeric attribute and make a new percentile plot for that attribute. Sketch the plot and explain what its shape means for the attribute's distribution.

INVESTIGATE

Box Plots

A box plot is also useful in showing the shape of a distribution and whether or not there are outliers.

5. Select the graph of *Age* and *Place* that you made in step 4 and choose **Graph | Remove Y Attribute: Place**.

6. Choose **Box Plot** from the pop-up menu. A box extends from the 25th percentile (or Q_1) to the 75th percentile (or Q_3) and is cut by a line at the median.

7. To see exactly the median and Q_3, hold the cursor over the right side of the box and read the status bar in the lower-left corner of the Fathom window as shown.

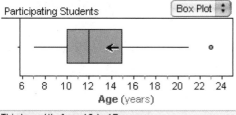

Q7 Using the box plot, find the median, Q_1, and Q_3. Compare your answers to your answers for Q1 and Q2. Were your estimates using the percentile plots fairly accurate?

Q8 The box plot shows an outlier. What is the value of that outlier? Verify that it is an outlier using the $1.5 \cdot IQR$ rule.

8. Split the graph again by dragging *Place* to the vertical axis.

Q9 Using the box plots, describe each country's distribution of *Age* in terms of shape, center, and spread.

Q10 The combined box plot was skewed right a bit and had an outlier, but the box plots from step 8 are all approximately normal with no outliers. Explain why this is so.

With the graph selected, choose **Object | Add Filter**. Type Age<15yr and click **OK**.

9. Add a filter to the graph to examine the distribution of *Age* for the three countries for students less than 15 years old.

Q11 Compare the three distributions of *Age* for students less than 15 years old.

Q12 Contrast the information you can learn from a box plot with that from a histogram. List the advantages and disadvantages of each.

Box Plots and Transformations—CensusAtSchool
continued

Sliders and Transformations

To experiment with transformations, you'll look at a new distribution: the heights of students from Queensland, Australia.

10. Make a new graph and plot *Height* on the horizontal axis.

Place = "Queensland"

11. Filter the *collection* to show only Queensland students.

Choose **Graph | Plot Value** and type mean(). Click **OK**.

12. On your new plot, plot the mean. Then plot the values mean()+stdDev() and mean()−stdDev().

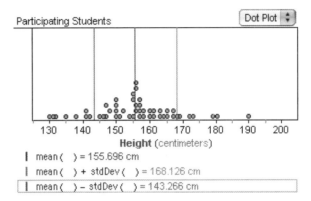

Q13 Do the two values mean()+stdDev() and mean()−stdDev() capture the majority of the dots?

How would the distribution of heights change if you added a constant value to every height measurement?

13. Drag a slider from the shelf. It will be named *V1*. Rename the slider *a*. Click once to the right of 5.00 and type cm. Press Enter.

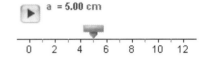

In the inspector, click once on the word *<new>*. Type in the name of the new attribute and press Enter. Double-click in the attribute's formula cell and type a+Height. Click **OK**.

14. In the collection's inspector, create an attribute named *Transformed_Height*. Define *Transformed_Height* with the formula *a* + *Height*.

15. Duplicate the *Height* graph by choosing **Object | Duplicate Graph** and replace *Height* with *Transformed_Height*. Use the slider to change the value of the constant *a* and observe the effects.

Q14 Compare the dot plot you made in step 12 to the ones you made in step 15 as you changed the value of the constant *a*. Does the center of your data change? The spread? If so, by how much do the values change?

Q15 Slowly drag the slider thumb to the right, then back to the left. Describe what happens to the distribution of transformed heights. Create formulas for the transformed mean and spread.

Box Plots and Transformations—CensusAtSchool
continued

Q16 When you add a constant to every value, does the range change? Explain.

Next you'll rescale the data by multiplying each height measurement by a constant.

16. Double-click in the formula cell for *Transformed_Height* and change the formula as shown.

> You will need to change the scale on your axis to see the data. Experiment with different ranges for the horizontal axis.

17. Observe the effects on the plot, mean, and spread as you change the slider this time.

Q17 Compare the original dot plot you made in step 12 to the ones you made in step 17 as you change the value of the constant *a*. Does the center of your data change? The spread? If so, by how much do the values change?

> Try some negative values of *a* as well.

Q18 Slowly drag the slider thumb to the right, then back to the left. Describe what happens to the distribution of transformed heights. Create formulas for the transformed mean and spread.

Q19 When you multiply every value by a constant, does the range change?

EXPLORE MORE

Using the techniques you've learned, explore the data to answer these questions or make up your own.

1. Redo the Sliders and Transformations section of this activity using the median as your measure of center and the interquartile range as your measure of spread. For each transformation, answer these questions: Does the center of your data change? The spread? If so, create formulas for the transformed center and spread.

2. Using box plots, compare the distributions of *Age* split by *TravelMethod*. Then use a filter and find out if there are any differences in method of travel between older students and younger students.

3. Using box plots, compare the distributions of *Time_to_Travel* split by *Place*. Which country's students have the longest travel times to get to school? Is there any explanation for the outliers?

Box Plots and Transformations—CensusAtSchool   Activity Notes

Objectives
- Understanding that a percentile is a measure of position and that the quartiles and median are also percentiles
- Learning how to describe or compare distributions, with either box plots or percentile plots
- Finding rules for how summary measures are affected when adding or multiplying each data value by a constant

Activity Time: 40–50 minutes

Setting: Paired/Individual Activity or Whole-Class Presentation (use **CensusAtSchool.ftm** for either)

Statistics Prerequisites
- Comparing distributions
- Familiarity with quartiles
- Familiarity with the mean, median, and spread (standard deviation) in a general sense

Statistics Skills
- Comparing distributions using percentile plots and box plots
- Transformations and their effects on summary measures
- Percentiles and using a percentile plot
- Box plots and the five-number summary
- Finding outliers using the $1.5 \cdot IQR$ rule
- Comparing types of plots

AP Course Topic Outline: Part I A, B (2–5), C, E

Fathom Prerequisites: Students should be able to make and split graphs and plot values.

Fathom Skills: Students make box plots and modified box plots, calculate five-number summaries, use sliders to transform attributes, work with percentiles and reading percentile plots, create formulas, split graphs using categorical variables, and use filters.

General Notes: This activity uses the CensusAtSchool data to introduce students to Fathom. It requires little to no knowledge of Fathom or statistics. This activity focuses on two different topics: measures of position, including percentile plots and box plots, and the effects of transformations on summary measures. Fathom makes it easy for students to work with a large data set and quickly compare distributions. Additionally, students can dynamically change transformation parameters and watch how the distributions change.

Procedure: Try working with the document **CensusAtSchool.ftm** before giving it to your students. This file has 500 cases. If it's too slow on your computers, select about half the cases and delete them. To select many cases, click on the first row number in the case table. Then scroll halfway down and Shift-click on another row number. All the cases in between will be selected. Choose **Edit | Delete Cases.**

EXAMINE DATA

Q1–Q3 The 25th and 75th percentiles are Q_1 and Q_3 respectively. The median is the 50th percentile. The age at the 25th percentile is 10 yr. The age at the 50th percentile is 12 yr. The age at the 75th percentile is 15 yr. The oldest 10% of the population is in the 90th percentile. The age at the 90th percentile is 17 yr, and the age at the 10th percentile is 8 yr. To find these, find points as close as possible to 90 and 10.

Q4 When the percentile plot is steep over an interval, that means there is a concentration of data values in that interval, whereas when it is shallow, that means there are fewer values in that interval. The percentile plot is steep at first and then gradually becomes shallower, so the age distribution is skewed right.

Q5 The split percentile graph shows that the narrowest range of ages is for the Queensland students; the South African students span the greatest range and have all the oldest students; and the youngest students are from the United Kingdom. The percentile plot for South Africa starts out shallow and gets steeper in the middle and then levels out at about 20 years old. The plot for the UK is, from the beginning, fairly steep and looks almost linear, then at the very last two dots it levels out. The plot for Queensland is mostly steeper still but has a small shallow part around 15.

Box Plots and Transformations—CensusAtSchool
continued

Activity Notes

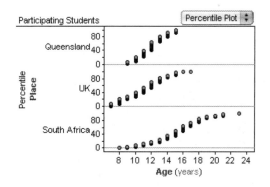

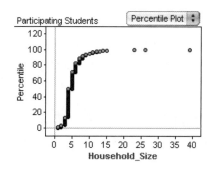

Q6 The percentile plot for *AgeMonths* is naturally similar to the plot for *Age*, but it shows a more detailed distribution. The plot for *Foot_Size* is steep in the center with a shallow part at either end, meaning that most of the data are concentrated between about 19 cm and 25 cm. The plot for *Height* doesn't have such a steep middle section, indicating that the heights are more spread out. The plots for *Household_Size*, *SchoolAgedMales*, and *SchoolAgedFemales* are all quite steep at the beginning, with long shallow ends, meaning that household sizes tend to be small. Similarly, the plot for *Time_to_Travel* has a steep early part and a long shallow end.

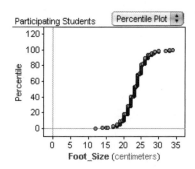

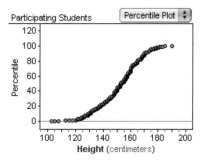

INVESTIGATE

Q7 Median = 12, $Q_1 = 10$, and $Q_3 = 15$. These values should be similar to, if not exactly the same as, the estimates in Q_1 and Q_2.

Q8 The outlier is at 23 years. The *IQR* is 5, so $1.5 \cdot IQR = 7.5$. Then $Q_3 + 1.5 \cdot IQR = 22.5$, so 23 is definitely an outlier.

Q9 The plot again shows that the narrowest range of ages is for the Queensland students; the South African students span the greatest range and have all the oldest students; and the youngest students are from the United Kingdom. The box plots clearly show, however, that all three are fairly mound-shaped and that the students from South Africa are definitely older—75% of the South African students are 14 or older, whereas for both the UK and Queensland, 75% of the students are younger than 14. The spreads (*IQR*) for both South Africa and the UK are the same (4 yr), while Queensland has a spread (*IQR*) of 2 yr. The box plot tells a great deal more about the shape, center, and spread than both the histogram and the dot plot in this case.

Q10 The outlier (23) is from South Africa. When all of the ages are lumped together into one distribution, 23 years old is quite a bit older than everyone else; however, compared to only the other South African students, the outlier is not that much older. For the whole group, $Q_3 = 15$, but for South Africa, $Q_3 = 18$. So, $Q_3 + 1.5 \cdot IQR = 18 + 1.5(4) = 24$ instead of 22.5.

Q11 The distributions of *Age* for both South Africa and the UK are skewed toward the younger values. The middle half of the ages of the younger Queensland students are between 11 and 13. The spreads for the younger UK and South African students are a bit larger: The middle half of the ages are between 9 and 12 for the UK and between 11 and 14 for South Africa. Half of the younger Queensland students have ages above 12 and half 12 or below. Half of the younger South African students have ages above 13 and half 13 or below, and half of the younger UK students have ages above 11 and half 11 or below. Note that for South Africa there is no upper whisker and 50% of the students are between 13 and 14.

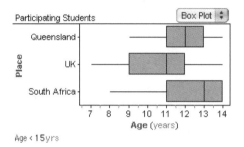

Q12 From a box plot, you can see the five-number summary exactly and outliers are clearly marked. These must be estimated from a histogram, which can be difficult. From a histogram, you can estimate the mean by estimating a balance point for the distribution. You cannot do this with a box plot. A histogram will reveal the frequency of the data within an interval. You do not know the exact values but you know how many are within the given boundaries. You know a lower bound and an upper bound, but not necessarily the exact least and greatest values. You know where there are clusters of data and where there are gaps. With a box plot, you get a sense of the basic shape of the distribution but you cannot see clusters or gaps. You cannot see the frequency but you can see the proportions.

10. This group of students was chosen because there are relatively few cases and it is easier to work with the transformed data than with the whole group.

Q13 The majority of dots are captured between the two values—38 out of 52 are captured.

Q14–Q16 When adding the constant *a* to each height, the whole distribution shifts by that constant. So the range and spread stay the same because the same constant is added to all values in the range (including the minimum and maximum values). The mean changes to *original mean + a*.

Q17–Q19 When multiplying height by the constant *a*, all three values change. Students will need to resize their plots to see their data. The new mean will be *original mean · a*. The new range and the new spread will be the original values multiplied by $|a|$.

EXPLORE MORE

1. When adding the constant *a* to each height, the whole distribution shifts by that constant. So the *IQR* stays the same but the median changes to *original median + a*. When multiplying height by the constant *a*, both values change. Students will need to resize their plots to see their data. The new median will be *original median · a*. The new *IQR* will be *original IQR · $|a|$*.

2. Taxis have the largest median age associated with them at 16; cars and cycles tie for the lowest median age at 10; walkers have the largest *IQR* at 5 yr. The largest range of older students take taxis, walk, or ride the bus, while the largest range of younger students walk, ride in a car or the bus, or take an unlisted mode of transportation.

3. It appears that the South African students have the longest travel time: 50% have a travel time of at least 15 min. The UK and Queensland students both have median travel times of 10 min. This could be due to the large number of students who walk to school in South Africa. The outliers could represent students that live in very rural areas served by regional schools or students on the last stop on a bus route.

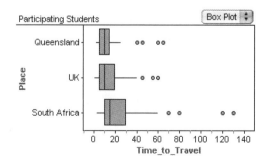

Deviation from the Mean—Hand Spans

You will need
- centimeter ruler

There are various ways you can measure the spread of a distribution around its mean. This activity gives you a chance to create a measure of your own.

COLLECT DATA

1. Spread your hand on a ruler and measure your hand span (the distance from the tip of your thumb to the tip of your little finger when you spread your fingers) to the nearest tenth of a centimeter.

2. In a new Fathom document, make a case table with two attributes: *Name* and *HandSpan*. Enter the names and measurements for your group.

	Name	HandSpan
1	Andrea	19.4 cm
2	Maya	18.2 cm
3	Pedro	24.0 cm
4	Kelly	21.0 cm
5	Jason	19.7 cm

INVESTIGATE

> You can identify the graph's points with the mouse: Select the *Name* column in the case table and choose **Table | Use As Caption**. Point the cursor at any point on the graph and the name will show up in the lower-left corner of the Fathom window.

3. Make a dot plot of the hand-span measurements for your group.

4. Plot the mean hand span: Select the graph, choose **Graph | Plot Value**, type mean() in the formula editor, and click **OK**.

Q1 Give two sources of variability in the measurements. That is, give two reasons why the measurements aren't all the same.

> You can enter attribute names and functions into formulas by typing them or by using the lists in the formula editor.

5. In the case table, create an attribute named *Deviation*. A *deviation* is the difference of a data value from the mean. Choose **Table | Show Formulas**. Double-click the shaded cell below *Deviation* and enter the formula shown.

Hands

	Name	HandSpan	Deviation
units		centimeters	centimeters
=			HandSpan − mean (HandSpan)
1	Andrea	19.4 cm	−1.08735 cm

6. Make a dot plot of the differences from the mean, or *Deviation*.

Q2 How far is your hand span from the mean of your group? How far from the mean are the hand spans of the others in your group? Use your case table and graphs to answer these questions.

Q3 What is the mean of *Deviation*? Plot this value on the dot plot of *Deviation*.

Exploring Statistics with Fathom

Deviation from the Mean—Hand Spans
continued

Q4 Experiment with dragging points in the dot plot of *HandSpan*. What is the effect on the dot plot of *Deviation* and on the mean of *Deviation* when you drag the points in the dot plot of *HandSpan*? Does it look like the mean of *Deviation* is a good measure of spread?

When you're done experimenting, repeatedly choose **Edit | Undo Drag Point** to return to your original data values.

Q5 Using the idea of differences from the mean, invent at least two measures that give a "typical" distance from the mean.

Q6 Compare your group's measures with those of the other groups in your class. Discuss the advantages and disadvantages of each group's method.

You can also use a Fathom summary table to compute the measures of typical distance from the mean that you invented in Q5.

7. Drag a summary table from the object shelf. Drag the attribute *HandSpan* from the case table to the summary table. Drop the attribute on the down arrow. Choose **Summary | Add Formula**. Enter one of your formulas from Q5 into the formula editor.

Hands	
HandSpan	20.48735 cm
	-1.4210855e-15 cm
S1 = mean( )	
S2 = mean(Deviation)	

8. On the dot plot of *HandSpan,* choose **Graph | Plot Value** and plot the values mean *plus* your measure of spread and mean *minus* your measure of spread.

Q7 Do the two new values capture most of the dots between them? Should they? Why or why not?

EXPLORE MORE

You might need to change the scale on your axis to see the data.

1. To add a constant value to every hand-span measurement, drag a slider from the shelf. It will be named *V1*. Click once to the right of 5.00 and type in cm. Press Enter. In the case table, create an attribute named *Recenter*. Define *Recenter* with the formula V1+HandSpan. Make a dot plot of *Recenter* and plot the mean, mean *plus* spread, and mean *minus* spread. Change the values of the constant with the slider and observe the effects. Does the center of your data change? The spread?

2. To rescale the data by multiplying each hand-span measurement by a constant, create an attribute named *Rescale* and define it with the formula V1·HandSpan. Observe the effects on the graph, mean, and spread as you change the slider. Does the center of your data change? The spread?

Deviation from the Mean—Hand Spans

Activity Notes

Objectives

- Becoming familiar with the concept of a measure of variation from the mean
- Recognizing that several measures of the "typical" distance from a mean may be defined (the measure most often used, the standard deviation, is by no means intuitive)
- Understanding that the mean of the deviations, because it is zero, is *not* a valuable statistic

Activity Time: 40–50 minutes

Setting: Group Activity (collect data, analyze)

Materials

- One centimeter ruler for each student or each group of students

Statistics Prerequisites

- Making dot plots
- Familiarity with the mean

Statistics Skills

- How changing data values affects the mean
- Variation from the mean
- Deviation
- Mean deviation and sum of the deviations is zero
- *Optional:* Rescaling, recentering (Explore More 1 and 2)
- *Optional:* Standard deviation (Extension 1)

AP Course Topic Outline: Part I B (1, 2, 5)

Fathom Prerequisites: Students should be able to make graphs and case tables and enter data with units.

Fathom Skills: Students define attributes with formulas, plot values on graphs, make summary tables, and use formulas in summary tables. *Optional:* Students make sliders (Explore More 1 and 2).

General Notes: Fathom relieves the computation burden in this activity and allows students to experiment with more and different measures of spread. The Explore More section and Extensions explore other concepts such as standard deviation, interquartile range, recentering, and rescaling. Students should be in groups of four or five for this activity.

INVESTIGATE

Q1 The major source of variability is that everyone in the group has a different hand span, depending on hand size and flexibility. The other source of variability is the uncertainty of the measurement of a given hand. That is, if different people measure the same hand, they may get different values. If the same hand is measured at different times, the hand span may be a little different each time.

Q3 The mean of *Deviation* is zero.

Q4 The deviations will move around the plot, but the mean of *Deviation* will stay at zero. While dragging points, students may notice that the numerical value of mean(Deviation) is changing. If they look closely, they will notice that any change in the mean is so miniscule that it still is considered zero, and in fact is zero. The error is introduced through calculation at several decimal places. The mean of the deviation is *not* a good measure of spread.

Q5 Groups will tend to invent the mean absolute deviation,

$$MAD = \frac{\sum |x - \bar{x}|}{n}$$

or in Fathom mean(|Deviation|). The *MAD* is easily understood and is a reasonable measure of spread for describing a distribution of data, but it has almost no importance in statistical theory. Sometimes a group will invent a statistic defined as the number of group members whose hand span is not equal to the mean (the number of "misses"). Other groups may suggest the largest difference from the mean. Here are a few other measures that students might invent: median(|Deviation|), $\sqrt{\text{median}(\text{Deviation}^2)}$, median(Deviation), max(Deviation), and the mode of residuals (which is not a built-in function in Fathom).

7. This table shows the calculation of the *MAD* with a formula equivalent to mean(|Deviation|).

Hands	
HandSpan	20.46 cm
	1.632 cm

$S1 = \text{mean}(\ )$
$S2 = \dfrac{\text{sum}(|\text{Deviation}|)}{\text{count}(\ )}$

Deviation from the Mean—Hand Spans
continued

Activity Notes

Q7 Depending on the measure of spread, the majority of values may or may not be captured.

DISCUSSION QUESTIONS

- What did you discover about the mean of the deviations?
- Why would we be interested in a typical distance from the mean?
- How did you compute a typical distance from the mean? What advantages or disadvantages did you find for these various measures?

EXPLORE MORE

1. When recentering, the mean will shift with the constant. The new mean will be mean(*HandSpan*) + *V1*. The spread will not change.

2. When rescaling, the mean will shift with the constant and the spread will change as well. The new mean will be *V1* · mean(*HandSpan*) and the new standard deviation will be | *V1* | · stdDev(*HandSpan*).

EXTENSIONS

1. Have students use their summary table to compute the standard deviation of *HandSpan*. (Choose

Summary | Add Formula. Enter **stdDev(HandSpan)**.) What is your standard deviation and how does it compare to the value of your measure of spread? Make a new dot plot of *HandSpan* and plot the values mean()+StdDev() and mean()−StdDev(). Do the two new values capture most of the dots? Does this measure of spread capture more or fewer dots than your measure of spread? [About two-thirds of values should be captured.]

2. Prove algebraically that the mean of deviations must be zero.

$$\bar{x} = \frac{\sum x_n}{n}, \text{ so}$$

$$\frac{\sum(x_n - \bar{x})}{n} = \frac{\sum x_n - (n \cdot \bar{x})}{n}$$

$$= \frac{\sum(x_n - (n\sum x_n/n))}{n}$$

$$= \frac{\sum(x_n - x_n)}{n} = 0$$

3. Redo this activity using the median as the measure of center and the interquartile range as the measure of spread. Which measures of center and spread produce the best results?

Exploring the Normal Distribution

You will need
• Normal.ftm

This activity has three parts: first you'll build a simulation, then you'll use your simulation to solve some problems, and finally you'll use your simulation to explore some data about different countries. The simulation is based on the function randomNormal(*mean, SD*), which gives you a random number from a normal distribution of the given mean and standard deviation.

GENERATE DATA

1. Open the Fathom document **Normal.ftm.** In this document you will find a collection of data from various countries around the world that you will use for comparison. But first, you need to make your simulation.

   To define an attribute, click once on <new> and type Simulated_Normal. Press Enter.

2. Drag a new collection from the shelf. Double-click the collection to show its inspector. Give it one attribute called *Simulated_Normal*. Choose **Collection | New Cases,** and add 40 cases.

3. Define *Simulated_Normal* with the formula randomNormal(0,1). The cases should fill with random numbers taken from a normal distribution with mean 0 and standard deviation 1.

INVESTIGATE

Simulated Data

4. Make a histogram of *Simulated_Normal*.

5. On your histogram, plot the mean, mean()+StdDev(), and mean()−StdDev().

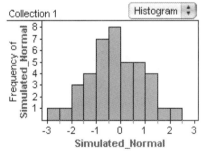

6. Make a summary table of *Simulated_Normal*. The mean is computed by default. Add the standard deviation.

*You can also rerandomize by pressing Ctrl+Y (Win) ⌘+Y (Mac) or by dragging the lower-right corner of the collection icon until you can see the **Rerandomize** button.*

7. Select the collection and choose **Collection | Rerandomize.** Do this ten times and observe the values for your plotted values.

Q1 What is the smallest value for the mean that you got in your ten randomizations? The largest?

Q2 What is the smallest value for the standard deviation that you got in your ten randomizations? The largest?

Q3 About what proportion of the cases are within one standard deviation of the mean? Two standard deviations of the mean?

*You may want to choose **Graph | Scale | Relative Frequency** to answer this question.*

How many standard deviations do you have to go from the mean to enclose half of the cases? You can use a slider to figure that out.

Exploring Statistics with Fathom
© 2007 Key Curriculum Press

1: Distributions of Data   37

Exploring the Normal Distribution
continued

8. Drag a new slider from the shelf and rename it *c*. Press Enter.

9. Double-click the slider to show its inspector. Change *Lower* to 0 and *Upper* to 2.

> You can insert *c* without typing the multiplication sign.

10. On your graph, change the values mean()+StdDev() and mean()−StdDev() to mean()+c·StdDev() and mean()−c·StdDev().

    Now you need a way to count the number of cases captured between the two new values. You'll do this in two steps. First you need to find each case's deviation from the mean and compare that to the value of c·StdDev(). Then you need to create a formula to find the proportion of cases that are between the two new values.

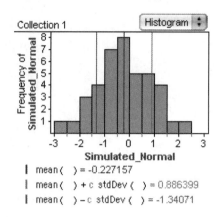

> For the absolute value, type abs().

11. Go to the collection inspector and define a new attribute, *Captured*. Define the attribute *Captured* with the formula

    |Simulated_Normal−mean(Simulated_Normal)|<c·stdDev(Simulated_Normal)

12. Now, in your summary table, add a new formula: Proportion(Captured).

Q4 Explain in your own words the formula in step 11 and what "true" and "false" mean in this context.

> With your slider set at 1.00, your plotted values are exactly one standard deviation away from the mean.

Q5 What proportion of the cases are within one standard deviation of the mean? How does your answer compare to your answer to Q3?

Q6 Using your slider, find out roughly how many standard deviations you have to go from the mean to enclose half of the cases. What value for *c* did you find? What are the two plotted values with this *c*?

> You can double-click the slider value to enter the exact proportion.

Q7 You may know that 95% of all cases are said to fall within 1.96 standard deviations of the mean. Change *c* to 1.96. What proportion of your cases are within 1.96 standard deviations of the mean?

Q8 Using your slider, find out roughly how many standard deviations you have to go from the mean to enclose 90% of the cases. What value for *c* did you find? What are the two plotted values with this *c*?

Exploring the Normal Distribution
continued

Countries Data

Now you'll look at some data from countries around the world and decide whether the data are normal. The *death rate* for a country is the proportion of the total deaths to the total population.

13. Make a graph of *Death_Rate* for the Countries collection. Drag *Region* to the graph and hold down the Shift key as you drop *Region* on the vertical axis of your graph.

Q9 Which region appears to have an approximately normal distribution for the attribute *Death_Rate*? Where are the countries in this region?

14. Make a filter for the collection so that your plot shows only the region you picked in Q9.

Q10 Using your plot or a summary table, find the mean and standard deviation for this region.

Choose **Graph | Scale | Density.**

15. Change your plot to a histogram with a density scale on the vertical axis. Then choose **Graph | Plot Function.** Enter the formula

    normalDensity(x,mean(Death_Rate),stdDev(Death_Rate))

Q11 How does the histogram compare to the normal density function? Do you think that this region's death rate is approximately normal?

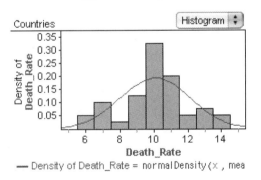

To better answer Q11, you can look at *z*-scores. A distribution of *z*-scores from an approximately normal population should look like an approximately normal population with mean 0 and standard deviation 1—just like your Collection 1.

16. In the Countries collection, define a new attribute, *z_Score*, with the formula

$$\frac{\text{Death\_Rate} - \text{mean}(\text{Death\_Rate})}{\text{stdDev}(\text{Death\_Rate})}$$

17. Make a new plot of *z_Score*.

Exploring the Normal Distribution
continued

Q12 If you rerandomize Collection 1, can you generate a group of data in Collection 1 that looks roughly like your plot of *z-scores*? Rerandomize Collection 1 until you find a plot that looks like your plot of *z-scores* or until you decide it is not possible to find one. Explain.

Q13 Based on your experimenting in Q12, do you think this region's death rate is approximately normal? Explain your answer.

EXPLORE MORE

1. Another way to answer Q13 is to make a new collection of 40 cases using the randomNormal function with the mean and standard deviation of *Death_Rate* instead of changing to *z-scores*, then plotting the new *Simulated_Normal* attribute, and rerandomizing the new collection like you did in Q12. Do this and see if your answer to Q13 stays the same.

2. Add 460 new cases to Collection 1 and redo Q1–Q8. Are the ranges for the values of your new means and standard deviations smaller and closer to 0 and 1 respectively? Compare the values of *c* that you got in Q6 and Q8 to the values 0.675 and 1.645. How close to these values did you get when you had 40 cases? 500 cases?

3. Make a new attribute, *twiceSimNorm*, with the formula 2·Simulated_Normal. What are the mean and standard deviation of *twiceSimNorm*?

4. Make another attribute, *b*, with the formula randomNormal(0,1). What is the standard deviation of *Simulated_Normal* + *b*? How many attributes like *Simulated_Normal* do you have to add up to get a standard deviation of 2?

5. Make the distribution of the *product* of two attributes like *Simulated_Normal* (for example, make a new attribute in which you multiply *Simulated_Normal* · *b*). Make its histogram. It does not look normally distributed. Describe in words the way in which it does not look normal.

6. Use Fathom to make a convincing argument that the product of two standard normal distributions is not normal.

Exploring the Normal Distribution — Activity Notes

Objectives

- Comparing actual and simulated normally distributed data with the normal distribution and seeing the variability in simulation that you don't see if you stick to the curve
- Understanding that a collection randomly generated from a normal distribution of mean 0 and standard deviation 1 does not necessarily look normal, and that the mean and standard deviation could vary quite a bit if the sample is small
- Recognizing the connection between a distribution of z-scores and the standard normal distribution
- *Optional:* Seeing how, when you add random variables, you do not add their standard deviations and seeing how multiplying normally distributed variables doesn't yield a normal distribution (Explore More 3–6)

Activity Time: 40–50 minutes

Setting: Paired/Individual Activity (use **Normal.ftm** and build simulation) or Whole-Class Presentation (use **NormalPresent.ftm**)

Statistics Prerequisites

- Some familiarity with z-scores and the standard normal distribution
- Familiarity with the mean and standard deviation

Statistics Skills

- The standard normal distribution
- Capture rate and central intervals
- Deviation from the mean and z-scores
- Comparing the ideal curve with an approximately normal distribution
- Normal density function
- Analyzing a data set to see if it is appropriate to use a normal model
- *Optional:* Sum and product of two normally distributed variables and their distribution (Explore More 3–6)

AP Course Topic Outline: Part I B (3, 5); Part III C

Fathom Prerequisites: Students should be able to make collections, graphs, attributes, and filters; and use summary tables or plotted values.

Fathom Skills: Students define an attribute with the randomNormal function, rerandomize to simulate a situation, work with sliders and plotted values to calculate proportions, use formulas to count cases between two values, and plot a normal density function.

General Notes: Using Fathom allows students to easily view many different normal distributions and note the differences in them, and to dynamically explore their properties, such as the relationship between proportion of cases captured and number of standard deviations from the mean.

Procedure: The **Normal.ftm** file is small enough that exploration in it is doable on almost all school computers, and the region the students will work with has only 40 cases. For that reason, the simulation that is set up in the first part of the activity is with 40 cases as well. If you don't wish to have students build the simulation, you can have a whole-class discussion using the document **NormalPresent.ftm**. This file has the simulation made in the Simulated Data section, and the first graph made in the "Countries Data" section.

INVESTIGATE

Q1–Q2 The range of reasonably likely means for 40 cases is from −0.3 to 0.3 and for standard deviations about 0.75 to 1.25. Here are the histograms for each for 1000 rerandomizations.

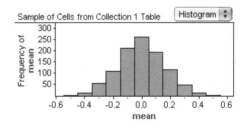

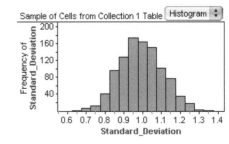

Q3 Theoretically, 68% are within one standard deviation and 95% are within two standard deviations.

Exploring the Normal Distribution
continued

Activity Notes

Students' values will vary; however, they should be somewhat close to these numbers in most cases. For example, for within two standard deviations, values will be between about 0.9 and 1.

Q4 The formula finds the absolute value of the distance from the mean, then compares it to the value of the standard deviation multiplied by the constant *c*. If the distance from the mean is less than this value, the formula returns "true." If it is outside the plotted values, it returns "false."

Q5 The values should be close to 68%.

Q6 The value of *c* should be close to 0.675 although with only 40 cases, this can vary widely. (With 500 cases as is done in Explore More 1, it will be close to this value.)

Q7 The value should be close to 95%.

Q8 The value should be close to 1.645.

13. *Death_Rate* was chosen for this activity because it is the only attribute with a normal distribution. You might want to have students explore the other attributes as well.

Q9 Region 3; Europe

14. The filter should be **Region=3**.

Q10 Mean: 10.06875; standard deviation: 2.0683

Q11 The histogram doesn't exactly fit the normal density function, but it is quite reasonable to think it is approximately normally distributed. It is fairly easy to rerandomize Collection 1 and find a histogram that looks roughly the same as the *z*-scores for *Death_Rate*.

Q12 If you rerandomize enough, you can get a graph with a shape similar to the *z*-score graph.

Q13 Experimentation in Q12 would lead most students to believe that the region's death rate is normal. Several of the rerandomized graphs from Collection 1 looked similar to the *z*-score graph.

DISCUSSION QUESTIONS

In these questions, *Simulated_Normal* and *b* are normally distributed attributes that students make during the activity (*b* is made in Explore More 4).

- When you multiplied *Simulated_Normal* by 2, the standard deviation doubled. Why?

- When you added *Simulated_Normal* to *b*, it did not double. Why not?

- You saw how the percentage of cases within one standard deviation varied around 68% but wasn't right on. What could you do to the simulation to reduce that variability? [Add more cases.]

- Suppose you looked at the distribution of *Simulated_Normal*^{2}. Would that be normal? [No!] Why not? [For one thing, it would always be positive.] What would the distribution look like? [It's skewed right.]

- (Advanced students.) You saw that the distribution of a product was not normal. But look:

$$e^{-x^2} \cdot e^{-x^2} = e^{-2x^2}$$

It certainly looks normal—it will just have a different standard deviation. What's going on? [The problem is not that you used *x* twice, though that's pretty bad. You can't just multiply distributions to get the distribution of the product of their random variables. Consider two disjoint distributions—their product would always be zero!]

EXPLORE MORE

1. The answer to Q13 would probably change. The *Simulated_Normal* histogram does not look like the *z*-scores anymore, no matter how many times it is rerandomized.

2. More cases should bring the experimental values closer to the theoretical values.

3. In general, the mean and standard deviation will be doubled: 0 and 2, respectively.

4. The standard deviation of *Simulated_Normal* + *b* will be about 1.4 times that of *Simulated_Normal*. If the cases truly had standard deviations of 1, then you would need to add four attributes to get a standard deviation of 2.

5. There is a spike in the middle of the distribution and it tails off more slowly than a normal distribution does.

6. Students could show that for multiple randomizations, the distributions of *Simulated_Normal* · *b* always show a spike in the middle.

2

Relationships Between Data

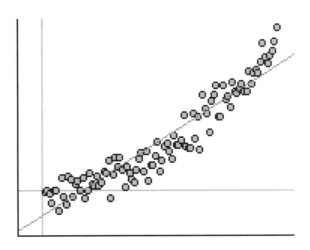

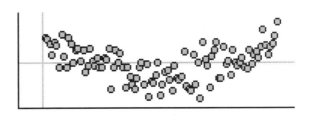

Least-Squares Lines—Pinching Pages

You will need
- millimeter ruler
- your statistics text

How thick is a single page of your statistics text? One page alone is too thin to measure accurately, but you could measure the thickness of 100 pages together, then divide. This method would give you an estimate of the thickness, but no information about the precision of your measurement. The approach in this activity lets you judge precision as well as thickness.

COLLECT DATA

1. In a new Fathom document, make a case table with attributes *Pages* and *Thickness*.

2. You are going to take measurements in millimeters. With the case table selected, choose **Table | Show Units.** Type millimeters in the units row of the *Thickness* column.

Pinch 50 sheets of paper, not up to page 50.

3. Pinch together the front cover and first 50 pages of your book. Measure the thickness to the nearest millimeter, and record the number of pages and thickness in your case table.

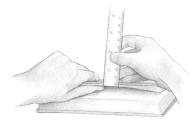

Page Thickness

units	Pages	Thickness
		millimeters
1	50	6.0 mm
2	100	11.0 mm
3	150	12.5 mm
4	200	17.0 mm
5	250	21.0 mm

4. Repeat step 3 for the cover plus 100, 150, 200, and 250 pages.

INVESTIGATE

Finding a Line of Fit

5. Drag a new graph from the shelf. Drag *Pages* to the horizontal axis and *Thickness* to the vertical axis. You get a scatter plot where each dot represents one pair of your measurements.

Q1 Does the plot look linear? Should it? Discuss why or why not. Make your measurements again if necessary.

6. Place a movable line on the scatter plot by choosing **Graph | Add Movable Line.** Run the cursor along this line, noticing how the cursor changes shape. Experiment with using different cursors to drag the line.

Least-Squares Lines—Pinching Pages
continued

The equation of the movable line appears below the graph and changes as you adjust the line.

7. Adjust the line to best fit the points by first dragging one end and then the other.

   This line is an eyeball fit; different people will fit the line somewhat differently. Fathom can also compute several standard lines of fit. One of these is the least-squares line. You'll find out what that means.

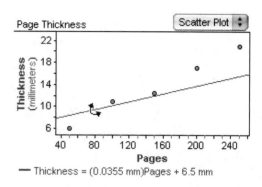

8. With the graph selected, choose **Graph | Show Squares**.

   Fathom draws a square constructed from each point to the line, a square whose side length is equal to the difference between the actual value and predicted value for the point—the *residual*. Below the graph, it reports the sum of the areas of all the squares. This sum is one way to judge how well a line fits data.

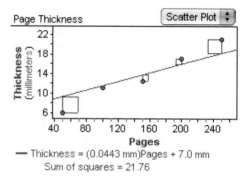

9. Drag the line around. Watch the squares and their sum change. Then adjust the line to make the sum of the squares as small as you can make it.

Q2 What is the equation of the line you got in step 9? What does the line's *y*-intercept tell you? What does its slope tell you?

Q3 What is your estimate of the thickness of one page?

Q4 Use the information in your graph to discuss the precision of your estimate in Q3.

Q5 How would your line have changed if you hadn't included the front cover?

Least-Squares Line

The line that minimizes the sum of the squares of the residuals is the *least-squares line*.

10. Choose **Graph | Least-Squares Line**. The new line is computed from the data and is not movable.

Q6 How closely did you manage to adjust the movable line to match the least-squares line?

Least-Squares Lines—Pinching Pages
continued

Q7 What is the sum of the squares for the least-squares line? How close did your measurement for the sum of the squares of the movable line come to this value?

You can use this line to predict likely values when given only one of the measurements. How well does this method predict the thickness given the number of pages?

11. Choose **Graph | Remove Movable Line.** Then click and hold your mouse over the least-squares line. The cursor changes and a red dot appears on the line. With the red dot showing, drag the cursor around in the graph to trace the line.

Q8 Use this feature to predict the thickness of 125 pages. Measure with your ruler the thickness you get for 125 pages of your book. How far off is the prediction from the thickness you measured?

Q9 Use the plot or the equation of the line to predict the thickness of 500 pages. Measure with your ruler the thickness you get for 500 pages of your book. How far off is the prediction from the thickness you measured?

You might need to adjust the scales on the horizontal and vertical axes to get this prediction.

Q10 Is there much difference between the prediction error you got in Q8 and the prediction error you got in Q9? If so, why do you think that is so?

Using a Residual Plot

The predictions you got in Q8 and Q9 were probably a bit off. How much off might you *expect* them to be?

12. With the graph selected, choose **Graph | Make Residual Plot.** A plot of the residuals appears below the main scatter plot. Each point in the original plot has a corresponding residual plot. The *y*-value for the residual point is the difference between the predicted value and the observed value (its vertical distance from the line).

13. Slowly drag one of the points in the top graph. Notice how much effect each point has on the line's equation and how its residual changes in the residual plot. Also notice that the least-squares line is changing in response to dragging the point, so the other residual points are changing as well.

When you are done experimenting, use **Undo** to return the data to its original values.

*Or select the collection and choose **File | Revert Collection.***

Least-Squares Lines—Pinching Pages
continued

Patterns in a residual plot suggest that the data aren't linear.

Q11 Examine the residual plot for any patterns or curvature that might suggest a linear model is not appropriate. Based on the residual plot, is a linear model appropriate for these data? Explain.

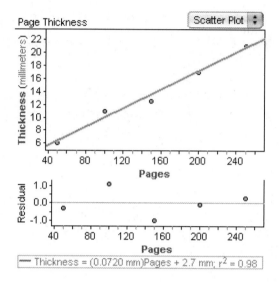

Q12 The residuals in the above plot appear to lie roughly in a band from -1 to $+1.1$. Between what two values do the residuals in your plot lie?

Q13 In Q8, how far off was the prediction of the book's thickness for 125 pages? Did it lie within the band you got in Q12? What about your prediction for 500 pages in Q9?

14. Add your two measurements for 125 and 500 pages to the case table.

Q14 How much does including these two measurements affect the slope and y-intercept of the least-squares line?

Least-Squares Lines—Pinching Pages

Activity Notes

Objectives
- Interpreting the slope and *y*-intercept of a line in context
- Recognizing two likely sources of error: measurement error and error in fitting a line through the points
- Understanding that the line of best fit is the line that minimizes the sum of the squares of the residuals
- Discerning the difference between errors that result from predicting values within the data's range versus those that result from predicting values outside the data's range

Materials
- One millimeter ruler for each pair of students
- One student text for each pair of students

Activity Time: 40–50 minutes

Setting: Paired Activity (collect data or use **Pinching Pages.ftm**) or Whole-Class Presentation (use **Pinching Pages.ftm**)

Statistics Prerequisites
- Introduction to linear data (elliptical cloud)
- Familiarity with the equation of a line
- Slope and *y*-intercept

Statistics Skills
- Slope and intercept in context
- Sources of error
- Sum of squares
- Least-squares line
- Residual and residual plots
- Examining residual plots
- Extrapolation versus interpolation
- Predicted value versus observed value

AP Course Topic Outline: Part I D (1–4)

Fathom Prerequisites: Students should be able to make graphs and case tables, and enter data.

Fathom Skills: Students make scatter plots, fit movable lines and the least-squares regression line to data, use residual squares, make residual plots, and use the least-squares line to make predictions.

General Notes: This activity provides a good opportunity to use Fathom's dynamic features with bivariate data—movable lines, residual squares, least-squares regression line, and residual plots. Students should work in pairs.

Procedure: Collecting the data for this activity can be done quickly and requires little equipment. To set up the activity, tell students that the thickness of one page of a book cannot be measured with a ruler but that looking for a trend in the thickness of larger, measurable groups of pages can help them estimate the thickness of one page. Note that "page" refers to a sheet of paper. Students who subtract *page numbers* instead of counting sheets will get estimates that are half the thickness of a sheet (because there are two "pages" per sheet). During or after the activity, discuss student responses to Q1–Q5. If you don't wish to have students collect data, you can use the document **PinchingPages.ftm.**

INVESTIGATE

Q1 The measurements should produce a scatter plot with a strong linear trend.

Q2–Q5 Be sure students correctly interpret the meaning of the slope and *y*-intercept. Slope measures change in *y*, per unit change in *x*, and because a unit change in *x* is one page and *y* is thickness, this slope is an estimate of the thickness of one page. The *y*-intercept is the thickness if no pages are included, so it is the approximate thickness of the cover of the book.

Q4 Sample answer: The points lie very close to the line, so we believe the measurements are fairly accurate. However, because we measured to only the nearest 0.5 millimeter, we expect some inaccuracy in each measurement—some will be too small and others too large. By using the line to estimate the slope, we are averaging out those errors.

Q5 If the front cover is not included, the *y*-intercept should be zero.

Q6–Q7 The two lines and the sums should be very similar.

Q8 The first value is within the range of their original data and students' predictions should be rather close to the estimate they get by actually measuring the 125 pages.

Least-Squares Lines—Pinching Pages
continued

Activity Notes

Q9–Q10 The second value is quite a bit outside the range of original values, and students' predictions will probably have a much larger error. You can use this discrepancy to introduce or discuss why extrapolating outside the range of the observed data is usually risky and needs to be qualified with warnings.

Q11 The residual plot should not have any curvature or pattern, which suggests that the line is an appropriate model for these data.

Q12–Q13 The range of the residuals will vary, but if students were measuring carefully, their first prediction should fall within the band of values, while the second prediction will probably fall well outside the band of values.

Q14 Adding the estimate for 500 pages will probably affect the equation of the line considerably and will illustrate how one point can influence the line and the slope.

DISCUSSION QUESTIONS

- Compare your estimate for the thickness of one page and the thickness of the cover with the estimates of your classmates. How can you explain the variation between the different approximations?
- How did you go about estimating the precision of your estimates of page thickness and cover thickness?
- How would you use the equation of the movable line to predict the total thickness of your book, including both covers? Calculate a prediction and measure the total thickness. How close is your prediction?
- How much did the range of the residuals increase when you added your measurement for the thickness of 500 pages?

Least-Squares Lines and Correlation—Was Leonardo Correct?

You will need
- Leonardo.ftm

Leonardo da Vinci (Italian, 1452–1519) was a scientist and an artist who combined these skills to draft extensive instructions for other artists on how to proportion the human body in painting and sculpture. Three of Leonardo's guidelines were

- height is equal to the span of outstretched arms
- kneeling height is three-fourths the standing height
- the length of the hand is one-ninth the height

In this activity you'll test Leonardo's rules.

EXAMINE DATA

1. Open the Fathom document **Leonardo.ftm**. In this document you will find a case table with data about some students' heights, kneeling heights, arm spans, and hand lengths.

   To make a scatter plot, drag a new graph from the shelf. Drag *Height* to the horizontal axis and one of the other attributes to the vertical axis.

2. Check Leonardo's three guidelines visually by plotting the data on three scatter plots.

   Q1 For each scatter plot, describe the overall shape of the relationship. Is the pattern linear (scattered about a line) or curved? Are there any clusters or outliers?

   Q2 For each scatter plot, describe the trend. Is there a positive or negative trend?

   Q3 For each scatter plot, describe the strength of the relationship. Is the association strong or is it weak? Does the relationship vary in strength or does it have constant strength?

INVESTIGATE

Correlation

To adjust a movable line, drag its ends to rotate it or drag its middle to move it up and down.

3. For each plot, choose **Graph | Add Movable Line**. Adjust the movable line so that it fits through each cloud of data points.

A *residual* is the distance of a point to a line measured parallel to the *y*-axis.

4. For each plot, choose **Graph | Show Squares**. Fathom draws a square constructed from each point to the line, a square whose side length is equal to the difference between the actual and predicted values for the point—the *residual*. Below the graph, it reports the sum of the areas of all the squares, or the Sum of Squared Errors.

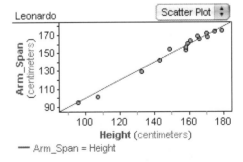

Least-Squares Lines and Correlation—Was Leonardo Correct?
continued

5. Adjust the movable lines so that the sum of squares is as small as you can make it for each plot. Write down the equation of each of your final lines.

6. The line that minimizes the sum of the squares of the residual is the *least-squares line.* Choose **Graph | Least-Squares Line** to display the least-squares line, the regression equation, and the squared correlation, r^2.

Q4 How close are the least-squares lines to the movable lines that you found in step 5? Write down the equation of each of these lines.

Q5 For each equation, interpret the slope of the regression line in context. How does each slope compare to Leonardo's guidelines?

Q6 For each of the plots, interpret the coefficient of determination in context. Then compute the correlation, the square root of r^2, and interpret it in context.

Q7 Do the three relationships described by Leonardo appear to hold? Do they hold strongly?

Residual Plots

Some of the relationships suggest curvature. To look into that more closely, examine the residual plots.

7. For each plot, choose **Graph | Remove Movable Line.** Then choose **Make Residual Plot.** A plot of the residuals appears below the main scatter plot. Each point in the original plot has a corresponding residual plot. The *y*-value for the residual point is the difference between the predicted value and the observed value (its vertical distance from the line).

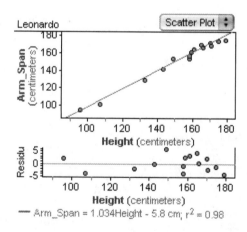

8. Drag points in either the original plot or the residual plot to explore the residual plot. When you are through, restore the original data (by undoing or reverting).

Q8 Examine each residual plot for any patterns or curvature that might suggest a linear model is not appropriate. Based on the residual plots, is a linear model appropriate for these data? Explain.

Least-Squares Lines and Correlation— Was Leonardo Correct?

Activity Notes

Objectives
- Interpreting the slope of a line and the coefficient of determination in context
- Visually representing the Sum of Squared Errors
- Developing a sense of the difference between strong and weak correlation
- Understanding that the least-squares line is just *one* way of fitting a line to data

Activity Time: 40–50 minutes (80 minutes if students collect their own data)

Setting: Paired/Individual Activity (use **Leonardo.ftm** or collect data) or Whole-Class Presentation (use **Leonardo.ftm**)

Materials
- *Optional:* One measuring tape, yardstick, or meterstick for each pair of students

Statistics Prerequisites
- Attributes of linear data (shape, trend, and strength)
- The equation of a line: slope and y-intercept
- Familiarity with the least-squares regression line

Statistics Skills
- Slope and intercept in context
- Sources of error
- Sum of squares and its relationship to the least-squares regression line
- Residual and residual plots
- Examining residual plots
- *Optional:* Influential points (Extension 1)

AP Course Topic Outline: Part I D (1–4)

Fathom Prerequisites: Students should be able to make scatter plots and case tables and, enter data (if students collect their own data).

Fathom Skills: Students use a movable line to explore the Sum of Squared Errors, fit data using a least-squares line, show and use residual squares, and make residual plots. *Optional:* Students create filters (Extension 1).

General Notes: In this activity, students use Fathom to quickly graph the least-squares line and calculate correlation for body proportion data. Make sure students realize that neither Leonardo's guidelines nor the regression lines should be regarded as a human ideal.

Procedure: A set of sample data is provided in the document **Leonardo.ftm.** However, if you have time, students enjoy collecting this data, though be aware that some students may be sensitive about their height or other body measurements. Have them work with a partner to measure their height, kneeling height, arm span, and hand length. You may want to collect data on one day, enter the data on a single computer, then distribute the saved Fathom document. Or you can split the data gathering and analysis across two classes.

EXAMINE DATA

2. Students should notice approximately linear trends in the scatter plots.

Q1–Q3 In the sample data, there does appear to be some curvature in the scatter plots of *Arm_Span* versus *Height* and *Hand_Length* versus *Height,* although in the latter plot, the appearance of curvature is largely due to the two lowest data values (outliers). The relationships are fairly strong with a positive trend and tend to be constant in strength.

INVESTIGATE

3.–5. These steps allow students to see a visual representation of the Sum of Squared Errors that cannot be done simply without technology. Looking at the changes in squares will solidify the concepts of the least-squares line and strength of correlation.

6. The least-squares line in Fathom automatically displays the value for r^2 below the graph. However, this activity also asks for the correlation, or the positive square root of r^2. A simple way to calculate r in Fathom is to use a summary table as a calculator.

Least-Squares Lines and Correlation—Was Leonardo Correct?
continued

Activity Notes

Make a summary table, choose **Summary | Add Formula,** and enter the square root of the value for r^2.

[summary table image: S1 = √0.98, value 0.98994949]

Q4 Arm_Span = $-5.8128 + 1.03439$Height;
Kneeling_Height = $2.1943 + 0.726933$Height;
Hand_Length = $-2.9696 + 0.124118$Height

Q5 *Arm_Span* vs. *Height:* The slope is 1.03, which means that for every 1 cm increase in height, there tends to be a 1.03 cm increase in arm span. Leonardo predicted a 1 cm increase.

Kneeling_Height vs. *Height:* The slope is 0.73, which means that for every 1 cm increase in height, there tends to be a 0.73 cm increase in kneeling height. Leonardo predicted a 0.75 cm increase.

Hand_Length vs. *Height:* The slope is 0.12, which means that for every 1 cm increase in height, there tends to be a 0.12 cm increase in hand length. Leonardo predicted a 1/9 cm, or 0.11 cm, increase.

Q6 *Arm_Span* vs. *Height:* $r = 0.99$. About 98% of the variation in arm span between these students can be attributed to their height.

Kneeling_Height vs. *Height:* $r = 0.989$. About 98% of the variation in kneeling height between these students can be attributed to their height.

Hand_Length vs. *Height:* $r = 0.96$. About 92% of the variation in hand length between these students can be attributed to their height.

Q7 In each case the correlation is quite high, greater than 0.96, and the points are packed tightly about the regression line, so there is a strong correlation. Leonardo's claims hold reasonably well. The slopes are about what Leonardo predicted and the *y*-intercepts are close to 0 in each case.

Q8 Although the plots suggest curvature, each of the residual plots is fairly randomly scattered about the line $y = 0$, and there is little pattern or curvature in the residuals. This suggests that a linear model is appropriate for each plot.

DISCUSSION QUESTIONS

- Do any of the relationships suggested by Leonardo's guidelines appear to be nonlinear?
- What does the intercept mean for each relationship? Does it make any sense?
- If a set of bivariate data has perfect correlation ($r = 1$ or $r = -1$), what would the squares of the residuals from the least-squares line look like visually? What if the set of data has a very weak correlation?

EXTENSIONS

1. In the case table or in the collection, eliminate any cases that are outliers. How does removing these outliers affect the slope of each least-squares line? The correlation? Are these outliers influential points? [To remove the outliers, students can select each case and choose **Edit | Delete Case** or they can use a filter. To create a filter, select the collection, then choose **Object | Add Filter.** In the formula editor, type Height>110 cm. Students can then choose **Graph | Rescale Graph** to rescale their plots.]

2. Art from different periods and cultures may show human proportions that are different from Leonardo's guidelines. Choose a period and culture from art resources available in your library or online, and investigate.

3. Devise a human-proportion guideline of your own based on data that you collect. Can you find a relationship that holds more strongly than any of Leonardo's?

Exploring Correlation

You will need
- Correlation.ftm

In this activity you'll experiment with a set of data to discover properties of the correlation coefficient.

EXAMINE DATA

1. Open the Fathom document **Correlation.ftm.** You'll see a table and a graph. The graph has a line in it (the least-squares line). Below the graph is the equation of the line and the square of the correlation coefficient, r^2.

2. Drag the points in the plot with your mouse and observe how the least-squares line moves with the point you are dragging. Experiment with dragging points close to the line and points far away from the line.

Q1 Which action influences the line more (that is, moves the line more): dragging a point close to the line toward the line or dragging a point far from the line toward the line? Explain.

Q2 Which action influences the line the more (that is, moves the line more): dragging a point close to the line away from the line or dragging a point far from the line away from the line? Explain.

INVESTIGATE

z-Scores

Now you will drag points in your scatter plot but concentrate on the changes in the correlation coefficient instead. In each question you will set up a situation to examine the effect on r^2.

Q3 Drag the points in the plot with your mouse so that they line up in a straight line, slanted upward. What's the value of r^2?

Q4 Make the points line up slanted downward. Now what's the value of r^2?

Q5 Put the points in a circle. Now what's the value of r^2?

Double-click in the formula cell to show the formula editor.

3. Make a new attribute, *Zx*. This will be the *z*-score in the *x*-direction. A *z*-score in the *x*-direction is the number of standard deviations between the *x*-value and the mean of *x*. Its formula is the one shown here. Enter this formula for *Zx*.

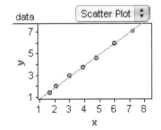

Formula for Zx

$$\frac{(x - \text{mean}(x))}{\text{stdDev}(x)}$$

4. Create another new attribute, *Zy*, where *Zy* will be a *z*-score in the *y*-direction. Give it an appropriate formula.

Exploring Correlation
continued

5. Make a new graph with Zy on the vertical axis and Zx on the horizontal axis. Make sure you can see both the x-y graph and the Zx-Zy graph.

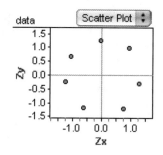

6. Experiment by dragging a few points on the x-y graph. Observe what happens on the other graph.

Q6 Describe what happens to the scatter plot of z-scores when you drag your points in the x-y graph.

Q7 One of the differences between the two graphs is that the one for the z-scores contains the origin. It has to. Why?

Product of the z-Scores

7. Make a new attribute, *Product*. Give it the formula $Zx \cdot Zy$. That is, it's the product of the z-scores.

8. Make a graph—a simple dot plot—of *Product*. Now you have three graphs: two scatter plots and a dot plot. Make sure you can see them all.

Q8 What has to be true of a point for its *Product* to be positive? What has to be true for its *Product* to be negative? Move points around in the original graph to explore this.

Q9 What must you do to make a scatter plot where all of the *Products* are positive but none of the points in the original graph fall on the line?

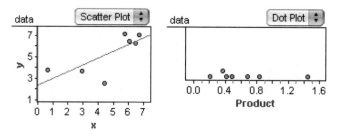

Q10 What must you do to make a scatter plot where all of the *Products* are negative but none of the points in the original graph fall on the line?

Average Product

9. Double-click the collection and go to the **Measures** panel in the inspector. Notice there is a measure n that counts the cases.

Exploring Correlation
continued

Click once on <new> and type Average_Product. Press Enter. Then double-click in the Formula column to show the formula editor.

10. Define a new measure, *Average_Product*, that is the average product (dividing by $n - 1$) of the z-scores.

Measure	Value	Formula
n	7	count( )
Average_Product	0.739788	$\frac{\text{sum}(Zx \cdot Zy)}{n - 1}$

11. Now select the *Product* graph and choose **Graph | Plot Value.** In the formula editor, type Average_Product. Close the editor.

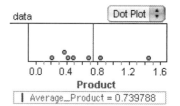

Remember, you can adjust only the points in the original graph. (Why is that?)

Q11 Once again, make all the points line up slanting upward. What is the average of the *Products*? Using the original graph, drag points so that you can fill in this table.

Point Slant	r^2	Average Product
Upward	1.00	
Upward	0.64	
Upward	0.25	
Upward		0.3
Downward	1.00	
Downward	0.49	
Downward		−0.5
	0.0625	−0.25

Q12 Generalize: What's the relationship between r^2 and *Average_Product*?

Q13 Describe in words what *Average_Product* tells you about the original graph of the points.

EXPLORE MORE

Suppose your data are a complete population and not a sample. How would the formulas for the z-scores, Zx and Zy, change? How would the formula for *Average_Product* need to change so that the relationship you discovered in Q11 still holds? Use your Fathom document to find out if you are right.

Exploring Correlation

Activity Notes

Objectives
- Understanding that correlation can be positive or negative, between -1 and $+1$, and that it measures how well the points are arranged in a line
- Realizing that the product of a point's coordinates—relative to the means—is that point's contribution to the correlation. If that point is in the first or third quadrant—relative to the means—the contribution is positive; if it's in the second or fourth, the contribution is negative.
- Seeing that z-scores are useful coordinates because they move the origin to the means of the two coordinates
- Realizing that when you move a point in the original graph, *the origin moves* in the z-graph—because you're changing the mean

Activity Time: 20–30 minutes

Setting: Paired/Individual Activity or Whole-Class Presentation (use **Correlation.ftm** for either)

Statistics Prerequisites
- Some familiarity with z-scores
- Familiarity with the least-squares line
- Familiarity with averages

Statistics Skills
- Correlation coefficient
- Spread in bivariate data
- Working with z-scores
- Mean and average

AP Course Topic Outline: Part I D

Fathom Prerequisites: Students should be able to define attributes in a case table, make scatter plots and dot plots, and use the formula editor.

Fathom Skills: Students define and plot measures, and use a sum formula.

General Notes: This activity demonstrates that the correlation coefficient is the average product of the z-scores. *Students should have seen z-scores before this activity.* Students first experiment with dragging points in a scatter plot to see how their action influences the least-squares regression line. They then define the average product of the z-scores (dividing by $n-1$) using the formula

$$\text{Average\_Product} = \frac{1}{n-1}\sum(Z_x \cdot Z_y) = r$$

EXAMINE DATA

Q1 When you move a point farther away from the line toward the line, it has more effect on the line than a point closer in. The line moves more dramatically.

Q2 The line moves more dramatically when you move a point close to the line away from the line than if you move a point farther out away from the line.

INVESTIGATE

Answers to Q3–Q5 depend on how close students can come to arranging the points in a straight line or a circle.

Q3 $r^2 = 1$

Q4 $r^2 = 1$; $r = -1$

Q5 $r^2 = 0$

Q6 When you drag a point in the *x-y* graph, all the points in the *Zx-Zy* graph move. This is because when you drag one point in the *x-y* graph, you are changing the mean, which affects all the z-scores of the other points. For example, if you drag your point horizontally to the right with no vertical change, the mean shifts to the right as well, so the z-scores of the other points will move to the left because, although they haven't moved, their position relative to the mean has moved left.

Q7 The "center" of the original plot is $(\bar{x}, \bar{y})$. Transforming to z-scores translates the mean of the *x*'s and the mean of the *y*'s to 0, so the center of the new plot is $(0, 0)$.

Q8 Points in Quadrant I have positive z-scores for both *x* and *y*, so their product contributes a positive amount to the calculation of the correlation. Points in Quadrant III have negative z-scores for both *x* and *y*, so their product contributes a positive amount to the calculation of the correlation. Points in Quadrants II and IV contribute negative amounts to the calculation for the correlation because one z-score is positive and the other is negative.

Exploring Correlation
continued

Q9 To make all the *Products* positive, all points in the original plot need to be moved so that their corresponding points in the *Zx-Zy* graph are all in Quadrants I and III.

Q10 To make all the *Products* negative, their corresponding points need to be in Quadrants II and IV.

Q11–Q13 If students line up the points in a perfect line sloping up, the average product will be 1. More than likely, if they drag the points, they will get an average product very close to 1. As shown in the formula for *Average_Product*, the average product of the *z*-scores is the correlation coefficient, *r*. If it is positive, the trend is positive and if *r* is negative the trend is negative.

Point Slant	r^2	Average Product
Upward	1.00	**1.00**
Upward	0.64	**0.80**
Upward	0.25	**0.50**
Upward	**0.09**	0.3
Downward	1.00	**−1.00**
Downward	0.49	**−0.70**
Downward	0.25	−0.5
Downward	0.0625	−0.25

DISCUSSION QUESTIONS

Make sure you discuss the questions posed on the student worksheet. Here are some additional questions:

- If you mapped the origin in the *z*-graph back into the original graph, where would it be? [At the coordinates of the means.]
- What's the relationship between the correlation and the slope? [Only the sign.]
- What would be a good name for this quantity, the average product? [You could let students name it— but as it is the square root of r^2, *r* is not a bad name.]

EXPLORE MORE

If you are working with a population, then instead of dividing by the sample standard deviation, you'd divide by the population standard deviation for the *z*-scores and you'd divide by *n* for the average product of the *z*-scores.

EXTENSIONS

1. Have students explain clearly and convincingly, in writing, why the correlation coefficient works the way it does. That is, how is it that the product of the *z*-scores is a number that tells us how well the points are lined up?

2. Have students explain—possibly with algebra—why the average of the product of *z*-scores cannot leave the interval $[-1, 1]$.

Creating Error

You will need
- CreatingError.ftm

In this activity you will look at how errors affect the way you interpret linear fits to data. First you'll look at some simulated data, then at data from various countries around the world.

GENERATE DATA

1. Open the Fathom document **CreatingError.ftm.** In this document is a collection of data from various countries. But first, you need to make your simulation. Drag a new case table from the shelf and add 100 new cases.

The function randomNormal (100,10) gives you a random number from a normal distribution mean 100 and standard deviation 10.

2. Create two attributes—*Cause* and *Effect*—with these formulas: For *Cause*, use randomNormal(100,10). For *Effect*, use **Cause**.

3. Make a scatter plot of *Effect* versus *Cause*, and choose **Graph | Least-Squares Line**.

Q1 What is the slope of the least-squares line? What is the value of r^2? Is this what you would expect?

INVESTIGATE

Error in Linear Data

You have set up a very simple situation. Now let's complicate it. So far, *Effect* has been perfectly related to *Cause*. Now you'll add error to *Effect* and see what happens.

4. Drag a slider from the shelf and name it *SD* (for standard deviation).

5. Make a new attribute, *Error*, and define it with the formula randomNormal(0,SD).

6. Now change the formula for *Effect* to **Cause+Error**. This adds a random, normally distributed amount to *Cause*. That random amount has mean 0 and standard deviation *SD*, in this case, 5.

Q2 Adjust the slider. What happens when *SD* is near 0?

You can increase the scale of SD by dragging on the axis in the slider.

Q3 As *SD* gets larger, what happens to r^2, the square of the correlation coefficient? What happens to the slope of the least-squares line? Explain why this happens.

7. Sometimes we get cause and effect mixed up. Make a second scatter plot. Put *Effect* on the horizontal axis and *Cause* on the vertical axis. Add the least-squares line.

Q4 Move the slider some more. What happens to the slope and the correlation coefficient in the new graph? Explain why the slope and the correlation coefficient behave differently in the other scatter plot.

Creating Error
continued

Error in Curved Data

So far you've been looking at basically linear data. Now you'll explore how measurement error can obscure the true shapes of functions. You will also see why residual plots are useful.

> Type time^2. The ^ will not appear—it just lets you type the exponent.

8. Make two new attributes, *Time* and *Value*. Give *Time* the formula $\frac{caseIndex}{100}$. Give *Value* the formula Time².

9. Graph *Value* as a function of *Time*. The graph you see should be familiar: It's the $y = x^2$ parabola between 0 and 1. It looks curved, obviously curved—for the moment. Now you will try to fit a line to it.

> You may want to stretch the graph vertically to make more room for the residual plot.

10. Add the least-squares line to the graph. Then choose **Graph | Make Residual Plot.**

Q5 Describe the trend in the function and in the residual plot between *Time* = 0 and *Time* = 0.5. Explain why this happens.

Right now, everything is nice and smooth. To shake things up, you'll add randomness as you did with the line.

> To change the axis to go from 0 to 1.2, change *Lower* to 0 and *Upper* to 1.2.

11. Double-click the *SD* slider near the axis to show its inspector. Change its axis to go from 0 to 1.2. Set the slider value to about 1.00.

12. Now change the formula for *Value* to Time²+Error. Choose **Graph | Rescale Graph Axes** to show all the data.

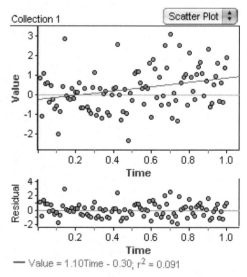

If everything has gone according to plan, you probably see that the graph seems to increase somewhat, but that *you can no longer tell that it's curved*. The randomness is so great that the curve is overwhelmed. You can confirm this by looking at the residual plot. It looks basically scattered and randomly distributed about the horizontal axis. That is the sign that the model—the straight line—is a good model for your data. Yet you know that the "real" data are curved. The question you will explore is: How much randomness can that curved model take before it no longer looks curved?

62   2: Relationships Between Data

Creating Error
continued

You can also rescale the graph by rechoosing **Scatter Plot** from the pop-up menu.

Q6 Adjust the *SD* slider to near 0 and rescale the graph. What do you notice about the original graph? About the residual plot?

Q7 Gradually increase the slider, rescaling the graph as necessary. What is the smallest value of *SD* that makes the residual plot look randomly scattered about the horizontal axis with no curvature?

Q8 Keep the slider at the value you found in Q7. Now repeatedly choose **Collection | Rerandomize.** (This makes Fathom reassign all the error values.) What changes in your graph when you rerandomize? Does the graph generally appear curved—or flat and random?

Q9 Readjust the slider, repeatedly rerandomizing, to find the largest value for *SD* you can use and *still see the curvature in the residual plot most of the time.* What value did you find?

Curvature and Error in Countries Data

Now you'll look at data on the birth rate and life expectancy for various countries.

13. Scroll down to the Countries collection. Make a scatter plot of *BirthRate* versus *LifeExp*. Add the least-squares line and a residual plot.

Q10 Describe the plot in terms of shape, trend, and strength. Describe the residual plot.

14. Select the collection and choose **Object | Add Filter.** Type Region=1. Decide whether a linear model seems suitable for Region 1.

You will need to rescale your graph each time.

Q11 Change your filter to examine Region 2, then Region 3. Which region's plot is most suited to a linear model? Which region seems the most curved?

15. Examine the scatter plot of the region you chose as most appropriate for a linear model. There appears to be an influential point. Select that point and choose **Edit | Delete Case.**

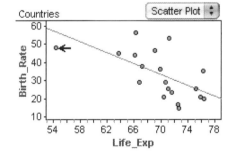

Q12 What effect did removing this point have on the slope of the line? On the correlation? In your opinion, is this point an influential point?

You might need to rescale your graph again.

16. Make a new attribute called *BR_Error*. Define this new attribute with the formula BirthRate+randomNormal(0,SD). In your plot, replace *Birth_Rate* with *BR_Error*.

Creating Error
continued

Q13 Adjust the slider, repeatedly rerandomizing. Try to find the smallest value for *SD* that introduces curvature in the residual plot most of the time. Is this possible? If so, what value for *SD* did you find?

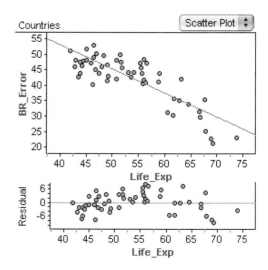

Q14 Can you find a value for *SD* that makes the slope of the line almost 0? If so, what is that value?

17. Readjust your slider back to 0 and change your filter to the region you said showed the most curvature back in Q11.

Q15 Adjust the slider, repeatedly rerandomizing, to find the largest value for *SD* you can use and *still see the curvature in the residual plot most of the time.* What value did you find, if any?

Q16 Discuss what you have learned in this activity.

EXPLORE MORE

1. Set *SD* to the critical value you found in Q9. Add 10 to the formula for *Time*. Sketch the new graph (including the residual plot) and explain why it looks so different.

2. Change the formula for *Time* so that the positions of the points are random within the interval [0, 1] instead of evenly spaced. What formula did you use? Does that make any difference in the amount of error you can put in and still detect the curvature?

3. Change the number of cases (by adding and deleting them) without changing the range of *Time* values. Does that make any difference in the amount of error you can put in and still detect the curvature?

4. If you're trying to detect an effect (for example, curvature), you have a better chance if your measurements have less error. This activity helped quantify the amount of error you can have and still detect the effect. Give some suggestions as to how you can improve your chance of detecting an effect if you're looking for one that's right on the edge of detectability.

Creating Error

Activity Notes

Objectives

- Understanding a common construction in statistics: that the observations consist of a signal plus an error term
- Understanding that adding error reduces the correlation coefficient. If the errors are much larger than the scale of the function, r^2 can appear to go to 0.
- Seeing that horizontal errors reduce the slope of the least-squares line, but vertical errors do not change it. This is because the residuals we look at to make the least-squares line are vertical. We can think of the least-squares line as trying to go through the mean value in each vertical stripe. That mean will be the same no matter how much error is added in.
- Realizing that random error can make it impossible to see an effect—in other words, that the shape of the original curve will be completely lost when the error is too great
- Testing the effect of error by simulating it; seeing that detecting the curvature in the data requires measuring to within about 0.1

Activity Time: 60 minutes

Setting: Paired/Individual Activity or Whole-Class Presentation (use **CreatingError.ftm** for either)

Statistics Prerequisites

- Familiarity with the least-squares regression line and correlation
- Familiarity with residual plots and how they help in determining the appropriateness of a linear model
- Familiarity with normal distributions, their center, shape, and spread

Statistics Skills

- Describing and comparing scatter plots, in terms of shape, trend, and strength
- Analyzing models using the residual plot
- Analyzing curvature and spread in bivariate data
- Finding an appropriate model for different groups of data
- Influential points and their effect on the slope and correlation
- Outliers in bivariate data

AP Course Topic Outline: Part I D

Fathom Prerequisites: Students should be able to make case tables, graphs, and sliders; and create attributes using formulas.

Fathom Skills: Students add least-squares lines and residual plots to plots, use the randomNormal function, rerandomize the collection, set the range for a slider, rescale a plot, and use filters.

General Notes: In this activity, students learn how error affects analysis by controlling the amount of error in their data and observing what features are visible in the data with different amounts of error. Finally, students will examine the relationship between the birth rate and life expectancy for various countries around the world. When the students use all the data, the scatter plot shows curvature, so they will use filters to find a region where a linear model looks appropriate and another region with obvious curvature. They will then introduce "error" to see if they can change the linear plot to make it show curvature and if they can change the plot with curvature into one where a linear model seems appropriate.

GENERATE DATA

Q1 The slope of the least-squares line is 1. The value of r^2 is 1.00. This is what you expect, because the graph is $y = x$ (Effect = Cause).

INVESTIGATE

Q2 When SD is near 0, the data are nearer the line $y = x$.

Q3 As SD gets larger, the error increases and points become more scattered about the original line, decreasing the r^2 value. The slope varies more because of sampling variation, but on average the values remain the same. On any given vertical swath of the graph, the mean value of Effect is the same.

Q4 The slope and r^2 both go to 0 as SD increases. The correlation coefficient is symmetrical, but the least-squares line is not. As SD increases, there is an increase in the mean value for Cause in a vertical swath where Effect is low. The reverse happens for large values of Effect, lowering the slope of the regression line.

Creating Error
continued

Activity Notes

Q5 Between *Time* = 0 and *Time* = 0.5, the function is increasing, but the residual plot is decreasing. This is because the points are decreasing relative to the least-squares line—they start out above it and end up below it.

Sample Graphs: All three have the same data—the function $y = x^2$—but each has a different amount of error added (see graphs below). The standard deviation (*SD*) of the first graph's error is 1.0, and the curvature is completely washed out. In the second, *SD* is only 0.15, and you can detect the curvature if you have a good imagination. Finally, at *SD* = 0.04—amazingly small—the curvature is obvious, especially in the residual plot. Notice the vertical scales on these graphs.

Q6 When *SD* is very close to 0, the graph and the residual plot will show the curve. Students must get the slider extremely close to 0 (less than 0.05) to see this. If they do not get the slider very close, they may see the curve as they adjust the slider, but lose it when they rescale the graph. This is a good opportunity to discuss the effects of scale on models of data. See Explore More 1 for more on this idea.

Q7 Students may be able to see the residual plot that is fairly randomly scattered about the horizontal axis at values of *SD* from 0.04 to 0.15. Students may have higher values at this step, but after rerandomizing they will need to lower their values.

Q8 The graph will change shape as the student rerandomizes. Some graphs may now appear flat and random, and students may need to adjust their values from Q7.

Q9 From 0.04 to 0.15, depending on the person viewing the graph.

Q10 The shape is basically curved with a downward trend. The strength is moderate and there looks to be some outliers in *Birth_Rate*. There appears to be clusters, both on the high end and on the low end. The residual plot shows curvature and also shows the clusters more clearly.

Q11 Region 1 shows the most curvature and does not appear to be linear. Region 2 is the region that is most appropriately modeled by a linear model because there is no obvious curvature in either the scatter plot or the residual plot. The residual plot also shows random scatter about the horizontal axis. Region 3's relationship is basically flat and nonexistent, with slope 0.0035 and correlation coefficient 0.004.

Q12 The slope changes from -1.50 to -2.06 and the correlation stays the same. So the point definitely influences the least-squares regression line but has no influence on the strength of the trend.

Q13–Q14 It is unlikely they'll find an *SD* that will make these data show curvature under repeated randomizations or that will make the slope flat all the

Q5

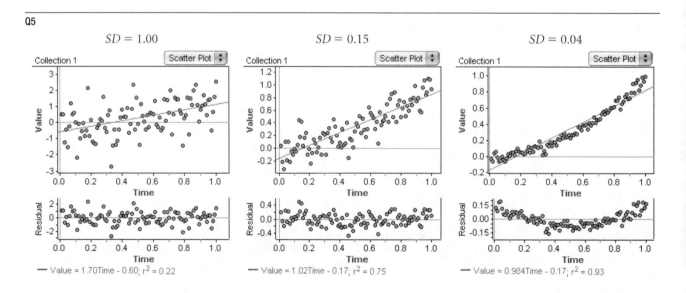

Creating Error
continued

Activity Notes

time. Once the *SD* is quite large (over 30), the line will flip back and forth from a positive slope to a negative one and will occasionally have a slope close to zero. But under repeated randomizations, it will not stay there.

Q15 Students will probably come up with values between 6 and 10 for the *SD*.

Q16 One conclusion is that it is very hard, if not impossible, to take data for which a linear model is appropriate and make them appear curved, whereas it is fairly easy to do the opposite, given a big enough error term. An interesting question is, then, what happens if you do a log or a log-log transformation and repeat this experiment? If the transformation straightens the data, will an "error" term unstraighten it?

DISCUSSION QUESTIONS

You should project **CreatingError.ftm** while you discuss these questions.

- Why is the correlation coefficient *r* the same no matter which attribute (*Cause* or *Effect*) is on which axis? [It's fundamentally different from the slope of the least-squares line. It's a measure of association between the attributes, not an analysis of how one depends on the other. The formula for the correlation coefficient is completely symmetrical.]

- When you see the two graphs and the two least-squares lines in step 7, do the lines go through the same points? [Here are two strategies for finding out: One is to select points in one graph and see that, for example, points on the line in one graph are off the line in the other; the second is to do the algebra on the two equations and show that when you solve for effect, the graphs are not the same.]

EXPLORE MORE

1. The scale of the graph changes, so the range of *y*-values is now 100 to 121, whereas the earlier graph had *y*-values from 0 to 1. The graph looks straight, but the residual plot will look similar.

2. Students can use the formula random(0,1) for *Time*. It shouldn't make any difference in the *SD* value.

3. By adding more cases, the curve can endure more error. The value for *SD* can be increased and the curvature will still be visible.

4. Including more cases can help reduce the effect of error on the curve. If you suspect curvature, increasing the range of the independent variable is very effective.

Nonlinear Transformations—Equilateral Triangles

You will need
• Triangles.ftm

How does the area of an equilateral triangle relate to the length of a side? You may know the formula for this relationship, but even if you do, your job in this activity is to derive it from measurements. Measurements always have some "measurement error" attached to them, so you will not get an exact rule from your data; however, you should get a good approximation of that rule.

COLLECT DATA

1. Open the Fathom document **Triangles.ftm.** You will see eight equilateral triangles on a square grid.

2. By counting squares, find the approximate length of a horizontal or vertical side of each triangle. Enter your values into the case table.

3. Find an approximation for the area of each triangle. Do this by counting the number of squares of the background grid that fall within each triangle. Make your own rule about what to do with partial squares. Enter these values.

INVESTIGATE

Fitting a Function

You may want to hide the picture of the triangles to make room for graphs. With the picture selected, choose **Object | Hide Picture.**

4. Make a scatter plot in which $(x, y) = (Side, Area)$. Add the least-squares line to the graph, then add a residual plot by choosing **Graph | Make Residual Plot.**

Q1 Describe the overall shape of the relationship (linearity, clusters, and influential points).

Q2 Is fitting a regression line appropriate for the pattern displayed in the scatter plot? Explain by inspecting the original plot as well as the residual plot.

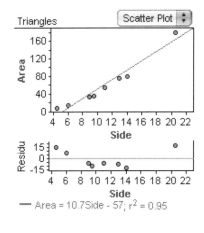

You should notice that your scatter plot is curved, or nonlinear, and that fitting a regression line is not appropriate for these data. You'll use Fathom to plot a nonlinear function through the points of the scatter plot. A slider will help you adjust the parameters of your function.

5. Drag a new slider from the shelf and rename it *a*.

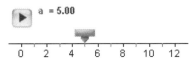

Nonlinear Transformations—Equilateral Triangles
continued

Some examples:
Side^a, a^Side,
a·Side^2,
a·Side^0.5,
a·Side^3.

You can also double-click on the slider to bring up the slider's inspector and change the Lower and Upper values.

6. Now you'll define a function. Make a new scatter plot in which $(x, y) = (Side, Area)$. Choose **Graph | Plot Function.** Type the expression for the right side of your function, using a as a parameter.

7. Adjust the slider to get a function that fits the raw data. If you want more precision with the slider, drag the numbers on the axis of the slider to change the range. If you want to try a different function, double-click the function name below the graph and edit its expression.

You can make a residual plot for a function plot to check how well your model fits the data.

Q3 What function did you plot, and what was the value of your slider parameter?

Transforming the Data

Now you'll experiment with some nonlinear transformations of the data.

To enter a formula into your case table, double-click the shaded cell below the attribute name and type the formula into the formula editor.

8. Define each new attribute using a formula and create a scatter plot for the relationship.

   a. *Area_Div_2* is *Area* divided by 2. Plot *Area_Div_2* versus *Side*.

   b. *Area_Root* is the square root of *Area*. Plot *Area_Root* versus *Side*.

   c. *Area_Squared* is the square of *Area*. Plot *Area_Squared* versus *Side*.

   d. *Side_Squared* is the square of *Side*. Plot *Area* versus *Side_Squared*.

Area_Div_2	Area_Root	Area_Squared	Side_Squared
$\frac{Area}{2}$	$\sqrt{Area}$	$Area^2$	$Side^2$

Q4 What effect does each of the transformations in step 8 have on the shape of your plot?

9. Pick the scatter plot(s) for which you think a line gives an appropriate summary. Fit a line through those points and make a residual plot.

Q5 Using the scatter plot as well as the residual plot, for which plot is fitting a regression line appropriate for the pattern displayed in the scatter plot? Explain your choice using both the scatter plot and the residual plot. What is the equation of the line you chose?

Q6 From the equation of the line of best fit in Q5, write a statement summarizing the relationship between the length of the side of an equilateral triangle and its area. Does this relationship correspond to the nonlinear function you plotted in step 6?

Nonlinear Transformations—Equilateral Triangles
continued

Q7 How does the slope of your line of fit in Q5 compare with the value of your slider in Q3? How close did you get?

Q8 There is more than one transformed data set in 8a–d for which a line fits well. Explain.

Powers and Logarithms

10. Drag a new slider from the shelf and rename it *b*.

11. Now you'll define another function. Make a new scatter plot in which $(x, y) = (Side, Area)$. Choose **Graph | Plot Function**. Enter the expression a·Side^b.

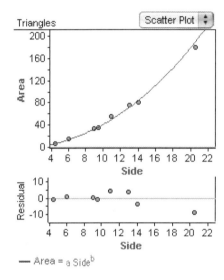

Again, you can make a residual plot for a function plot to check how well your model fits the data.

Q9 What happens when you move the slider for parameter *b*? Adjust both of the sliders to get a function that best fits the raw data. What are the values of your parameters?

12. Define new attributes with formulas ln(Area) and ln(Side) and plot *ln_Area* versus *ln_Side*.

Q10 Find the equation of the line of best fit, and write a statement summarizing the relationship between the length of the side of an equilateral triangle and its area. Does this relationship correspond to the nonlinear function you plotted in step 11?

Q11 How does the slope of your line of fit in Q10 compare with the value of parameter *b*? How does the constant of your line of fit in Q10 relate to the value of parameter *a*?

EXPLORE MORE

1. Use geometric reasoning to derive the relationship that should exist between *Side* and *Area*. How closely does this derived relationship correspond to the one you observe?

2. Scroll down in the Fathom document until you see the Black Crappies collection. Use what you learned in this activity to find an appropriate transformation for these data. (Black crappies are a type of fish.)

Equilateral Triangles

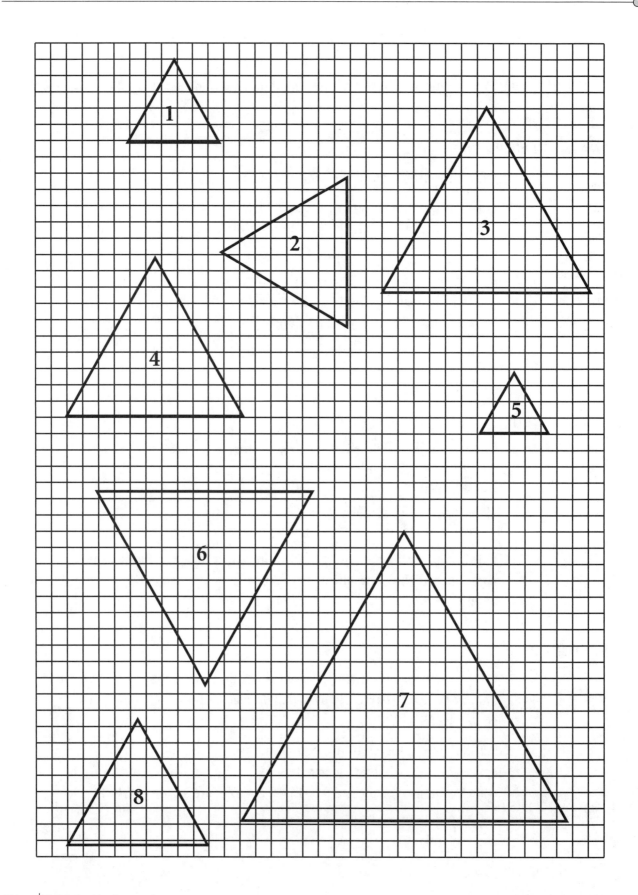

Nonlinear Transformations—Equilateral Triangles   Activity Notes

Objectives

- Seeing that nonlinear transformations, such as squaring, *do* change the basic shape of a plot and learning that this can sometimes lead to simplified statistical analyses
- Recognizing that area is a square measure and length is a linear measure, so one can be expressed as a linear function of the other by either squaring the side length or taking the square root of the area

Activity Time: 40–50 minutes

Setting: Paired/Individual Activity or Whole-Class Presentation (use **Triangles.ftm** for either)

Materials

- *Optional:* Equilateral Triangles worksheet

Statistics Prerequisites

- Attributes of linear data (shape, trend, and strength)
- The equation of a line: slope and *y*-intercept
- Familiarity with the least-squares line

Statistics Skills

- Comparing transformations of data
- Power transformations
- Log-log transformations
- Using residual plots to choose the best model
- Working with logarithms

AP Course Topic Outline: Part I D (1–5)

Fathom Prerequisites: Students should be able to make scatter plots and add least-squares lines.

Fathom Skills: Students use sliders, graph functions, transform data using formulas, and make residual plots.

General Notes: In this activity students fit power functions to raw and transformed data. They do this first in a geometric context and then to real-world data (Explore More 2). Fathom helps students quickly try and test many different functions and transformations.

Procedure: The Fathom document **Triangles.ftm** includes a picture of the triangles, but if students have trouble counting squares on the computer screen, use the Equilateral Triangles worksheet.

If your students are familiar with Fathom and with modeling data, you may want to have them explore the central problem of the activity on their own rather than use the student worksheet. Pose the problem and have them take measurements for *Side* and *Area*. They should first look at a linear model, and then use one or two sliders to find a power model. They should next experiment with transforming the data in various ways, particularly squaring *Area* and *Side*, taking the square root of *Area*, and doing a log-log transformation. They should explain which transformations work and why, how the transformations change the shape of the plot, and how the coefficients of the lines of fit for the transformed data correspond with the parameters of the fitted function. They can also look at the problems posed in the Explore More section on the student worksheet.

INVESTIGATE

Q1 The relationship appears slightly curved up. Triangle 7 is an influential point.

Q2 A line is not an appropriate model for these data—both the scatter plot and the residual plot show distinct curvature.

Q3 The function should be $a \cdot Side^2$. Students may need to move the slider around to actually see the function with their data. The parameter's value should be close to 0.433.

Q4 Taking the square root of *Area* (8b) and squaring *Side* (8d) each result in linearized data. Dividing *Area* by 2 (8a) does not change the shape, it merely rescales the vertical axis. Squaring *Area* (8c) increases the curvature.

Q5 Either 8b or 8d is appropriate. Sample equations: (8b) $Area\_Root = -0.21 + 0.67\,Side$; (8d) $Area = -0.75 + 0.43\,Side\_Squared$. The true geometric relationship is given by $Area = 0.5 \sin(\pi/3) \cdot Side^2$, or $Area \approx 0.5 \cdot (0.866) \cdot Side^2 = 0.433 \cdot Side^2$. So the approximate relationship found by measuring and fitting a line is close to the truth.

Q6 The area of an equilateral triangle is approximately 0.433 times the square of the length of a side, or the square root of the area is approximately 0.667 times the length of a side. The constant terms in the

Nonlinear Transformations—Equilateral Triangles
continued

Activity Notes

regression lines should be zero but do not turn out that way because of the variation in the sample data. You could force a regression line through zero in these cases (select the line and choose **Graph | Lock Intercept At Zero**) and see what happens to the slope.

Q7 The values should be fairly close to each other.

Q8 Because area is a square measure and length of a side is a linear measure, one can be expressed as a linear function of the other by either squaring the side or taking the square root of the area. So a line fits both 8b and 8d well.

Q9–Q11 The slope of the fitted line is the power of the power model, b, and should be close to 2. The y-intercept of the fitted line is related to the coefficient a, namely, $a = e^{\wedge}(y\text{-intercept of the fitted line})$.

There will be differences between the model students get using 8b or 8d and the model they get using logs. Encourage them to think about why this might be so.

DISCUSSION QUESTIONS

- Were there any outliers? What constitutes an outlier in this situation?
- How did you choose a function to fit the raw data? How did you use the slider as a parameter, and what value did you get for it?
- Which combinations of attributes produced roughly linear graphs? Why does that make sense?
- Is there any difference between fitting the raw data with a curve and transforming the data and fitting with a straight line?

EXPLORE MORE

1. The true geometric relationship is given by
$$Area = \left(\frac{1}{2}\right)\sin\left(\frac{\pi}{3}\right) \cdot Side^2 \approx 0.433 \cdot Side^2$$

2. The untranslated data clearly have curvature. Both the log transformation and the log-log transformation straighten the pattern considerably. There are so few points that it is difficult to choose between them on the basis of the residual plots. However, because there is reason from the physical situation to believe the relationship is close to cubic, students should use the log-log transformation. The regression equation is $\ln(weight) = -6.17 + 2.51 \cdot \ln(length)$. The estimated slope is 2.51, which gives the power model $weight = e^{-6.17} \cdot length^{2.51}$.

EXTENSION

Have students repeat the entire activity using the radii and areas of circles. They should be able to come up with an estimate for a famous constant.

Modeling Data—Copper Flippers

You will need
- **CopperFlippers.ftm**

A certain insect, the "copper flipper," has a life span determined by the fact that there is a 50-50 chance that a live flipper will die at the end of the day. (The name of the insect comes from its remarkable resemblance to a penny.) So, on average, half of any population of copper flippers will die during the first day of life. Of those that survive the first day, on average half die during the second day, and so on.

In this activity you will create a Fathom simulation of the life and death of 200 copper flippers. You'll then find an appropriate model for their survival rate.

GENERATE DATA

In this document there is a collection of data about different countries that you will explore later.

1. Open the Fathom document **CopperFlippers.ftm** and make a new case table.

2. In the case table, create the attribute *Survives* and name the collection Copper Flippers.

3. You want the original collection of copper flippers to contain 200 cases. Choose **Collection | New Cases** and add 200 cases.

You can enter a formula by showing the formula row (**Table | Show Formulas**), by selecting the attribute and editing the formula (**Edit | Edit Formula**), or by double-clicking the collection and entering it on the **Cases** panel of the inspector.

4. Define *Survives* with the formula randomPick("Yes", "No"). The value of each case now indicates whether or not the particular copper flipper survived to the end of the day.

5. Make a new graph with *Survives* on the horizontal axis.

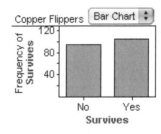

6. To find the number of copper flippers that survive, move the cursor over the "Yes" bar and look in the lower-left corner of the Fathom window.

    107 cases (53.5%) are Yes.

7. You'll record the results from each day in a second collection. Make a new collection called Results. Make a case table for Results and define two attributes, *Day* and *Survivor*. Enter the number of Day 1 survivors.

Modeling Data—Copper Flippers
continued

Now it's time to simulate how many copper flippers survive on Day 2, Day 3, and so forth.

8. Click on the "No" bar of the bar chart to select those cases. Choose **Edit | Delete Cases.** This leaves only the survivors from Day 1. Fathom will automatically rerandomize the data. You should see two bars again. These represent the survival outcomes of Day 2.

9. Enter the Day 2 survivors in your Results collection.

10. Repeat steps 8 and 9 until you have fewer than 5 survivors, but stop before you get to 0 survivors.

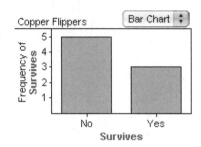

Q1 Make a scatter plot with *Day* on the horizontal axis and *Survivor* on the vertical axis. Describe the overall shape of the relationship (linearity, clusters, and influential points). Does the pattern look linear?

Q2 Add the least-squares line to the graph and make a residual plot. Is fitting a regression line appropriate for the pattern displayed in the scatter plot? Explain by inspecting the original plot as well as the residual plot.

INVESTIGATE

Transforming the Flippers

11. Define two new attributes in Results, *Ln_Survivor* and *Ln_Day*, using formulas as shown.

12. Make each new scatter plot and its residual plot.

    a. A log transformation: *Ln_Survivor* versus *Day*.

    b. A log-log transformation: *Ln_Survivor* versus *Ln_Day*.

Q3 What effect does each of these transformations have on the shape of your plot?

Q4 For which transformation is fitting a regression line appropriate for the pattern displayed in the scatter plot? Explain your choice using both the scatter plot and the residual plot. What is the equation of the line for the plot you chose?

Q5 From the equation of the line of best fit in Q4, write a statement summarizing the relationship between *Survivor* and *Day* for your data.

Modeling Data—Copper Flippers
continued

Finding an Exponential Model

13. Drag two new sliders from the shelf and rename them *a* and *b* respectively.

Choose **Graph | Plot Function.** Type a·exp(b·Day).

To get more precision, drag the slider axis to change the range.

14. On your original scatter plot of *Survivor* versus *Day*, plot the function $a \cdot e^{b \cdot Day}$.

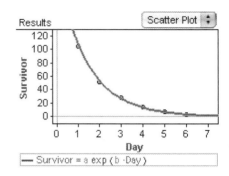

Q6 Adjust both of the sliders to get a function that best fits the raw data. What are the values of your parameters?

Q7 How does the slope of your line of fit in Q4 compare with the value of parameter *b*? How does the *y*-intercept of your line of fit in Q4 relate to the value of parameter *a*?

EXPLORE MORE

1. Scroll down until you see the collection labeled Countries. Experiment with pairs of different attributes. Pick two attributes from the collection and use what you learned in this activity to find an appropriate transformation for the attributes you picked.

2. Start with 1000 copper flippers instead of 200. Predict how many days you think it will take before there are fewer than 5 left. Then use Fathom to test your prediction.

3. Figure out a way to redo the simulation, assuming that only 30% of the copper flippers die each day. Predict beforehand how many days it will take to get down to fewer than 5. This incomplete formula will help.

$$\text{if}(\text{random}() < 0.3) \begin{cases} ? \\ ? \end{cases}$$

4. Figure out a way to transform the graph of *Survivor* versus *Day* so that the data lie on a straight line. Find the regression line and compare it with your nonlinear function in Q6.

Modeling Data—Copper Flippers

Activity Notes

Objectives

- Exploring nonlinear data and transformations, specifically exponential data
- Seeing that an exponential relationship can be straightened by transforming the y-values by taking the logarithm
- Working with different types of log transformations and exploring which transformation fits the data based on the residual plot
- Experimenting with sliders to find the connection between the slope and y-intercept of the fitted line and an exponential function's coefficient and exponent

Activity Time: 40–50 minutes

Setting: Paired/Individual Activity (use **CopperFlippers.ftm** and build simulation) or Whole-Class Presentation (use **CopperFlippersPresent.ftm**)

Materials

- *Optional:* 200 pennies for each group of students

Statistics Prerequisites

- Attributes of linear data (shape, trend, and strength)
- The equation of a line: slope and y-intercept
- Familiarity with the least-squares line
- Residual plots

Statistics Skills

- Comparing transformations of data
- Log and log-log transformations
- Using residual plots to choose the best model
- Working with logarithms

AP Course Topic Outline: Part I D

Fathom Prerequisites: Students should be able to make plots and add least-squares lines and residual plots; and make case tables and enter data.

Fathom Skills: Students use formulas to simulate a random event, work with categorical variables, use sliders, plot functions, and transform data using formulas.

General Notes: This activity introduces students to the use of Fathom for simulation. It helps students learn how to use random value generators as an alternative to hands-on probability experiments and allows them to try and test many different models and transformations.

Procedure: Students create a Fathom collection of 200 copper flippers whose survival is determined by a random function. The instructions on the student worksheet are step by step and self-contained. If you don't wish to have students build the simulation, you can use **CopperFlippersPresent.ftm** for a whole-class discussion. This document contains everything students make in the activity. Students could also use 200 coins as described below and then compare those results to their simulated results.

1. Place your pennies in a cup, shake them up, and toss them on a table. Record the number of heads.

2. Set aside the pennies that came up tails. Place the pennies that came up heads back in the cup and repeat step 1.

3. After each toss (day), set the tails aside, collect the heads, count them, and toss them again. Repeat this process until you have fewer than 5 heads left, but stop before you get to 0 heads.

GENERATE DATA

Q1 The pattern appears to be exponential, not linear. It will appear that the data could be modeled as exponential decay. Any one copper flipper, chosen at random, has a certain probability of death in a given time period. If this probability remains relatively constant across time periods, then the number of copper flippers "dying" in a time period will be a fixed proportion of the number "alive" at the beginning of the time period. In this example, approximately half of the copper flippers "die" in any one time period.

Q2 The residual plot should show curvature and a pattern, which suggests that a linear model is not appropriate for these data. We are looking for a residual plot with random scatter about the line $y = 0$.

INVESTIGATE

Q3 Although some students' log-log transformation might look like it straightens the data, the residual plot for that transformation will still show a pattern.

Modeling Data—Copper Flippers
continued

Activity Notes

The log transformation straightens the data and shows a residual plot with little pattern, provided the students' data are relatively well-behaved. Sample plots:

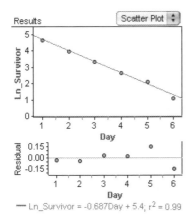

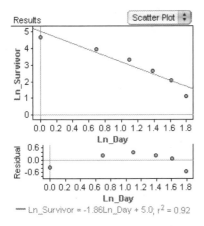

On the top scatter plot it looks like the transformed points lie very close to a straight line. The residual plot reveals, however, that there is a regular oscillation around that line, as is often the case with time-series data. The exponential decay model is still very good, however, and about the best we can do with these data. The bottom scatter plot shows definite curvature, in both the scatter plot and the residual plot. This curvature show that the log-log transformation is not appropriate for these data.

Q4 The equation for the students' fitted line should have a slope around ln(0.5), or -0.6931, and a y-intercept close to ln(200), or 5.2983.

Q5 Their equation should be close to $\hat{y} = 200 \cdot (0.5)^x$. For the graph in Q3, the equation is $Survivor = (e^{5.36})(e^{-0.6868x}) = (212.72)(0.503)^{Day}$. In the situation of exponential decay, the 0.502 is (1 − the rate of decay). Therefore, the rate of "death" for the copper flippers is estimated to be 0.498, or 49.8%. About half die in each time period.

Q6 The parameter a needs to be as close as possible to the value $e^{y\text{-intercept}}$, and the parameter b needs to be as close as possible to the slope of the students' fitted line. So, for the sample data, $a \approx e^{5.36} \approx 212.72$ and $b \approx -0.6868$.

DISCUSSION QUESTIONS

- Why does the plot of *Survivor* versus *Day* have the shape it does?
- What parameters determine the shape of this curve?
- What equation should fit through these points?

EXPLORE MORE

1. The Countries collection has many different possible relationships to explore. Most of the relationships between any two attributes are not linear and will require a transformation. Students could also explore how their models change if the data are restricted, for example, first looking at GNP versus Life Expectancy, then looking at GNP versus Female Life Expectancy or Male Life Expectancy.

3. The complete formula is

$$\text{if(random}(0, 1) < 0.3) \begin{cases} \text{"No"} \\ \text{"Yes"} \end{cases}$$

3

Collecting Data

Importing Data—U.S. Census

You will need
- access to the Internet

Since 1790, the U.S. Census Bureau has conducted a thorough survey of the United States population once every ten years. The first censuses were primarily concerned with counting the number of people so that the federal government could make decisions about representation and taxation. Today, the U.S. Census asks people to self-identify their age, sex, race, national origin, marital status, and education, among other things. The most detailed information published by the U.S. Census Bureau is called *microdata,* or data about individuals.

Using Fathom, you can import samples of census microdata from 1850 to 2000. The attributes you get depend on the questions asked in a particular year. You can use these data to explore many political and demographic trends in the United States. In this activity you'll look first at racial diversity, then at school attendance.

GENERATE DATA

1. In a new Fathom document, choose **File | Import | U.S. Census Data.** Read the help message in the inspector. Select Attributes and read the help for that, too.

The bottom section of the panel shows the current request: By default, this is a sample of people from all over the United States (because no state or metropolitan area is selected) from the 2000 census. Note the attributes requested.

2. Expand the Attributes list. Click Year and Location, and read the attributes.

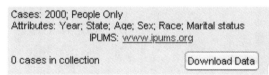

3. Move your cursor over *Urban or Rural.* A pop-up description and the status bar at the lower left of the Fathom window both show information about the attribute that describes the attribute and tells the years for which it is available:

> Urban is a place with >2500 people. 1850-1920, 1960, 1970, 1990.

Expand the Person list. Skim through the list of attributes, clicking on headings of interest in the left pane and reading about the attributes in the right pane. For now, you'll keep the default request but add one more attribute.

4. In Education, check *School Attendance,* then click **Download Data.**

Fathom connects to the Internet and submits your request to IPUMS (Integrated Public Use Microdata Series, at the University of Minnesota), which has a searchable database of census microdata samples. Fathom decodes and imports the results into a collection. (If left coded, all data would be in the form of numbers, rather than, for example, "male" and "female.")

Importing Data—U.S. Census
continued

INVESTIGATE

Racial Diversity

First you'll explore the racial diversity of the United States as a whole and compare that with diversity in specific regions. You'll begin with this first sample.

*Go to the **Cases** panel of the inspector to see the attribute names.*

5. Make a graph of *Race_General* by dropping it on the vertical axis (it has long values, so it will be easier to read that way).

Q1 Describe the distribution of *Race_General* for your sample.

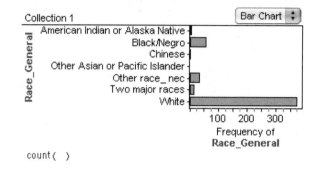

There are many ways to quantify racial diversity. One way is to look at the ratio of cases in the sample with particular races to the total number of cases, and to compare those ratios. Another is to look at the number of counted categories of race that have members appearing in the sample.

6. Make a new summary table, and drop the collection's name in it.

The collection has been connected to the summary table, so it "knows" about the collection and its attributes.

You can find Proportion in Functions: Statistical: One Attribute.

7. Double-click the formula to show the formula editor, and delete the existing formula. Enter a formula similar to the one shown here.

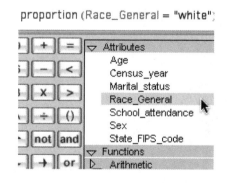

You have told Fathom to calculate the ratio of cases with a particular race to the total number of cases. This expression illustrates a common formula in Fathom, where you specify for *which* category you want something calculated.

Now you want Fathom to calculate how many counted categories of race are in this sample.

8. With the summary table selected, choose **Summary | Add Formula.** Enter the formula uniqueValues(Race_General).

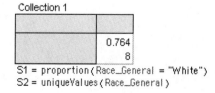

84 | 3: Collecting Data

Importing Data—U.S. Census
continued

Q2 What can you conclude from the proportions in your sample? How many categories do you have in your sample?

Your numbers may be a bit different from those of others in your class. When IPUMS has more cases available than you ask for, you get a simple random sample. Try downloading data several times to see how much the numbers change.

You now have a rough idea of the racial diversity for the United States as a whole. To compare this to the diversity in one area, you need to replace the data. First, you need a record of the current values.

9. Make the summary table a good size—as small as possible but still showing all the information. Then select the summary table and choose **Edit | Copy As Picture.** Click in a blank place in the document, then choose **Edit | Paste Picture.**

This isn't a live summary table and won't change when you change the data. (You might also want to have a picture of the graph of race.)

Now you will change the request from its default of all of the country.

10. In the inspector, go to the **Microdata** panel. Expand the Choosing Cases list and click States or Metropolitan Areas. Pick one area by clicking its checkbox and click **Download Data.** The data fill whatever graphs or other objects you have (except pictures) with the new sample.

Q3 Is this area more or less racially diverse than the country as a whole?

School Attendance

You could continue the exploration of racial diversity, looking at different states and metropolitan areas to find the most and least racially diverse places in the United States. But let's move on and look at some of the other attributes.

11. Make a graph with *Age* on the horizontal axis and *School_attendance* on the vertical axis.

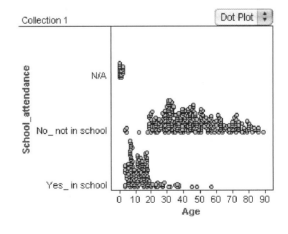

Q4 Who's the oldest person in school now? What does N/A mean for this attribute?

Let's look at the age range of students and the rate of school attendance over the years.

Importing Data—U.S. Census
continued

12. Go to the **Microdata** panel of the inspector. Click the Years heading in the Choosing Cases list. Check the boxes for 1850, 1900, 1940, 1970, and 2000, and submit the request.

13. When the data come in, graph *Census_year*. Notice that you don't get a dot plot; you get a bar chart, instead. Fathom is treating this attribute as categorical.

14. Rearrange the bars in chronological order by dragging labels to the correct position. Now, whenever you graph *Census_year*, the years will be in chronological order. Change the graph to a ribbon chart.

15. Make another graph of *School_attendance*. Select the "Yes_in school" bar. Notice the pattern of selection in the other graph. Within each vertical slice representing year, the proportion of people in school is colored red.

16. Drop *School_attendance* in the middle of the ribbon chart of year. You now have a time series display. The vertical bands are the census years, and the legend patterns show changes in proportions of the population that are in school for that census year.

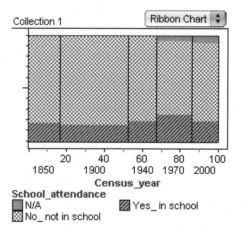

17. Make a new graph of *Census_year* versus *Age* and change it to a histogram.

Q5 What do these graphs tell you about schooling over time in the state you chose?

If you select groups of cases in one graph, they will highlight in the other graph.

Importing Data—U.S. Census

Activity Notes

Objectives

- Choosing an appropriate and meaningful way to display data
- Learning how to work with distributions, either with proportions or with shape, center, and spread, based on their plots
- Discovering some of the differences between categorical and quantitative data and how to work with the differences

Activity Time: 30–50 minutes depending on Internet connection

Setting: Paired/Individual Activity or Whole-Class Presentation (download data and analyze for either)

Materials

- Access to the Internet

Statistics Prerequisites

- Making and analyzing one-variable graphs
- Familiarity with the mean and median
- Familiarity with describing and comparing graphical distributions

Statistics Skills

- Working with categorical variables
- Recognizing different representations of data
- Calculating appropriate summary statistics
- Sampling
- Making appropriate plots
- Cases versus variables

AP Course Topic Outline: Part I; Part II A, D

Fathom Prerequisites: Students should be able to make graphs and summary tables.

Fathom Skills: Students use the uniqueValues function, calculate proportions and column proportions, copy and paste summary tables and plots for future comparisons, create a new collection by random sampling, download data from the Internet, and choose attributes of interest.

General Notes: This activity is adapted from Tour 3 of the Fathom *Learning Guide*. It is an introduction to using Fathom to import samples of census microdata from 1850 to 2000 from the U.S Census Bureau. Census microdata (data about individual Americans), has great immediacy for students, and Fathom makes it easy to explore and analyze this data. Little to no knowledge of Fathom or statistics is required. *Note:* For this activity to work, your computers must be connected to the Internet. If you've done the normal installation of Fathom, you will have the files you need in place. (If not, make sure the Fathom application folder contains the file **IPUMS_USA_InterfaceSpec.xml** in the **Helpers | ImportSpecs** folder.)

Procedure: Try the first part of this activity before giving it to your students so that you will know, based on the speed of your machines and Internet connection, how long downloading will take for your students. If it's too slow for your students to download 500 cases, have them change the sample size before they download the data. On the **Microdata** panel, expand the Choosing Cases list, click Sample Size, and change it to something smaller. You may need to experiment beforehand to find a sample size that works best with your computers and Internet connection. Then students can proceed with the activity as written.

INVESTIGATE

Q1–Q2 Whites will make up a substantial majority of the cases in this sample of the country as a whole. The other races counted in the census will vary in terms of their proportions.

Q3–Q5 Answers will vary depending on the state chosen and random sample selected.

ADDITIONAL QUESTIONS

This activity is aimed mainly at introducing your students to downloading samples from the Internet. Here are some other questions you can pose to your students with these data.

1. Analyze the distribution of *Age* for your sample.

    a. Describe the shape of the graph and any important features.

    b. How many people in your data set are between 15 and 20 years old?

    c. How old is the oldest person in your file? What kind of graph did you use?

Importing Data—U.S. Census
continued

Activity Notes

   d. What is the median age of the people in your file? Again, what kind of graph did you use?

   e. What is the median age for males? For females?

2. Consider the distribution of the categorical variable *Marital_status*.

   a. What observations can you make about the people in your collection who have never married?

   b. How many people in your data set are married?

3. Make a graph of *Wage_and_salary_income*.

   a. Are the richest people predominantly from one group? That is, are they mostly male? Are they mostly older? Look for patterns. What do you find?

   b. Decide how much income will define what you mean by *rich*. How many "rich" people are there in your data set? What percentage of the population is "rich"? How do you know?

   c. What is the mean income for elderly women? For elderly men?

   d. Filter out retired people as well as children. What is the median *Wage_and_salary_income*? The mean *Wage_and_salary_income*?

4. For your state, what is the most common method of travel to work? How does that compare to the country as a whole?

DISCUSSION QUESTIONS

- Census data are collected to help people make decisions. Why might someone be interested in looking at school attendance over time? What are some questions that might be informed by an analysis of racial diversity?

- What proportions did you use to look at racial diversity? What are some other proportions that could be used? Other measures?

- The census has changed the questions and categories it uses over the years. It has also been criticized for not having methods that result in a complete count of the population. Did anything surprise you about these data? What concerns might you have in drawing conclusions from them?

EXTENSIONS

For more project ideas using census microdata, see *Data Are Everywhere: Project Ideas for Fathom*.

1. Have students download data for California in the years 1850 and 2000, then make a ribbon chart with *Census_year* on the horizontal axis and *Sex* in the middle. What's going on here? (*Hint:* Who came to California before 1850? *Note:* The 1850 census did *not* count Native Americans.) They can verify their hypothesis by getting more data, such as *Occupation*. How long did it take for the sexes to even out? (Find out by getting some years in between.)

2. Students could explore the idea that people now move around more than did people in the past. (One way to do this is to make an attribute with a formula that compares people's current state with their birthplace, such as

$$\text{if (Birthplace\_General = State\_FIPS\_code)} \begin{cases} \text{"Same State"} \\ \text{"Moved"} \end{cases}$$

3. You can access many other data sources on the Internet by dragging the URL from your Web browser into a Fathom document or by choosing **File | Import | Import From URL**. However, some data will not import correctly or will require tweaking to format properly. For example, try importing data from www.statsci.org/data/oz/sydrain.txt. These data are from OzDASL, the Australasian Data and Story Library. This is a library of data sets and associated stories intended as a resource for teachers of statistics with emphasis given to data sets with an Australasian context. This particular data set is the annual maximums of daily rainfall recorded over a 47-year period in Turramurra, Sydney, Australia. For each year, the wettest day was identified (that having the greatest rainfall). The data show the rainfall recorded for the 47 annual maximums.

Sampling Bias—Time in the Hospital

You will need
• HospitalSim.ftm

In this activity you'll estimate the average length of stay in a five-bed hospital ward. You'll use a pre-made Fathom simulation of this hospital ward. In the simulation, each bed gets filled with a patient whose length of stay is chosen at random from between 1 and 10 days. You'll estimate the average length of stay from a sample of five patients.

EXAMINE DATA

1. Open the Fathom document **Hospital Sim.ftm.** You'll see a case table that is similar, but not identical, to the one shown here. Each column (attribute) corresponds to one bed in the ward. Look, for example, at *Bed2Stay.* The first 2 in that column stands for a person who comes to that bed on the 1st day and stays for 2 days. The second 2 in that column represent a continuation of that person's stay. Continuing down, we have seven 7's representing a person who stays 7 days.

Days

Day	Bed1Stay	Bed2Stay
1	1	5 d	2 d
2	2	5 d	2 d
3	3	5 d	7 d
4	4	5 d	7 d
5	5	5 d	7 d
6	6	9 d	7 d
7	7	9 d	7 d
8	8	9 d	7 d
9	9	9 d	7 d
10	10	9 d	10 d

2. To see this visually, make graphs for two beds, as shown.

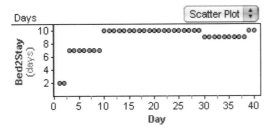 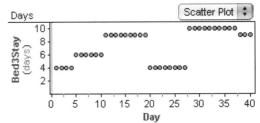

You can also choose **Collection | Rerandomize.**

3. To get a different set of data, drag the lower-right corner of the Days collection until you can see the **Rerandomize** button. Click **Rerandomize** several times and notice what happens in your plots.

Q1 Choose one of your plots ("beds") and estimate the average length of stay for that particular bed.

INVESTIGATE

A "Random" Choice of Day

One way to estimate the average length of stay is to compute the average length of stay for the five patients in the beds on each day, then choose a random day as typical.

Exploring Statistics with Fathom 3: Collecting Data 89
© 2007 Key Curriculum Press

Sampling Bias—Time in the Hospital
continued

Choose **Table | Show Formulas.** Double-click the shaded formula cell.

4. In the case table, define a new attribute, *AverageStay*, with the formula

$$\frac{Bed1Stay+Bed2Stay+Bed3Stay+Bed4Stay+Bed5Stay}{5}$$

Choose **Line Plot** from the pop-up menu.

5. Make a line plot of *AverageStay*.

Q2 What does each dot in the plot represent?

You can also rerandomize by pressing Ctrl+Y (Win) ⌘+Y (Mac).

6. Choose **Collection | Rerandomize.** Notice that the case table changes and you get a new plot of *AverageStay*. Do this several times to see how the randomized data change.

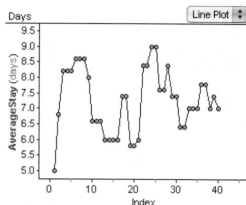

You probably notice that there is nothing special about any given day. In fact, you are going to treat the 40th day as a typical day and record the value of *AverageStay* over and over again as Fathom rerandomizes.

7. To record the value of *AverageStay* on the 40th day, you need a measure. Double-click the Days collection to show its inspector. Go to the **Measures** panel and define a new measure, *AnAverageStay*, using the formula last(AverageStay). You should see that the value of *AnAverageStay* is, in fact, the last value in the *AverageStay* column of the case table.

Measures from Days

8. Select the Days collection and choose **Collection | Collect Measures.** By default, Fathom rerandomizes Days five times and places the values for *AnAverageStay* in a new collection named Measures from Days.

9. Show the inspector for Measures from Days. Go to the **Collect Measures** panel and enter these settings. Click **Collect More Measures.**

☐ Animation on
☒ Replace existing cases
☐ Re-collect measures when source changes
⊙ 100 measures

10. Make a dot plot of the 100 averages: Make a new graph, go to the **Cases** panel of the inspector for Measures from Days, and drag *AnAverageStay* to the horizontal axis of the graph.

Q3 Sketch the dot plot. Where is the distribution centered? Is there much variability in your estimates?

Q4 If each patient's length of stay is chosen at random from between 1 and 10 days, what is the expected average length of stay for the whole population?

Q5 Compare your results from Q3 and Q4. Are your estimates clearly too low, clearly too high, or about right?

Sampling Bias—Time in the Hospital
continued

Q6 You are trying to estimate the average length of stay of a patient. In this simulation, did every possible patient's length of stay (1–10) have an equal chance of being in the sample? If so, explain. If not, which patients and their lengths of stay had the greater chance of being chosen?

A Random Choice of Patient

Another approach is to compute the average length of stay for any five patients, each assigned a length of stay chosen at random from 1 to 10.

11. Make a new collection and name it Patients. Add five cases to the collection by choosing **Collection | New Cases.** Type 5 and click **OK.**

12. Double-click the Patients collection to show its inspector. On the **Cases** panel, define an attribute with the formula shown.

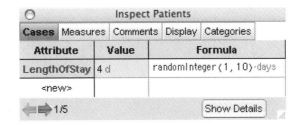

13. On the **Measures** panel, define a measure with this formula.

Q7 Rerandomize the Patients collection a few times and write down the values of *AverageLengthOfStay*. If you collect a lot of values, what do you think the distribution of these values will be (in terms of shape, center, and spread)?

14. Select the Patients collection and choose **Collection | Collect Measures.** By default, Fathom rerandomizes Patients five times and places the *AverageLengthOfStay* measure in a new collection named Measures from Patients. Use the inspector for Measures from Patients to change the settings to 100 and collect more measures.

To make your dot plot, drag *AverageLengthOfStay* from the Measures from Patients inspector.

Q8 Make a dot plot of the 100 averages and sketch your plot. Where is this distribution centered?

Sampling Bias—Time in the Hospital
continued

Q9 The average length of stay for the whole population is 5.5 days. (Is this what you got in Q4?) Are your estimates clearly too low, clearly too high, or about right?

Q10 Compare the dot plots from Q3 and Q8. Are there any similarities? What are the main differences? How do you account for these differences?

EXPLORE MORE

1. Your sampling method in steps 4–9 is biased because it tends to give samples that are not representative of the population. Would the method be more or less biased if the maximum length of stay were 20 days instead of 10? Explain your prediction, being careful to define what you mean by "more" or "less" biased. Look at the formulas used to define the attributes in the Days collection and see if you can modify the simulation to test your prediction.

2. Suppose that instead of using the patients on the 40th day as your sample, you use the patients on the 1st day. Predict the results, try a simulation in Fathom, and compare and contrast the results with those from the previous simulation.

Sampling Bias—Time in the Hospital

Activity Notes

Objectives

- Being introduced to the concept of *bias* and understanding that bias is in the method of the sampling, not the result
- Understanding that a biased sampling method tends to result in a nonrepresentative sample, but not always
- Predicting and explaining the *direction* of the bias—for example, that the method of sampling introduced in Random Choice of Day results in longer stays being chosen more often than shorter stays

Activity Time: 40–50 minutes

Setting: Paired/Individual Activity or Whole-Class Presentation (use **HospitalSim.ftm** for either)

Statistics Prerequisites

- Making dot plots
- Familiarity with the mean
- Familiarity with describing and comparing graphical distributions

Statistics Skills

- Sampling bias
- Different methods of sampling
- Different methods in finding averages
- Sampling
- Calculating summary statistics
- Comparing distributions

AP Course Topic Outline: Part I B (1, 2), C; Part II B

Fathom Prerequisites: Students should be able to create formulas in a case table and make graphs.

Fathom Skills: Students calculate averages, make line graphs, define attributes with the randomInteger function, create a new collection based on random selection, and define and collect measures.

General Notes: This Fathom activity uses a pre-made simulation. Using the pre-made document has the advantage of minimizing setup time. Another advantage is that the simulation can be repeated any number of times at the click of a mouse.

Procedure: Students work with the Fathom document **HospitalSim.ftm**. It has a collection, Days, with a case table that shows the first 9 days. The case table extends to the 40th day. The mechanics of how this simulation is constructed need not concern you or the students, but being convinced that the simulation works properly is important. The student worksheet attempts to do that in steps 2 and 3. (*Note:* Not all attributes are displayed in the case table, deliberately so, to keep things as simple as possible.)

EXAMINE DATA

Q1 The average length of stay for Bed 2 in the given plot is $\frac{2 + 7 + 10 + 10 + 9 + 2}{6 \text{ patients}}$, or 6.67 days.

INVESTIGATE

Q2 Each dot represents the average length of stay for a particular day.

6.–10. For some students, these steps may seem contrary to selecting a random day. You may need to point out that although they are always using the 40th day, rerandomizing the data in Fathom produces random results for this one day. It is equivalent to randomly choosing any day for a set of data that is static.

Q3 Plots will probably be centered around 7 with not much variability.

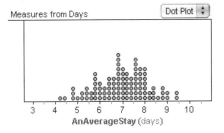

Q4–Q6 The population average is $\frac{(1+2+\cdots+10)}{10} = 5.5$ days, so the estimates are clearly too high. The sampling method in steps 6–10 is more likely to choose long stays than to choose short stays. A stay of 7 days had seven times the chance of being chosen as a stay of 1 day. This results in an estimate of an average length of stay that is too long.

Sampling Bias—Time in the Hospital
continued

Activity Notes

Q7–Q9 The dot plots for this sampling method should be centered around 5.5 with more variability. So these estimates are about right.

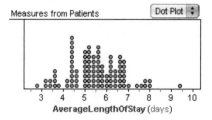

Q10 The most important difference between the plots is that the averages for the random samples of patients seen in the plot for Q8 tend to cluster near the population mean of 5.5, whereas the samples from the biased method, seen in the plot for Q3, tend to have averages that are too large. A second difference is that the averages from the random choice of patient have more variability than those from the biased method, because you are likely to get both short and long stays in a random sample.

DISCUSSION QUESTIONS

- Exactly what are you trying to estimate? Considering this, which sampling method (Choice of Day or Choice of Patient) gives the most accurate insight into the average length of stay?
- Suppose the hospital records the lengths of stays as patients are discharged from the ward. If you take a random sample of these numbers, which sampling method (Choice of Day or Choice of Patient) would produce similar results?

- What question does Choice of Day's simulation answer without bias?
- In Choice of Day, instead of using the 40th day as representative, would it make any difference to use the 30th day? The 10th? The 1st? (See Explore More 2 for a way to examine the 1st day in Fathom.)
- Computing the average length of stay of patients present on the ward on a given day is biased. In what sense? In which direction? What kinds of decisions might be wrongly made based on the biased estimate?

EXPLORE MORE

1. Answers will vary depending on what students think "more" or "less" biased means. To alter the simulation, students need to redefine the "hidden" attributes *Bed1, Bed2, Bed3, Bed4,* and *Bed5* to use randomInteger(1,20) instead of randomInteger(1,10). You can see the formulas for these hidden attributes by showing the Days collection's inspector.

2. Because all new values are generated on the 1st day, the lengths of stays can be any number between 1 and 10. Hence, using the 1st day removes the bias that is associated with using the 40th day (or any other day). To modify the simulation, students should redefine the measure *AnAverageStay* to use first(AverageStay) instead of last(AverageStay).

Choosing a Representative Sample—Random Rectangles

You will need
- **Random Rectangles.ftm**

In this activity you'll be working with a collection of 100 rectangles. Your goal is to choose a sample of 5 rectangles from which to estimate the average area of the 100 rectangles.

EXAMINE DATA

1. Look at the display of random rectangles in **RandomRectangles.ftm.** Without studying the display too carefully, quickly choose five rectangles that you think represent the whole population of rectangles. This is your judgment sample.

The areas are given below the rectangles.

Q1 Find the area of each rectangle in your sample of five and calculate the sample mean, that is, the average area of the rectangles in your sample.

2. In a new, empty Fathom case table, combine your sample mean with those of other students in the class. Name the collection Judgment Sample and the attribute *meanArea*.

*To plot the average, choose **Graph | Plot Value** and enter mean().*

3. Make a plot of the means. Plot the average of the sample means.

Q2 Here is a plot of another class's judgment sample. Compare your class's distribution with the one shown here in terms of shape, center, and spread.

Q3 Do you think the judgment method of sampling gives an estimate for the mean area that is too high, too low, or just about right? Explain your choice.

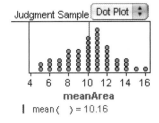

INVESTIGATE

Random Places

RandomRectangles.ftm is pre-made to demonstrate two kinds of sampling. The first demonstration chooses random places in the display of rectangles and, if there is a rectangle at that place, adds the area to the sample.

Random Place Mean Areas

4. To begin the demonstration, select the collection Random Place Mean Areas and choose **Collection | Collect More Measures.**

You can stop the process by pressing the Esc key.

You'll see a small green square move around inside the display of Rectangles. Each time it lands on a rectangle, that rectangle turns red. When five rectangles have been chosen, their mean area is calculated, moved into the Random Place Mean Areas collection, and plotted on the graph. The process stops when

Choosing a Representative Sample—Random Rectangles
continued

30 means have been collected. If you collect all 30 samples, you will end up with a histogram similar to this one:

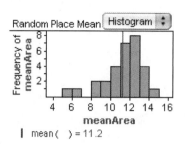

Q4 Compare the shape, center, and spread of the plots from judgment sampling and place sampling.

Q5 Do you think the random place method of sampling gives an estimate for the mean area that is too high, too low, or just about right? Explain your choice.

Random Sample

The second demonstration in **RandomRectangles.ftm** chooses five rectangles at random from the 100 in the collection. Fathom chooses a rectangle by a process similar to using a random number table or a random number generator.

5. Select the collection Random Sample Mean Areas and choose **Collection | Collect More Measures.**

Random samples of size 5 are chosen from the Rectangles collection using a random number generator. The mean area is calculated, moved into the Random Sample Mean Areas collection, and plotted on the graph. The process stops when 30 means have been collected.

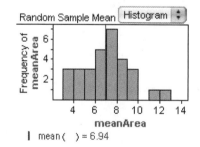

Q6 How does this distribution of mean areas compare with the previous two?

Q7 Do you think this method of sampling gives an estimate for the mean area that is too high, too low, or just about right? Explain your choice.

Q8 Which method of producing sample means do you think is better if the goal is to use the sample mean to estimate the population mean? Explain your reasoning.

The value of the population mean is in the summary table in the lower left of the Fathom document.

Choosing a Representative Sample— Random Rectangles

Activity Notes

Objectives

- Understanding that choosing your sample by chance is the only method guaranteed to be unbiased and that it is randomization that makes inference possible
- Seeing that sampling methods in which each unit does not have the same chance of being chosen are biased
- Understanding that most people mistakenly feel they can choose a judgment sample that is as good as a simple random sample

Activity Time: 40–50 minutes

Setting: Paired/Individual Activity or Whole-Class Presentation (use **RandomRectangles.ftm** for either)

Statistics Prerequisites

- Finding a mean
- Familiarity with describing and comparing graphical distributions

Statistics Skills

- Judgment and size bias
- Different methods of sampling
- Comparing distributions
- Simple random sampling
- Calculating summary statistics
- Sources of bias

AP Course Topic Outline: Part I A–C; Part II B–C

Fathom Prerequisites: Students should be able to make case tables and graphs.

Fathom Skills: Students plot the mean and collect measures.

General Notes: This activity gives students the opportunity to easily compare *three* distributions of sample means: one from judgment samples, one from random places, and one from simple random samples. In the case of random places, Fathom animates the selection and the collection of measures so that students can "see" these processes in action.

Procedure: Because the construction of the Fathom document **RandomRectangles.ftm** is fairly complex, the logic behind the sampling process is invisible, especially in the case of the simple random sample. If your students need experience using a random number table or random number generator, you may prefer to have them generate five distinct random numbers between 00 and 99. (The rectangle numbered 100 can be called 00.) The students should find the rectangles with case numbers corresponding to their random numbers and then that will be their random sample of five rectangles. Then follow steps 2–3 in the activity to plot the class's data.

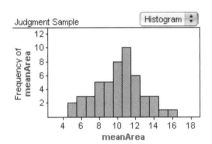

The first of the three distributions—sampling by judgment—requires pooling class data. You could collect these data one day prior to the activity. You or a student can enter the data into a new case table in the Fathom document **RandomRectangles.ftm** before distributing the document to the entire class.

EXAMINE DATA

Q1–Q3 The judgment sample is likely to be biased toward the high side. (You can point out that the population of rectangles has population mean 7.4 with standard deviation 5.2. It is calculated in a summary table in the lower left.) So, samples of size 5 should have a standard deviation of $5.2/\sqrt{5}$, or 2.34. The plot given in Q2 is fairly mound-shaped with mean 10.16 and SD 2.55.

INVESTIGATE

4. Students generate a second distribution, sampling by place. Fathom animates the choice of random places in the collection of rectangles. Depending on the speed of your computers, collecting 30 samples can take up to 10 minutes. If this is more time than

Choosing a Representative Sample—Random Rectangles
continued

Activity Notes

you want students to spend, they can stop the process after only a few sample means and refer to the sample histogram on the student worksheet.

Q4–Q5 By geometric probability, rectangles with larger area are more likely to be chosen, so this sample is biased toward the high side, too. The sample histogram in the handout has mean 11.2 and standard deviation 2.002.

5. Fathom gathers the third collection—simple random sampling—at a reasonable speed.

Q6–Q7 This distribution will be centered near the population mean and with a larger variability than the random place sampling but smaller usually than in judgment sampling.

Q8 In general, simple random sampling is not susceptible to bias and results in a good estimate of the population mean. On occasion, just by the luck of the draw, your class will get samples that are less representative. If this happens, you have a serendipitous opportunity to proselytize: "Although chance-based sampling methods give representative samples most of the time, there will be times when, just by chance, you get an unrepresentative sample. One of the great advantages of relying on chance is that you can use mathematical theory or simulations to calculate the probability of getting a bad, that is, unrepresentative, sample. No other sampling method can offer you protection anywhere near as good as that."

DISCUSSION QUESTIONS

- What similarities and differences did you find among the three distributions of mean areas? (Encourage students to consider bias.)
- Which sampling method did you conclude was the best for estimating the mean area of the rectangles in the original collection? What was wrong with the other two methods; that is, why were they biased?
- Can you describe another sampling method that would estimate the mean? If so, would your method be susceptible to bias?
- Using geometric probability, explain why the random place sampling is biased.

EXTENSIONS

1. If you have students who are becoming very proficient with Fathom, you may want to challenge them to dissect the formulas and processes that drive this simulation.

2. You can also encourage your students to scroll down to the Rectangles2 collection and ask them to describe how to take a stratified random sample from the population of rectangles. Then have them use Fathom and filters to do that. Compare their results to the simple random sampling case.

Randomization in Experiment Design

You will need
- one box
- equal-sized slips of paper

The students in your class are to be the subjects in an experiment with two treatments, A and B. Your task is to find the best of three ways to assign the treatments to the subjects.

COLLECT DATA

Be sure that you've completed the survey about gender, siblings, and reading habits and that the results have been entered into a Fathom case table.

INVESTIGATE

Pick Treatment

1. Choose two leaders for the class, one for treatment A and one for treatment B. The leaders should flip a coin to decide who goes first, then alternately choose class members for their teams.

2. Open the Fathom document that contains the survey results.

3. Enter the treatment assignments of each individual into a new attribute, *PickTreatment*.

4. Make summary tables, ribbon charts, and stacked bar charts to investigate the proportions of each gender, the proportion who have brothers and sisters, and the proportion who like to read novels in their spare time.

   When you place your cursor over a portion of a graph, the percentage displays in the lower-left corner of the Fathom window.

   a. To make a summary table, drag a new summary table from the shelf. Then drag *PickTreatment* and one other attribute from the case table to the arrows of the summary table.

   b. To make a ribbon chart, make a new graph and drag *PickTreatment* to the horizontal axis. Drag another attribute to the *interior* of the graph. Choose **Ribbon Chart** from the pop-up menu.

Students

		PickTreatment		Row Summary
		A	B	
Gender	Female	8	13	21
	Male	7	2	9
	Column Summary	15	15	30

S1 = count ( )

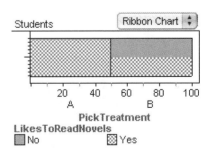

Randomization in Experiment Design
continued

c. To make a stacked bar chart, make a new graph and drag one attribute to the horizontal axis. Then drag a second attribute to the horizontal axis of the graph and drop the attribute on the plus. Then drag the attribute *Treatment* to the interior of the graph.

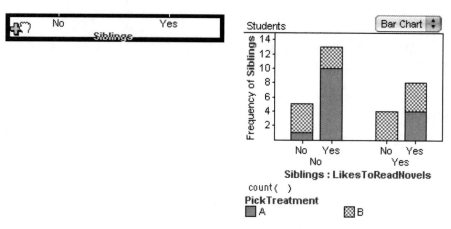

Q1 Based on your summary tables and graphs, do the two treatment groups look quite similar or quite different? Explain your reasoning.

Slip Treatment

5. Next, divide the class by writing your names on pieces of paper and putting them in a box; randomly draw out half of them (one by one) to be assigned treatment A. The names remaining in the box get treatment B. Record these assignments in a new attribute, *SlipTreatment*.

6. Repeat steps 3 and 4 for the *SlipTreatment* groups.

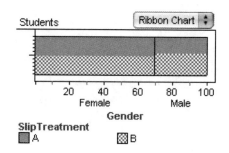

Q2 Based on your summary tables and graphs, do the two treatment groups look quite similar or quite different? Explain your reasoning.

Random Choice Treatment

*To add the formula, select the attribute in the case table and choose **Edit | Edit Formula**.*

7. Now you'll divide the class by using random assignment in Fathom. Define an attribute named *RandomChoiceTreatment* with the formula randomPick("A", "B").

Randomization in Experiment Design
continued

8. Repeat steps 3 and 4 for the *RandomChoiceTreatment* groups.

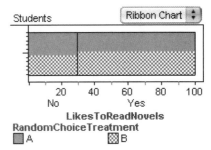

Q3 Based on your summary tables and graphs, do the two treatment groups look quite similar or quite different? Explain your reasoning.

Q4 What are the strengths and weaknesses of each method of assigning treatments to subjects? Which method is the least random? Which method do you think is best for this type of experiment?

Q5 Which method might have some confounded variables or lurking variables?

EXPLORE MORE

1. Divide the class into treatment groups by standing single-file and assigning treatment A to the first student in line, treatment B to the second student, treatment A to the third student, and so on. What are the advantages and disadvantages of this method? Is this a random assignment? Why or why not?

2. Use sampling in Fathom to divide your class into treatment groups. To do this, select the collection, then choose **Collection | Sample Cases.** Open the inspector to the **Sample** panel, uncheck With replacement, and type in your class size for the number of cases. Click **Sample More Cases.** Then make a new attribute, *Treatment2*, defined by the formula below. (Use half your class size in place of 15.)

Which treatment in the activity does this most resemble? Explain.

Surveys

Name: _____

What is your gender?	Male	Female
Do you have any brothers or sisters?	Yes	No
Do you like to read novels in your spare time?	Yes	No

— — — — — — — — — — — — — — — — — — — —

Name: _____

What is your gender?	Male	Female
Do you have any brothers or sisters?	Yes	No
Do you like to read novels in your spare time?	Yes	No

— — — — — — — — — — — — — — — — — — — —

Name: _____

What is your gender?	Male	Female
Do you have any brothers or sisters?	Yes	No
Do you like to read novels in your spare time?	Yes	No

— — — — — — — — — — — — — — — — — — — —

Name: _____

What is your gender?	Male	Female
Do you have any brothers or sisters?	Yes	No
Do you like to read novels in your spare time?	Yes	No

— — — — — — — — — — — — — — — — — — — —

Name: _____

What is your gender?	Male	Female
Do you have any brothers or sisters?	Yes	No
Do you like to read novels in your spare time?	Yes	No

— — — — — — — — — — — — — — — — — — — —

Name: _____

What is your gender?	Male	Female
Do you have any brothers or sisters?	Yes	No
Do you like to read novels in your spare time?	Yes	No

Randomization in Experiment Design

Activity Notes

Objectives
- Understanding the goal of random assignment of treatments: to have each group as nearly alike as possible
- Seeing that randomization tends to take care of balancing the groups on all variables, those that can be observed (such as gender) as well as those that cannot be observed (such as genetic conditions)
- Seeing that randomization protects against confounding by distributing any variables that might be confounded with the explanatory variables as evenly as possible among the treatment groups

Activity Time: 50–60 minutes if data are collected the day before

Setting: Paired/Individual Activity (collect data) or Whole-Class Presentation (collect data or use **Randomization.ftm**)

Materials
- One survey for each student (see Surveys worksheet)
- One box
- Equal-sized slips of paper with a different student's name on each

Statistics Prerequisites
- Comparing distributions with proportions
- Some familiarity with experimental protocol
- Some familiarity with lurking or confounded variables

Statistics Skills
- Randomization of data to protect against confounding
- Confounding and lurking variables
- Random assignment of treatments to subjects
- Comparing multiple distributions to make conclusions
- Simple random sampling
- Experimental protocol

AP Course Topic Outline: Part I E; Part II A–C

Fathom Prerequisites: Students should be able to make summary tables and graphs.

Fathom Skills: Students make and interpret ribbon charts and stacked bar charts (with multiple attributes), use summary tables with two variables, and use random generator formulas. *Optional:* Students sample without replacement (Explore More 2).

General Notes: Fathom adds convenience to this activity by using random value generators, ribbon charts, stacked bar charts, and summary tables. In this activity, students have a chance to compare proportions. A ribbon chart and stacked bar charts are good plots for this purpose, and it's important for students to get comfortable with how to read them.

As an alternative to using slips of paper in the second method, you could use Fathom's ability to sample from a collection. See Explore More 2 in the activity for how students can do this.

Procedure: The day before you do the activity, have each student fill out the brief survey and turn it in. You or a student should enter all of the information—name, gender, siblings, and novels—in a Fathom document in time for distribution the following day. The document **Randomization.ftm** contains sample data, with treatments assigned.

COLLECT DATA

1. Pick two leaders who are not alike, such as a male student and a female student, to increase the probability that there is more discrepancy between the treatment groups. While these two students are choosing their groups, someone should enter the treatment-group information into the Fathom document. Be prepared to distribute this document to the students shortly after the treatment groups are selected.

INVESTIGATE

Q1 The first method is not random and can lead to other variables confounded with the treatment. The two groups will probably look different because each leader will usually choose their friends or people with whom they have something in common. Another attribute you could add to the survey is something about study habits. Study habits within one group could be very similar within the group and very different between groups.

Randomization in Experiment Design
continued

Activity Notes

Q2 Randomly drawing names from a box should balance the groups nicely unless your class is small.

Q3 Using Fathom to assign treatments should balance the groups nicely unless your class is small. Assignment by this method may not lead to equal group sizes. It is not essential that the group sizes be equal, but most experiments assign the same number of subjects to each treatment group.

Q4–Q5 Both the second and third methods are random in the sense that each person has equal likelihood of being assigned to group A or B. An advantage of the second method is that you guarantee equal groups (or at most are off by one). The third method has the advantage that it works even when you don't know how many subjects there are in the experiment. The drawback to the third method is that there is no guarantee that the groups will have nearly equal sizes. It is possible that one group could have very few members and that, too, could lead to groups that differ on significant variables. The first method is bound to have confounded variables because of how the selection was made.

DISCUSSION QUESTIONS

- Were the characteristics of groups A and B unbalanced under any of the methods of assignment? If so, how do you explain the unbalance?
- If your goal were to evaluate the effectiveness of the two different treatments, which of the three methods of assignment would you choose?
- When might you choose the second method in preference to the third method, or vice versa?

EXPLORE MORE

1. This method could create similar treatment groups if students line up next to their friends, as you might expect. Also, this method will create groups of equal sizes. However, it is not random, so it could easily have confounded variables.

2. This method is most like the first method, as friendly students are more likely to line up next to each other.

Designing an Experiment, Part 1—Bears in Space

You will need
- launch ramp and launcher
- coin
- tape measure or meterstick
- supply of gummy bears
- four copies of your statistics text

This activity consists of two treatments: one book (flat ramp) and four books (steep ramp). The response variable is the launch distance. Pay close attention to the variation in launch distances within each of these treatments.

COLLECT DATA

1. Form launch teams and construct your launcher. Your class should have an even number of teams, at least four. Make your launcher from tongue depressor sticks and rubber bands. First, wind a rubber band enough times around one end of a stick so that it stays firmly in place. Then, place that stick and another together and wind a second rubber band tightly around the other end of the two sticks to bind them firmly together. Insert a thin pencil or small wooden dowel between the two sticks as a fulcrum.

How to build a bear launcher.

2. Randomize. Use random numbers or a random ordering of cards containing team names to assign each team to produce data for one or the other of the two treatments: steeper ramp (four books) or flatter ramp (one book). The same number of teams should be assigned to each treatment.

3. Organize your team. Decide who will do which job or jobs: Hold the launcher on the ramp, load the gummy bear, launch the bear, take measurements, and record the data. (When you launch, use a coin instead of your fingers, or you could end up with very sore fingers.)

Setting up the launch.

Designing an Experiment, Part 1—Bears in Space
continued

4. Gather data. Each team should do ten launches. After each launch, record the distance. Measure the distance from the *front* of the ramp, and measure only in the direction parallel to the ramp.

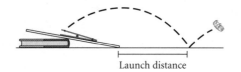

Measuring the launch distance.

5. Record data. Open Fathom. In a new document, drag a new case table from the shelf and record your data. So that your data can be combined with other teams' data later on, use these attributes.

> The values of *Team* and *Books* will be the same for all ten of your measurements.

Launch	Team	Books	Distance
1	1	1	15 in
2	2	1	1	24 in
3	3	1	1	48 in

INVESTIGATE

Within-Team Variability

6. Make a dot plot of your team's ten launch distances.

7. To summarize the center and spread, calculate the mean and standard deviation in a summary table. Drag a new summary table from the shelf. Drag *Distance* to either arrow in the summary table. The mean will automatically be calculated. To get the standard deviation, choose **Summary | Add Formula,** and enter stdDev().

8. Plot the center and spread on your graph. Select the graph, choose **Graph | Plot Value,** and enter mean(). Do the same to plot mean()+stdDev() and mean()–stdDev().

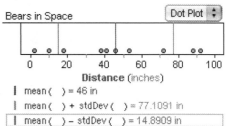

Q1 Why do the launch distances vary so much for your team? List as many explanations as you can, then order them, starting with the one you think has the most effect, and on down.

Q2 One important source of variability between launches is the effect of practice. Can you think of an alternative strategy that would reduce or come close to eliminating the variability due to practice?

Designing an Experiment, Part 1—Bears in Space
continued

Q3 Use your answers to write an *experimental protocol*, a set of rules and steps to follow to keep launch conditions as nearly constant as possible. Your protocol should include rules for deciding when, if ever, not to count a "bad" launch.

Between-Team Variability

9. Drag a new, empty case table from the shelf. Record each team's summary measures, using these attributes.

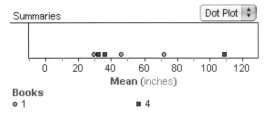

10. Make a dot plot with two kinds of dots, one for each launch angle. To do this, choose a summary measure, say *Mean*, and drop the attribute on the horizontal axis. Then drag *Books* and drop it in the interior of the graph while holding down the Shift key.

Q4 Which measure of center, mean or median, gives a better summary of a team's set of ten launches? Or are they about equally good? Give reasons for your judgment.

11. Make a summary table. Drag your choice of center to one arrow and *Books* to the other arrow. Hold down the Shift key while you drop *Books*. Double-click the formula for count() and replace it with the appropriate measure of spread.

Q5 Is there more variation between teams with the same launch angle or between launches for the same team (calculated in step 7)? Why do you think the team summaries vary as much as they do? List as many explanations as you can, then order them, starting with the one you think has the most effect, and on down. Here you summarized a team's results with a single-number measure of center—why not do only one launch per team and use that number?

Q6 What strategies would you recommend for managing the variability between the teams that had the same launch angle?

Designing an Experiment, Part 1—Bears in Space
continued

Difference Between Treatments (Day 2)

Now you'll use the data from the entire class to think about ways to use the data to measure and compare the sizes of the three kinds of variability: within-team variability, between-team variability, and the variability between treatments.

12. Open the document **PooledBears.ftm.** This document has each team's results.

13. Make the plot and summary table shown, using your measures of center and spread. Hold down the Shift key while dropping *Teams* and *Books*. Plot your choice for the measure of center. In your summary table, calculate a measure of center and spread.

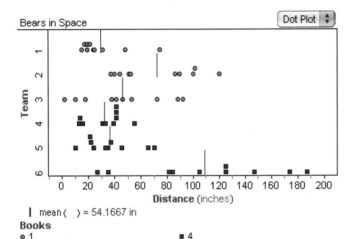

Q7 Can you say which kind of variability is the largest? The smallest? Explain.

Q8 Without doing a formal analysis, choose a tentative conclusion from this set of three choices, and give your reasoning.

   I. Launch angle clearly makes a big difference.

   II. Launch angle may very well make a difference, but there's too much variability from other sources to be able to isolate and measure any effect due to launch angle.

   III. There's not much variability anywhere in the data, so it's safe to conclude that the effect due to launch angle, if there is any, is quite small.

14. Make a side-by-side box plot of one-book launches and four-book launches. Show all individual launches, ignoring teams.

Q9 Compare the centers and spreads for the two sets of launches. Why might this analysis be misleading?

Designing an Experiment, Part 1—Bears in Space   Activity Notes

Objective: Understanding that too much variability among units that are given the same treatment (within-treatment variability) obscures any difference caused by the treatments themselves (between-treatment variability).

Activity Time: 100–120 minutes if data are collected by students

Setting: Small Group Activity (collect data, combine as class) or Whole-Class Presentation (use **BearsInSpace.ftm**)

Materials
- Launch ramps (such as thin books or pieces of wood)
- Tongue depressors
- Rubber bands
- Pencils or dowels
- Gummy bears
- Tape measures, metersticks, or yardsticks
- Coins
- Several copies of your statistics text

Statistics Prerequisites
- Using random numbers to assign treatments
- Some familiarity with experimental protocol
- Some familiarity with lurking or confounded variables
- Measures of center and spread and comparing distributions

Statistics Skills
- Randomization of data to protect against confounding
- Random assignment of treatments to teams
- Comparing multiple distributions to make conclusions
- Simple random sampling
- Experimental protocol
- Between-treatment variability
- Within-treatment variability

AP Course Topic Outline: Part I C; Part II A, C

Fathom Prerequisites: Students should be able to make summary tables, graphs, and case tables.

Fathom Skills: Students use summary tables with two variables, make split graphs and graphs with legend attributes, make "numerical" data categorical, and create new attributes to sort data using "if" statements.

General Notes: The main advantage to using Fathom in this activity comes when the data are combined and between-team and between-treatment variability are discussed. This activity is highly recommended, but it is time-consuming. If you choose not to do it as a class activity, bring in one bear launcher and demonstrate how the experiment is done. Then use the sample data in **BearsInSpace.ftm**.

Procedure: Although one person can generate data, a team approach works much better. As few as two and as many as eight on a team will work. This is a sketch of the various roles, but the main thing to keep in mind is that there is a lot of room for flexibility, and you can leave it to the teams to divide the tasks themselves. Ideally, two people are responsible for setting up each launch: placing books under the ramp, positioning the fulcrum, and positioning the launcher on the ramp, and so on. These two people can also be responsible for holding the ramp and launcher in place. One person should be designated to load gummy bears onto the launcher, and another does the actual launching. Students should use a coin instead of their fingers to release the spring, or they will have very sore fingers. Measuring is best done by two people, one at each end of the tape, and you may want to have two people record as protection against mistakes.

In this activity each team uses only one of the launch angles. Randomize using a table of random numbers or a random number generator on a calculator.

Because launch distances are quite variable, with a standard deviation of around 15 inches, it works best to make multiple launches and report averages. See the document **BearsInSpace.ftm** for some actual results of individual distances for sets of ten launches, along with some summary statistics.

For Day 2, you or a student should enter the groups' data into one Fathom document, called **PooledBears.ftm**.

Designing an Experiment, Part 1—Bears in Space
continued

Activity Notes

INVESTIGATE

Q1 The main sources of within-treatment variability in distance are variability in the force of the launch and the effect of practice. Other sources include the position of the fulcrum (dowel), the position of the bear on the launcher, how the launcher is held, the position of the launcher on the ramp, and the position of the books under the ramp. The plot below shows side-by-side dot plots for the launch distances from six teams. Teams 1–3 used one book and teams 4–6 used four books.

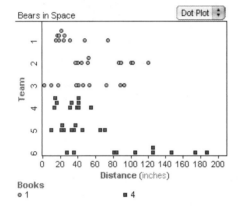

Q2 It is standard practice in lab-based research to practice procedures until trends disappear and variation stabilizes. Doing several practice launches is the analog for this activity.

Q3 The basic idea is to keep all the sources of variability listed in Q1 as constant as possible. Sample protocol:

- The book(s) under the ramp should always touch the ramp at the same place. Mark the ramp at this spot to make sure.
- Mark the position of the fulcrum on the launcher, and always reposition the fulcrum to its designated place before each launch.
- Always position the launcher at the back of the ramp. Mark the location to make sure it is held constant.
- The person who holds the launcher on the ramp should check the position of the book(s) and ramp, then check the position of the fulcrum, and then place the launcher on the ramp, holding it steady.
- Always position the bear the same way: Place the bear on its back on the launcher, on the stick rather than the rubber band, but touching the rubber band, and lying sideways so that its long direction (head to foot) is perpendicular to the direction of the launcher.
- The person doing the launching should try to keep the force of the launch constant from one launch to the next.
- To eliminate time trends, a team should practice several launches before beginning the experiment.

Q4 Either the median or the mean will ordinarily provide a reasonable summary. If the distances are strongly skewed toward extreme values or there are outliers, however, the median may be better.

Q5 The plot in the activity is a dot plot of team means. Sources of variability include those listed in Q1, to the extent that these vary from one team to another. Typically, the single biggest source of variability is the technique of the person doing the launching. In **BearsInSpace.ftm,** the variability between teams is about 31, which is larger than the variability between launches for teams 1, 4, and 5, about the same for teams 2 and 3, and much smaller than team 6's variability. If each team does only one launch, it might not be very representative of their launch distance.

Q6 Students may have ideas for improving the protocol to reduce differences between teams, but the best strategy (blocking) is to ask each team to do two sets of launches, one set with each ramp angle.

Q7 Here is an informal sense for measuring components of variation:

Launch angle: Find the overall average distance for each of the two launch angles, and compute the difference. For the sample data, the averages are 49.1 for the teams with one book and 59.2 for the teams with four books. The difference is 10.1.

Particular team: Start with the team averages, and compute one SD for each treatment. For the sample data, the standard deviation for teams 1–3, with one book, is 21.6. The standard deviation for teams 4–6, with four books, is 43.2.

Individual launches: Use the SD or *IQR* for each team's set of ten launches. For the sample data, the standard deviations for the six teams are 18.4, 30.5, 31.1, 14.0, 19.2, and 53.6.

Designing an Experiment, Part 1—Bears in Space
continued

Activity Notes

It is difficult to say whether the largest variability is within the teams (the individual launches) or between teams (the particular team). The smallest variability is between the two treatments (the launch angle).

Q8 If results from your class are typical, then II will be the best answer. Variability between teams tends to be large in comparison to the effect of launch angle, as it is with the sample data. This means that it is often not possible to tell whether launch angle has an effect.

Q9 There is a clear suggestion in the sample plot that the steeper angle leads to longer launches. However, ignoring teams gives up the essential information that team 6 alone accounts for the greater mean distance when four books are used. The plot also shows, however, that the data are fairly skewed, so perhaps using the medians and the *IQR* is better for comparison. It is hard to make a case for either launch angle.

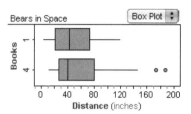

Designing an Experiment, Part 2—Block Those Bears

You will need
- launch ramp and launcher
- coin
- tape measure or meterstick
- supply of gummy bears
- four copies of your statistics text

Ordinarily, when teams carry out the bear-launch experiment (see Part 1), their results show that the variability between teams is so large that it is hard to tell how much difference the angle of the launch ramp makes. Following a careful protocol can reduce the variability somewhat, but there will still tend to be big differences from one team to another. In this activity you'll see that sometimes there are much better ways to manage variability.

COLLECT DATA

You're going to perform the experiment again, but you'll assign treatments differently.

1. Form teams and assign jobs. See steps 1–3 in Part 1.

2. Each team will conduct two sets of five launches, one set with four books under the ramp and one set with just one book. Each team should flip a coin to decide the order of the treatments: If the coin is heads, then use four books first; if tails, use one book first.

3. Carry out the launches, following the protocol your class developed in Q3 of Part 1. Record your data in a Fathom case table with the attributes shown. (It's necessary to record your team number so that your data can be merged with the data of other teams.)

   Blocked Bears

	Team	Books	Distance
1	1	1	64.5 in
2	1	1	70.3 in

4. Pool the results from all teams into a single Fathom collection.

INVESTIGATE

Hold down the Shift key while dropping *Team*.

Cells from Blocked Bears Table

5. Make a summary table as shown. The two summary formulas will give each team's averages for the two treatments.

6. Select the summary table and choose **Summary | Create Collection From Cells.** A new collection is created, Cells from Blocked Bears Table. Make a case table of the cells collection. The formulas *S1* and *S2* are now attributes of the cells collection.

Blocked Bears

	Team	
	1	53.12 in
		104.58 in
	2	69.5 in
		130 in
	3	71.06 in
		141.44 in
	4	71.88 in
		88.64 in
	5	76.7 in
		116.82 in
	6	48.28 in
		133.9 in
Column Summary		65.09 in
		119.23 in

$S1 = \text{mean}(\text{Distance}, \text{Books} = 1)$
$S2 = \text{mean}(\text{Distance}, \text{Books} = 4)$

Designing an Experiment, Part 2—Block Those Bears
continued

7. Define a new attribute, *Difference*, with the formula S2−S1. The results are the differences between the two treatments for each block.

Cells from Blocked Bears Table

Team		S1	S2	Difference
1	1	53.12 in	104.58 in	51.46 in
2	2	69.5 in	130 in	60.5 in

8. Make a dot plot of the *Difference* values. Compute or plot the center and spread of *Difference*.

Q1 Based on the differences between the averages, is it reasonable to conclude that launch angle makes a difference? Or is it impossible to tell?

Q2 What kind of experimental design is used in this activity? What are the blocks?

Q3 Remember that a unit is what the treatment gets assigned to. For the design of this activity, which is the unit: a single launch, a set of five launches, or a set of ten launches by a team?

9. Make the two plots shown. For each plot, hold down the Shift key while dropping the attribute on the vertical axis. For the left plot, also hold down the Shift key while dropping *Books* in the interior.

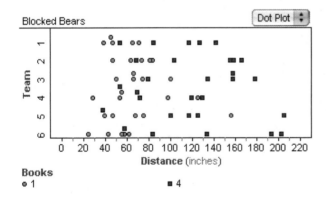
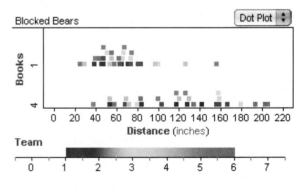

Q4 Based on your two new plots, is it reasonable to conclude that launch angle makes a difference? Or is it impossible to tell? Discuss the between-treatment variability and the within-treatment variability as shown by your plots.

EXPLORE MORE

1. Make a box plot of *Distance* split by *Team*, thus removing any blocking from the design. Comment on whether you would be able to draw a conclusion from these data.

2. Suppose you were to redo this activity. How could you improve on the design?

Designing an Experiment, Part 2—Block Those Bears

Activity Notes

Objectives
- Seeing that a randomized block design reduces variability, thereby allowing you to isolate the effect of the treatment
- Understanding that the effectiveness of blocking depends on how similar units are in each block and how different the blocks are from each other. Here, "similar" units are units that would tend to give similar values for the response if they were assigned to the same treatments. *The more similar the units in a block, the more effective blocking will be.*

Activity Time: 50 minutes if data are collected by the students

Setting: Paired/Individual Activity (collect data, combine as class) or Whole-Class Presentation (use **BlockThoseBears.ftm**)

Materials
- Launch ramps (such as thin books or pieces of wood)
- Tongue depressors
- Rubber bands
- Pencils or dowels
- Gummy bears
- Tape measures, metersticks, or yardsticks
- Coins
- Several copies of your statistics text

Statistics Prerequisites
- Using random numbers to assign treatments
- Familiarity with experimental protocol
- Familiarity with blocking
- Measures of center and spread and comparing distributions

Statistics Skills
- Randomization of data to protect against confounding
- Randomized block design
- Comparing multiple distributions to make conclusions
- Blocking to reduce within-treatment variability
- Experimental protocol
- Within-treatment versus between-treatment variability

AP Course Topic Outline: Part I C; Part II C

Fathom Prerequisites: Students should be able to make summary tables, graphs, and case tables.

Fathom Skills: Students use summary tables with two variables, make split graphs and graphs with legend attributes, make "numerical" data categorical, and collect measures from summary tables.

General Notes: This activity is meant to show that a randomized block design can reduce the variability in Part 1. Each team (block) is reused so that it can provide two sets of five launches. This procedure makes it possible to isolate the effects of the launch angle by making it a within-team factor. With Fathom, students will be able to work with the entire data set and take on the process of reducing it to differences between treatments within each block. Hopefully, the power and beauty of the blocked design will become clear.

Procedure: After students have formed teams and assigned jobs, they might want to do a few practice launches for each angle.

Each team (block) will do ten bear launches. Along with their measurement of distance, they will enter their team number and the number of books (the treatment).

You can make this activity more efficient by preparing a blank Fathom master document in advance, with the attributes *Team*, *Books,* and *Distance*. During class, open this master document on one computer that is accessible to all. As students finish measuring their launch distances, they should enter their data into the case table. After all data have been entered, distribute a copy of this master document to each student for completion of the activity.

A summary table provides a good tool for computing the means for each treatment by each team. However, as instructed on the student worksheet, you want to drag only the *Team* attribute to the summary table. Formulas are used to calculate the means of each treatment. Then a collection of measures is used to compute the differences. This process gives you a collection of differences that can be graphed and analyzed.

Designing an Experiment, Part 2—Block Those Bears

continued

Activity Notes

INVESTIGATE

Q1 In general, students should conclude that a four-book launcher results in a longer-distance launch than a one-book launcher. (You might have students check whether their results match the theory from calculus that a 45-degree launch angle maximizes the horizontal distance.)

Q2–Q3 This is a randomized block design or, more specifically, a randomized paired comparison design with repeated measures. The blocks are the teams. The unit is a set of five launches by a team. In this experiment the basic terms have the following interpretations:

Response variable: Launch distance

Treatment: Launch angle (flat or steep)

Unit: A set of five launches by a team

Block: A team (both treatments are carried out in each block)

Design: Randomized block design with the teams as blocks. For each team, one (randomly assigned) set of five launches is done with a steep angle, the other set of five with a flat angle. The randomization is done separately for each team. Each team can toss a coin to decide on the order of treatments.

Q4 The two additional plots display the within-treatment variability and also show that there is a difference in treatments. In the righthand plot on the student worksheet, you can see that the four books do make for better launches, but you can also see each individual team's variability as well. On the left, you see each team's performance and how their values for the four books are generally higher. You can also see clearly the within-treatment variability. For the sample data, the mean for one book is about 65.09 and the standard deviation is 27.41, the mean for four books is 119.23 and the standard deviation is 46.89. There is quite a bit more variability within the four-book treatment but the whole distribution is generally higher than the one-book launches. Here is the box plot for the different treatments.

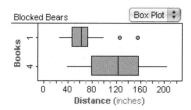

DISCUSSION QUESTIONS

- How is the experimental design of this activity different from the design in Part 1?
- Which activity do you think has the better design? Explain your reasoning.
- What were you able to conclude about the two treatments?

EXPLORE MORE

1. It would be difficult to draw a conclusion from these data; without being able to see the blocking, you can't tell much of anything except that the distributions are rather skewed.

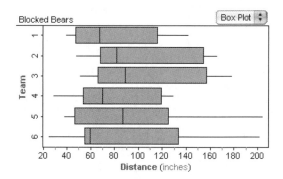

2. Instead of having the team do all five launches of one type and then all five launches of the other type, randomly decide how many books to use for each launch by drawing one of ten slips of paper from a box. Five of the slips of paper would say "one book" and five would say "four books."

Comparing Experiment Designs—Sit or Stand

You will need
- slips of paper
- one coin
- **SitOrStandT.ftm**

In this activity you will serve as a subject for three different versions of an experiment. The goal of the experiment is to see whether there is a detectable difference in heart rate measured under two conditions, or treatments: standing, with eyes open (treatment 1), and sitting, with eyes closed, relaxing (treatment 2).

You will take your pulse several times. Each time follow these steps:

a. Get ready, sitting or standing.

b. Your teacher will time you for 30 seconds. When your teacher says "go," start counting beats until your teacher says "stop."

c. Double your count to get your heart rate in beats per minute.

The goal of the activity is for you to think about the advantages and disadvantages of each version of the experiment.

COLLECT DATA

Experiment A: Completely Randomized Design

In Experiment A, you and the other subjects will be randomly assigned to one treatment or the other.

1. **Random assignment.** Your teacher will pass around a box with slips of paper in it. Half of the slips say "stand" and half say "sit." When the box comes to you, mix up the slips and draw one. Depending on the instruction you get, either stand with your eyes open or sit with your eyes closed and relax.

2. **Measurement.** Take your pulse.

3. **Record data.** Open **SitOrStandT.ftm.** Record your data in the Completely Randomized case table: your group (either Sit or Stand) and your pulse (in beats per minute).

4. **Summaries.** Plot the data using a split dot plot by first dragging *Pulse* to the horizontal axis and then dragging *Group* to the vertical axis.

You can add the mean to your plot by choosing **Graph | Plot Value.** Add the mean and standard deviation to your summary table by choosing **Summary | Add Formula.**

5. Use a summary table to compute the mean and sample standard deviation for each group.

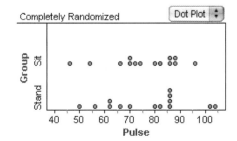

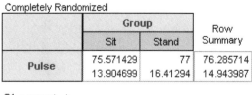

Exploring Statistics with Fathom
© 2007 Key Curriculum Press

Comparing Experiment Designs—Sit or Stand
continued

Q1 Based on your plot and summary statistics, do you think the treatment makes a difference? Explain.

Experiment B: Randomized Paired Comparison Design (Matched Pairs)

In Experiment B, you and your classmates will first be sorted into pairs based on an initial measurement. Then, within each pair, one person will be randomly chosen to stand and the other will sit.

1. **Initial measurement.** Take your pulse sitting with eyes open. (You will not record this measurement.)

2. **Form matched pairs.** Line up in order, from fastest rate to slowest, and then pair off, with the two fastest in a pair, the next two after that in a pair, and so on.

3. **Random assignment within pairs.** Either you or your partner should prepare two slips of paper, one that says "sit" and one that says "stand." One person should then mix the two slips and let the other choose a slip. Thus, within each pair, one of you randomly ends up sitting and the other one standing.

4. **Measurement.** Take your pulse.

5. **Record data.** Record your and your partner's pulses in the Matched Pairs case table. (*Difference* is calculated by the formula Stand_Pulse−Sit_Pulse.)

6. **Summaries.** Plot the set of differences in a dot plot. Then compute the mean and standard deviation of the differences in a summary table.

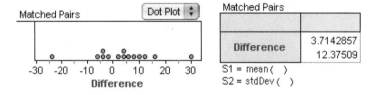

Q2 What should the mean be if the treatment makes no difference?

Q3 Do you think the observed difference is real or simply due to variation between individuals?

Experiment C: Randomized Paired Comparison Design (Repeated Measures)

This time each person will be his or her own matched pair: Each of you will take your pulse under both treatment 1 and treatment 2. You'll flip a coin to decide the order: sit first, then stand; or stand first, then sit.

1. **Random assignment.** Flip a coin. If it lands heads, you will sit first, then stand. If it lands tails, you will stand first.

Comparing Experiment Designs—Sit or Stand
continued

2. **First measurement.** Take your pulse in the position chosen by your coin flip.

3. **Second measurement.** Take your pulse in the other position.

4. **Record data.** Record your pulses in the Repeated Measures case table. (*FirstTreatment* tells whether you first took your pulse sitting or standing. *Difference* is calculated by the formula Stand_Pulse−Sit_Pulse.)

Repeated Measures

	FirstTreatment	Sit_Pulse	Stand_Pulse	Difference
=				Stand_Pulse − Sit_Pulse
1	Sit	60	64	4
2	Stand	70	72	2
3	Sit	72	76	4

5. **Summaries.** Display the set of differences in a dot plot. Then compute the mean and standard deviation of the differences in a summary table.

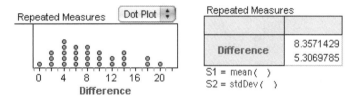

Q4 Do you think the treatment makes a difference?

Q5 Which design do you think is best for studying the effect of position on heart rate: completely randomized design, randomized paired comparison design (matched pairs), or randomized paired comparison design (repeated measures)? Explain what makes your choice better than the other two designs.

Q6 Which of the three designs do you think is least suitable for studying the effect of position on heart rate? Explain what makes it less effective than the other two. Make up and describe a scenario (choose a response and two treatments to compare) for which the least suitable design would be more suitable than the other two.

Q7 Describe the within-treatment variability for the design in Experiment A. How did the other two designs eliminate within-treatment variability?

INVESTIGATE

One way to compare designs is to plot the two Randomized Paired Comparison Designs on the same plot.

1. Drag a new collection from the shelf.

Comparing Experiment Designs—Sit or Stand
continued

2. Select the Matched Pairs collection and choose **Edit | Select All Cases.** Then choose **Edit | Copy Collection.** Select your new collection and choose **Edit | Paste Cases.**

3. Copy the Repeated Measures collection and paste those cases into your new collection as well.

4. Define a new attribute in your new collection called *Treatment* using an "if" statement: for example, if there were 14 cases in the Matched Pairs collection one possibility would be

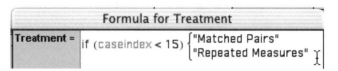

5. Display the set of differences in a split dot plot with *Difference* on the horizontal axis and *Treatment* on the vertical.

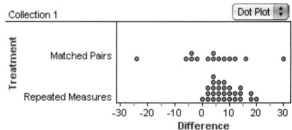

Q8 Compare these two distributions (graphically) in terms of shape, center, and spread.

6. Compute the mean and standard deviation of the differences in a summary table.

Q9 Based on this new plot and summary statistics, which design best supports the claim that the treatment does make a difference in pulse rates?

EXPLORE MORE

1. Investigate using box plots instead of dot plots. Which type of plot gives you a better summary of the data? Do your conclusions change when you change plots?

2. In the Repeated Measures design, you recorded whether the first pulses were measured sitting or standing. Explore whether or not *FirstTreatment* makes any difference.

3. Choose one of the designs and make a Fathom simulation that generates pulses and has one or more sliders that control how different one treatment is from the other.

Comparing Experiment Designs—Sit or Stand

Activity Notes

Objectives

- Emphasizing that blocking—placing similar units in the same block, then randomly assigning treatments within each block—reduces variability so that you can see how different treatments affect similar units
- Recognizing that more than one experimental design can apply to a given situation and that one design may have advantages over another

Activity Time: 40 minutes if done as a class, about 50–70 minutes if students do it separately, or 30–40 minutes if students use the sample data

Setting: Paired/Individual Activity (use **SitOrStandT.ftm** to collect data or use **SitOrStand.ftm**) or Whole-Class Presentation (use **SitOrStandPresent.ftm**)

Optional Document: SitOrStandRandom.ftm (Explore More 3 solution)

Materials

- Clock or watch with a second hand
- Slips of paper
- Coins

Statistics Prerequisites

- Using a random process to assign treatments
- Familiarity with experimental protocol
- Familiarity with blocking
- Measures of center and spread and comparing distributions

Statistics Skills

- Completely randomized design
- Randomized block design
- Randomized paired comparison design (matched pairs and repeated measures)
- Comparing experimental designs for the best and most appropriate design
- Comparing distributions to make conclusions
- Blocking to reduce within-treatment variability
- Experimental protocol
- Within-treatment versus between-treatment variability

AP Course Topic Outline: Part I; Part II C, D

Fathom Prerequisites: Students should be able to create summary tables and add formulas, make graphs and collections, and define attributes.

Fathom Skills: Students use summary tables with two variables, make split graphs, copy and paste cases to make a new collection, and use an "if" statement to define an attribute. *Optional:* Students use sliders to generate a random collection (Explore More 3).

General Notes: Collecting the data for this activity goes quickly and illustrates how blocking can reduce variability. Fathom helps students easily manipulate the data. If you aren't able to collect data with your class, your students can use the sample data in **SitOrStand.ftm**. Or you can have a whole-class discussion using **SitOrStandPresent.ftm** where everything is already created for you.

Procedure: As written, this activity uses the template **SitOrStandT.ftm** as a master document. During class, open this document on one computer that is accessible to all students. As students finish measuring their pulses, have them enter their data into the appropriate case table. After all data have been entered, you can use the document for a whole-class discussion by following the activity or you can distribute a copy of this master document to each student so that they can complete the activity.

COLLECT DATA

Experiment A: Completely Randomized Design doesn't use blocking. Students are randomly assigned to the treatments of sitting or standing. You could use Fathom to randomly assign students to treatment groups, but slips of paper with "sit" or "stand" on them may be quicker.

Q1 Students are likely to see a great deal of variability in the data, making it hard to tell whether the treatments made any difference. For the sample class, the dot plot and summary table are shown in the student's activity. Both distributions show a great deal of variability,

Comparing Experiment Designs—Sit or Stand
continued

but their means and medians are fairly close. Both are somewhat skewed left.

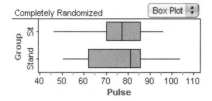

Experiment B: Matched Pairs assigns two students with similar heart rates to the same block. Then the treatments of sitting and standing are randomly assigned, and the difference in their heart rates is computed.

Q2–Q3 The center of the differences is likely to be very close to zero, which implies that the treatment makes no difference. For the sample class, the dot plot and summary table are shown in the activity. Here the mean and median are fairly close to zero, considering the high variability (SD 12.38, IQR 14.50). The observed mean difference of 3.71 could easily be due to variation between individuals and not due to a particular treatment. Note that the middle 50% contains 0, so 0 is reasonably likely.

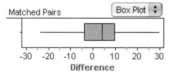

Experiment C: Repeated Measures assigns one student to a block, with two measurements taken on that student. The treatments of sitting and standing are assigned in random order to the student, and the difference in the heart rates is computed.

Q4 Here the center of the differences is likely to be larger than in Experiment B, and the variability smaller. This suggests that the treatment really does make a difference. For the sample class, the mean and median are larger than in Experiment B (8.357 and 8.0, respectively), and the variability is smaller (SD 5.3, IQR 8). The observed mean difference of 8.357 and lower variability suggest that treatment really does

make a difference. Note that here, the complete box plot is above 0, suggesting that 0 is a rare event.

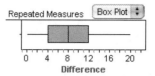

Q5 The unwanted variability is the range in individual heart rate, which can be 30 or even 40 beats per minute within a class of students. For this experiment—to see whether standing people have a higher heart rate than sitting people—the repeated measures design is best as it eliminates more of the unwanted variability than do the other two designs by looking at the *difference* between the sitting heart rate and the standing heart rate for each student.

Q6 The completely randomized design is the least suitable as all of the individual variability remains. As an example where the completely randomized design is best, suppose you are a theater owner and you want to determine the effect of giving moviegoers a flier advertising popcorn when they buy their tickets. Your response variable will be whether the ticket purchaser goes directly to buy popcorn or not. If you expect, say, 200 people at the movie where you are doing this experiment, you would have to use a completely randomized design, assigning each treatment randomly to the people as they enter. You cannot use a matched pairs design because you can't examine the people beforehand in order to match them. You cannot use repeated measures because once a moviegoer gets one method, you cannot use the other method on him or her.

Q7 The variability is the natural differences in heart rate between different individuals. By matching two students with approximately the same heart rate and taking the difference in their rates under the two treatments, the design in Experiment B made the fact that some students naturally have higher rates and some have slower rates largely irrelevant. By having the same student measure his or her heart rate under the two treatments, the design in Experiment C accomplished even more by eliminating individual differences almost entirely.

Comparing Experiment Designs—Sit or Stand
continued

Activity Notes

INVESTIGATE

Q8–Q9 Having both designs in the same plot makes it easier to compare distributions and easier to see the differences between the two distributions. It is much easier to see that the matched pairs design has considerably more variability and is centered at zero, while the repeated measures design has little variability and is definitely centered above zero. It is harder to see the shape of the distributions, but they both look basically mound-shaped. The Repeated Measures plot does look like it could be skewed right, but that could also be due to the scale of the graph. The summary statistics support these findings and both the summary statistics and plots support that the repeated measures design shows best that the treatment does make a difference.

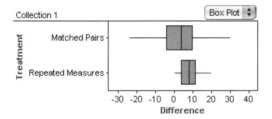

EXPLORE MORE

1. The box plots for the sample data are given in each individual section. Box plots illustrate the summary statistics better, showing the variability in each experiment. Conclusions should be the same.

2. Here is the box plot for the *Difference* split by *FirstTreatment*. The group that sat first is clustered more between 4 and 10, although the distribution is skewed right. Their variability is less than the standing first group, which is quite skewed with a lower median but a much higher variability. It would be interesting to collect more data and see if these findings are true in other samples.

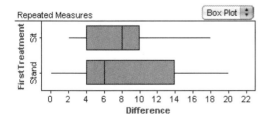

3. The document **SitOrStandRandom.ftm** has one possible simulation. You could use this simulation as a class presentation after the activity. Here is a step-by-step guide to creating the simulation for a completely randomized design.

   First, make a case table with attributes *Group* and *Pulse*. Choose **Collection | New Cases** and enter an even number for the new cases (students). Double-click the collection to show the inspector. Define *Group* with the formula

   $$\text{if(even(caseIndex))} \begin{cases} \text{"Sit"} \\ \text{"Stand"} \end{cases}$$

   This randomly assigns the students to either treatment group.

   Next, use sliders to adjust the pulses randomly selected for each student. Make three sliders: *BasePulse*, *Difference*, and *StDev*. Then define the attribute *Pulse* with the formula

   $$\text{round}\left(\text{randomNormal}\left(\text{if (Group = "sit")} \begin{cases} \text{BasePulse} \\ \text{BasePulse + Difference} \end{cases}, \text{StdDev}\right)\right)$$

   Spend a few minutes changing the slider values to understand how they change the values in your case table.

   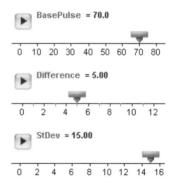

   Display the data in a split dot plot, and use a summary table to compute the mean and sample standard deviation for each treatment.

   Change the slider values and observe the effects each attribute has on the graph and summary statistics.

4
Sampling Distributions

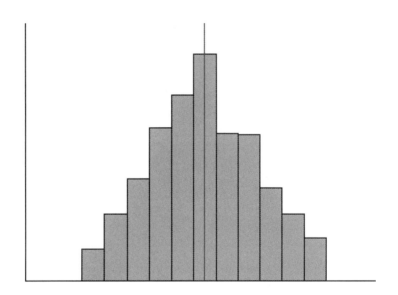

Introduction to Sampling Distributions—Random Rectangles

You will need
- Sampling Rectangles.ftm

In this activity you'll use Fathom to draw samples of size 5 from a collection of rectangles many times to explore the sampling distribution of various summary statistics.

EXAMINE DATA

1. Open **SamplingRectangles.ftm.** The collection named Rectangles contains 100 random rectangles.

   Q1 Make a plot of *Area* and describe the distribution of *Area* in terms of shape, center, and spread.

INVESTIGATE

Taking a Sample

Sample of Rectangles

2. With the collection selected, choose **Collection | Sample Cases.** By default, Fathom takes a sample of ten cases with replacement and places them in a new collection named Sample of Rectangles. You'll change this to five cases without replacement.

Notice that animation is on by default. You may want to change this later.

3. Double-click the sample collection to show its inspector. On the **Sample** panel, change the settings to match these. Click **Sample More Cases.**

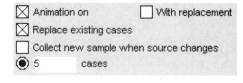

Remember that you can drag an attribute from the **Cases** panel of the inspector.

4. You now have a sample of 5 areas from the collection of all 100 rectangles. Make a dot plot of the attribute *Area* for your sample.

Choose **Graph | Plot Value** and type mean(). Then do the same for the other values.

5. You need to compute some summary statistics for your sample. Plot the values for the mean, median, and maximum of your sample.

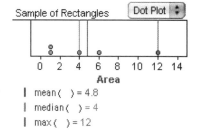

Q2 In a moment, you are going to repeat this sampling, say, 200 times, to create sampling distributions of these summary statistics (or measures). But first, sketch distributions to predict what you will get for the set of 200 mean, median, and maximum areas.

Introduction to Sampling Distributions—Random Rectangles
continued

Collecting Measures

Now you need to define the measures and collect them for several samples.

6. Double-click the sample collection to show its inspector. On the **Measures** panel, define the three measures shown. The values in the inspector should be the same as those plotted on your graph.

Measure	Value	Formula
MeanArea	4.8	mean(Area)
MedianArea	4	median(Area)
MaxArea	12	max(Area)

You can see the samples being taken by watching the dot plot.

7. Select the Sample of Rectangles collection and choose **Collection | Collect Measures.** You should see Fathom take five samples from the Rectangles collection. Each time, Fathom places the measures in a new collection named Measures from Sample of Rectangles.

8. Double-click the Measures from Sample of Rectangles collection to show its inspector. Go to the **Cases** panel. Confirm that the attributes are the measures you defined in step 6.

9. Make three histograms, one for each of the attributes in Measures from Sample of Rectangles.

> Notice that Replace existing cases is off by default, so you need only 195 more measures to make a total of 200 measures.

10. Five samples don't make a very good distribution. Show the inspector for the Measures from Sample of Rectangles collection. Go to the **Collect Measures** panel and change the settings to match these.

   - ☐ Animation on
   - ☐ Replace existing cases
   - ☐ Re-collect measures when source changes
   - ⦿ 195 measures

> You can also select the measures collection and choose **Collection | Collect More Measures.**

11. Click **Collect More Measures.** Even with the animation off, it will take some time to collect another 195 samples of size 5. You can see Fathom taking these samples from the Rectangles collection and calculating the measures by watching the histograms you created in step 9.

Q3 Sketch your histogram of the sampling distribution for each measure.

Introduction to Sampling Distributions—Random Rectangles
continued

Choose Graph | Plot Value.

12. Plot the mean, mean *plus* stdDev, and mean *minus* stdDev on each graph.

The standard deviation of the sampling distribution is often called the *standard error*.

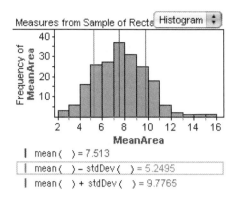

Q4 For the sampling distribution of *MeanArea*, describe the shape, mean, and standard error. How does your sampling distribution of *MeanArea* compare to your prediction in Q2?

Q5 How does your sampling distribution of *MeanArea* compare with the population's distribution of *Area* that you plotted in Q1 in terms of shape, mean, and standard deviation?

Q6 Is the sample mean a good estimate of the mean of the areas of all 100 rectangles?

Q7 Approximately what values of the sample mean for samples of size 5 would be reasonably likely? Which would be rare events?

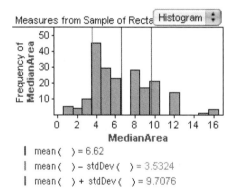

Q8 Repeat Q4–Q7 for the sample median. For Q6, you will need to compute the median of the population of 100 rectangle areas.

Q9 Repeat Q4–Q7 for the sample maximum. For Q6, you will need to compute the maximum of the population of 100 rectangle areas.

Q10 How does the sampling distribution of *MedianArea* compare to the sampling distribution of *MeanArea*?

EXPLORE MORE

1. Define two new measures in your Sample of Rectangles collection:

MaxArea	16	max(Area)
SampleSD	5.49545	stdDev(Area)
PopSD	4.91528	popStdDev(Area)
<new>		

Introduction to Sampling Distributions—Random Rectangles
continued

Collect 200 measures and make a histogram for both standard deviations. Plot the center on both graphs. Compare the center to the population standard deviation, which is 5.20. Which gives a better estimate for the population standard deviation?

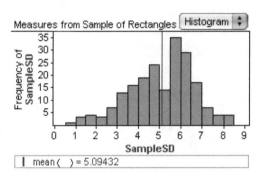

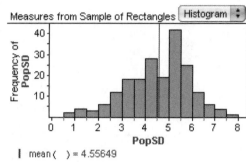

You'll want to turn animation off for both the sample and measures collections.

2. What will happen to the sampling distributions if you take 1000 samples rather than 200 (in the Measures from Sample of Rectangles collection)? Make your predictions and then try it.

3. What is the effect of choosing samples of size 10 instead of 5? Predict, do it, and write up the results.

4. In addition to the mean, median, and maximum, there are a lot of other summary statistics you could use. Pick one, then predict and generate its sampling distribution. Describe your results.

Introduction to Sampling Distributions— Random Rectangles

Activity Notes

Objectives

- Understanding the concept of a (simulated) sampling distribution—the distribution of summary statistics you get from taking repeated random samples
- Identifying the characteristics of sampling distributions: The sampling distribution of the sample mean is mound-shaped and approximately normal, and the mean is at the population mean, whereas the sampling distribution of the sample median is more spread out and less mound-shaped, and the median is near, but not always at, the population median.

Activity Time: 40–50 minutes

Setting: Paired/Individual Activity or Whole-Class Presentation (use **SamplingRectangles.ftm** for either)

Statistics Prerequisites

- Familiarity with taking a sample
- Comparing distributions graphically
- Measures of center and spread

Statistics Skills

- Sampling distributions of summary statistics: mean, median, and max (SD optional)
- Definition of a sampling distribution
- The mean and SE of a sampling distribution
- Biased vs. unbiased statistics
- Identifying characteristics of the sampling distribution in terms of shape, center, and spread
- Preview of the Central Limit Theorem
- The necessity of repeated sampling

AP Course Topic Outline: Part I; Part II B (4); Part III D (1, 2, 6)

Fathom Prerequisites: Students should be able to make graphs, plot values, and define attributes.

Fathom Skills: Students sample from a collection, define and collect measures, and use the population standard deviation function.

General Notes: This activity is essential to ensure that students understand the concept of a sampling distribution. Fathom allows students to easily sample repeatedly.

Procedure: Because this is the first activity in which students generate a sampling distribution, confusion is inevitable. There are three levels of abstraction in this simulation, each level represented by a Fathom collection: the population of rectangles, a sample of five rectangles, and a collection of measures (summary statistics) that result from repeated sampling. Keeping these three levels straight is not easy. If you can project the computer screen so that everyone can see it, you can use Fathom to help students understand these levels. Divide the Fathom window into three vertical areas, one for each level, as shown here. With animation on, collecting summary statistics will cause balls to move from the population to the sample, then from the sample to the measures collection. If you use this activity as a whole-class presentation, open the Sample of Rectangles collection so that students can see which rectangles have been randomly selected for each sample.

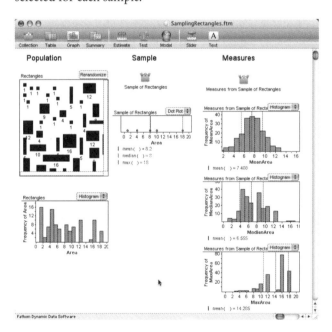

Before you begin the activity proper, you may want to have students open **SamplingRectangles.ftm** and explore the Rectangles collection. Looking in the case table, you'll see that "behind" the display are five attributes: *Height*, *Width*, *Area*, *Px*, and *Py*. *Height*, *Width*, *Px*, and *Py* are primarily used to create the illustrated display. *Area* contains the important data, and it is the attribute that students will use for collecting measures.

Introduction to Sampling Distributions—Random Rectangles

EXAMINE DATA

Q1 The population has mean 7.41 and SD 5.2. The shape is not normal and is closer to uniform although it has two values (1 and 4) that rise higher than the general pattern.

INVESTIGATE

Simulated answers will vary.

Q2–Q3 The three histograms in the right column on the preceding page should be similar to what the students draw here.

Q4–Q7 The simulated sampling distribution of *MeanArea* is approximately normal with mean 7.408 and SD about 2.08. The shape of the sampling distribution is completely different from the shape of the original population, but the mean is very close to the population mean. The SD of the sampling distribution is quite a bit smaller (as it should be—theoretically it is $5.2/\sqrt{5} = 2.33$). The sample mean is a very good estimate (unbiased estimator) of the population mean because it tends to give values very close to the mean, on average. (If you took all samples of size 5, you would get that the mean of the sampling distribution was exactly the population mean.) Because the sampling distribution is approximately normal, we can estimate that reasonably likely outcomes are those in the interval $7.408 \pm 2(2.08)$, or 3.428 to 11.568.

We can also estimate these values using the histogram. Rare events are those in the upper 2.5% of the distribution and the lower 2.5% of the distribution. Because there are 200 samples, this would be the largest five means and the smallest five means. We will have to approximate as we cannot isolate the five largest and five smallest from the histogram: about 12 or larger, or 3 or smaller. So, a mean larger than 12 or smaller than 3 from a sample of size 5 would be a rare event.

Q8–Q10 The typical approximate sampling distribution of the median has mean 6.62 and standard error 3.16. The mean is a little less than that of the population, and the spread is a lot less. It is more mound-shaped than the population, but we still would not call it approximately normal. The center of the distribution of medians should be below the center of the distribution of means. The distribution of sample medians is more spread out and less mound-shaped than the distribution of sample means.

The median of the population of 100 rectangle areas is 6, so it does appear that the median of the sampling distribution of medians, for random samples, is close to the population median. Here, in this sample, the median of the sampling distribution is 6. So, it appears to be a good estimator of the population median.

The sampling distribution is not very mound-shaped, so it is better to use the histogram to approximate the upper 2.5% and lower 2.5% of the 200 sample medians. Rare events would be the smallest five sample medians and the largest five. The first bar contains four medians and the second bar contains three, so we cannot isolate the smallest five. Sample medians less than 2 or more than 13 would be rare events. Medians between 2 and 13 are reasonably likely.

Q9 A typical simulated sampling distribution of the sample maximum for random samples of five rectangles is strongly skewed left. The maximum in the population is 18. The sample maximum underestimates the population maximum. A small sample is not very likely to contain the maximum value from the population, so the best estimate of the population maximum should be a little larger than the sample maximum. The sample maximum is a biased estimator of the population maximum and is biased in the direction of tending to be too small. The distribution is not close to being mound-shaped, so we will use the histogram to approximate the upper 2.5% and lower 2.5% of the 200 sample maximums. Sample maximums less than 6 would be rare events. Maximums between 6 and 18 are reasonably likely. Here is a box plot of the sample maximums:

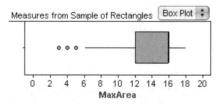

Introduction to Sampling Distributions—Random Rectangles

continued

Activity Notes

DISCUSSION QUESTIONS

- What does one case in the Rectangles collection represent? One case in Sample of Rectangles? One case in Measures from Sample of Rectangles?
- Why is the distribution of sample medians so much bumpier than the distribution of sample means?
- Why is the distribution of sample maximums skewed left?

EXPLORE MORE

1. The effect of dividing by 4 rather than by 5 makes the standard deviation larger. When you divide by $n - 1$, the center of the sampling distribution is much closer to the population standard deviation than when you divide by n. Note that, even dividing by $n - 1$, the sample standard deviation is a biased estimator of the population standard deviation—it tends to be a bit too small. The normal distribution is not a good model for sampling distributions of standard deviations unless the sample size is very large. For smaller samples, the distribution is skewed right.

2.–3. The values should remain very close to where they were for 200 samples. The sample median area will likely get closer to the mean, and the sample maximum will likely get slightly closer to the population maximum.

Sampling Distributions of the Sample Mean—Pocket Pennies

You will need
- 25 pennies collected from recent day-to-day change

Some of the distributions of data that you have studied have had a roughly normal shape, but many others were not normal at all. What kind of distribution tends to emerge when you create sampling distributions of the mean from these non-normal populations? Here you'll explore that question.

COLLECT DATA

1. Enter the dates on your random sample of 25 pennies into a Fathom case table, using the attribute *Year*. Create a second attribute, *Age*, with a formula that calculates the difference between the current year and *Year*.

   Pocket Pennies

	Year	Age
=		2005 – Year
1	1996	9
2	1985	20
3	1985	20
4	1993	12

   Q1 If you were to make a histogram of the ages of all the pennies from all the students in your class, what do you think the shape of the distribution would look like? Sketch your prediction.

   You can copy cases from one case table to another by choosing **Select All Cases**, **Copy**, and **Paste** from the **Edit** menu.

2. Combine everyone's data into one collection and make sure everyone has a copy of that Fathom document.

3. Using the complete collection, make a histogram of the ages of all the pennies in the class.

   Q2 How does the actual distribution compare with your prediction in Q1?

   Q3 Estimate the mean and standard deviation of the distribution. Confirm these estimates by computing the mean and standard deviation in Fathom. Either plot the values on the histogram or use a summary table.

INVESTIGATE

Building a Sampling Distribution

Next you'll take a random sample of size 5 from the ages of your class's pennies.

4. Select the collection, and choose **Collection | Sample Cases.** By default, Fathom takes a sample of ten cases with replacement and places them in a new collection named Sample of Pocket Pennies. You'll change this to five cases without replacement.

Sampling Distributions of the Sample Mean—Pocket Pennies
continued

> Notice that animation is on by default. You may want to change this later.

5. Double-click the Sample of Pocket Pennies collection to show its inspector. On the **Sample** panel, change the settings to match these. Click **Sample More Cases** to re-collect your sample.

   ☒ Animation on   ☐ With replacement
   ☒ Replace existing cases
   ☐ Collect new sample when source changes
   ⦿ 5   cases

> The *SampleSize* measure will allow you to compare different sample sizes later on.

6. Go to the **Measures** panel of the inspector and define these measures.

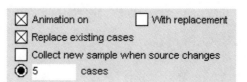

Q4 If you were to make a histogram of the mean ages from several samples, do you think the mean of the values in this histogram would be larger than, smaller than, or the same as the mean of the population of the ages of all pennies? Regardless of your choice, try to make an argument to support each choice. Estimate what the standard deviation of the distribution of the mean ages will be.

> You may want to turn off animation in the sample and measures collections.

7. Collect the mean ages from several samples by selecting the sample collection, then choosing **Collection | Collect Measures.** Show the inspector for the measures collection and change to these settings. Click **Collect More Measures.**

   ☐ Animation on
   ☒ Replace existing cases
   ☐ Re-collect measures when source changes
   ⦿ 100   measures

8. Make a histogram of *MeanAge*. Compute the mean and standard deviation of *MeanAge* by plotting values on the graph or using a summary table.

Q5 Which of the three choices in Q4 appears to be correct?

Changing the Sample Size

You'll now collect measures for samples of size 10 and size 25. You'll be able to create a split histogram to compare the effect of sample size.

9. Show the inspector for the sample collection. On the **Sample** panel, change the sample size to 10.

10. Show the inspector for the measures collection. On the **Collect Measures** panel, uncheck Replace existing cases. This allows you to put the measures from all the samples into one collection. Then click **Collect More Measures.**

Sampling Distributions of the Sample Mean—Pocket Pennies
continued

11. Repeat steps 9 and 10 to change the sample size to 25 and collect 100 more measures.

Holding down the Shift key tells Fathom to use the numerical values of the attribute as categories.

12. Drag the attribute *SampleSize* of the measures collection and drop it on the vertical axis of the histogram for *MeanAge* while holding down the Shift key. Your histogram should split three ways, showing distributions for each sample size (5, 10, and 25).

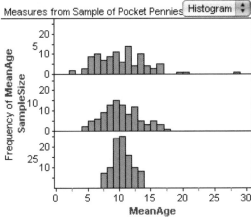

13. Compute the mean and standard deviation for each of the three sampling distributions, using a summary table. Again, hold down the Shift key when you drop *SampleSize* in the summary table.

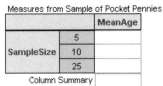

Q6 Look at the four histograms you constructed. As the sample size increases, what can you say about the shape of the histogram of sample means? About the center? About the spread?

Q7 Compare the values you got in step 13 for the mean and SD for the three sampling distributions with the values you got in Q3 for the whole population. Then figure out formulas for the mean and SD of a sampling distribution that relate them to the population mean and SD.

14. On your histogram, plot the values mean, mean *plus* 2SD, and mean *minus* 2SD.

    MeanAge
    | mean( ) = 10.2621
    | mean( ) + 2 stdDev ( ) = 16.3766
    | mean( ) − 2 stdDev ( ) = 4.14756

Q8 What percentage of sample means are within 2 SD's of the population mean for each sample size?

Q9 For which sample sizes would it be reasonable to use the rule stating that 95% of all sample means lie within approximately 2 SD's of the population mean?

EXPLORE MORE

Open the Fathom document **LifeExp.ftm.** In this file you will find data on the life expectancy for females in Asia and Africa. Discuss the shapes of the original population. Take 200 samples of size 5 from each population. Do your conclusions from Q6–Q9 still hold up?

Sampling Distributions of the Sample Mean— Pocket Pennies

Activity Notes

Objectives

- Understanding the concept of a sampling distribution of the sample mean and how to generate one
- Discovering the properties of the shape, mean, and standard deviation of the sampling distribution of the sample mean
- Recognizing that the mean of the sampling distribution of the sample mean is approximately the mean of the population
- Seeing that the standard deviation of the sampling distribution of sample means decreases as the sample size increases
- Being introduced to the Central Limit Theorem: The sampling distribution of the sample mean approaches the normal distribution as the sample size increases, regardless of the shape of the original population distribution.

Activity Time: 30–50 minutes (the shorter time is when data collection is done the day before the activity)

Setting: Paired/Individual Activity (collect data using **PenniesTemplate.ftm** or use **Pennies.ftm**) or Whole-Class Presentation (use **Pennies.ftm**)

Optional Document: LifeExp.ftm (Explore More)

Materials

- 25 pennies collected by each student from recent day-to-day change

Statistics Prerequisites

- Familiarity with taking a sample
- Comparing distributions graphically
- Measures of center and spread

Statistics Skills

- Sampling distributions of the sample mean
- Properties of the shape, center, and spread of the sampling distribution of the sample mean
- Introduction to a geometric distribution
- Normal distribution
- Reasonably likely sample means
- Central Limit Theorem

AP Course Topic Outline: Part II B (4); Part III C, D (2, 3, 6)

Fathom Prerequisites: Students should be able to make case tables and graphs, plot values, find statistics in a summary table, and define attributes.

Fathom Skills: Students sample from a collection, define and collect measures, and combine measures for different sample sizes. *Optional:* Students use the normal density function (Extension 2) and a normal quantile plot (Extension 3).

General Notes: In this activity, students discover the properties of the shape, mean, and standard deviation of the sampling distribution of the mean for samples taken from a distribution that is decidedly not normal. It involves a lot of repeated random sampling to create distributions of means for different sample sizes. Because Fathom automates the sampling process, students will spend less time on the busywork and more time examining the results.

Procedure: During the week before the activity, have each student collect the first 25 pennies that he or she receives in change from various purchases and emphasize that these should not be some collection of pennies stored from years long past. Students should bring in the pennies and a list of the 25 dates on the pennies.

You will need to collate all the data into a single Fathom document that students can work with. The student worksheet suggests that this be done using copy and paste. Alternatively, you could use the master document, **PenniesTemplate.ftm,** into which all students enter data. A third alternative is to collect a list of the dates of the pennies from the previous class session, type them into a Fathom document, and distribute this document to the class. Another possibility is to have students type in their data and then email their document to one person who will copy and paste the data into a new class document. If you don't have time to collect data, the document **Pennies.ftm** contains sample data.

Sampling Distributions of the Sample Mean—Pocket Pennies
continued

Activity Notes

COLLECT DATA

Q1–Q3 Few students realize that the shape of the distribution of the ages of all pennies will be roughly geometric. Many will believe that it should be normal ("A few pennies are new, a few pennies are old, most are lumped in the middle.").

In a number of places, students are asked to compute the mean and standard deviation of a distribution. Two ways of doing that are shown here. Plotting values for the mean and the mean plus or minus the standard deviation on top of a histogram is very satisfying, but it doesn't give you a direct value for the standard deviation. Using a summary table gives you the numbers but no visual context.

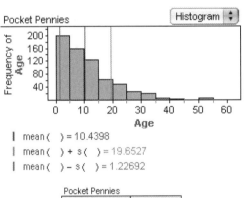

The mean age for the population tends to be between 7 and 8 years, with standard deviation about 8 years. If your results for Q3 are contrary to this standard, you may want to discuss potential causes.

INVESTIGATE

Q4–Q5 The mean will be the same. Most students will not realize this. A typical answer is to say it will be smaller. Most students will realize that the standard deviation will be smaller.

Q6–Q7 This split histogram shows the kind of results students are likely to get. The shape of the distributions becomes approximately normal as the sample size increases, the mean stays the same, and the standard deviation decreases. Specifically, the mean of the three sampling distributions should be approximately the same as the mean of the population of all pennies: $\mu_{\bar{x}} = \mu$. The standard deviation of the sampling distributions should approximately equal the standard deviation of the population divided by the square root of the sample size: $\sigma_{\bar{x}} = \sigma/\sqrt{n}$.

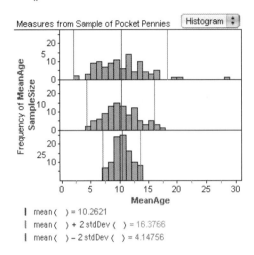

Q8 It is easiest to use the histogram. For both samples of size 5 and 10, using the sample statistics as shown in the above histogram will give nearly the same values as would plotting the theoretical values. For samples of size 5 in the sample document, 97% are within 2 standard deviations of the population mean (10.439). For samples of size 10, 97% are within 2 standard deviations of the population mean. For samples of size 25, 100% are within 2 SD's of the mean using the sample statistics as shown below. Using the population parameters and the Central Limit Theorem, the percentage within 2 SD's is exactly 95%.

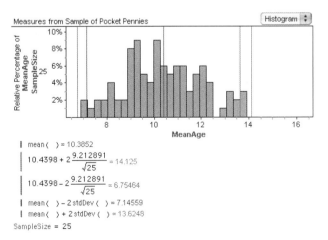

Sampling Distributions of the Sample Mean—Pocket Pennies
continued

Activity Notes

Q9 For this sample of pennies, it seems that we could use the rule for any of them. However, it would be safer to try sampling again to see if this holds up. Typically, it would not be a good idea to use this rule with samples of size 5.

DISCUSSION QUESTIONS

- What does one case in the Pennies collection represent? One case in Sample of Pennies? One case in Measures from Sample of Pennies?
- What did you observe as you increased the sample size from 5 to 10 to 25? Explain why this makes sense.
- For samples of size 25, what characteristics does the sampling distribution of the sample mean share with the normal distribution?

EXTENSIONS

1. Have students extend the sample sizes to 50 and 100. Does this confirm or modify the conclusions they made in Q6–Q9?

2. Have students figure out how to plot a normal curve on top of the histograms of *MeanAge* for each sample size. (Try searching Fathom Help for "normal distribution.") What do they observe about the fit of the histogram with the curve? [Students should plot the function 100•normalDensity(x, *population mean,* stdDev()). The 100 scales it vertically, and the population mean translates the center. Students should find that as sample size increases, the distribution approaches the normal curve—the Central Limit Theorem. This is especially visible if students try collecting measures for larger sample sizes, such as 100.]

3. Students can investigate using a normal quantile plot to determine closeness to normality. (Use Fathom Help to learn about normal quantile plots.) Which sample size produces a distribution closest to normal? [The normal quantile plot should show each sample size to be normal. The sample of size 25 should appear slightly more normal than the others, however.]

Sampling Distributions of the Sample Proportion—Seat Belts

You often hear reports of percentages or proportions: About 60% of automobile drivers in Kentucky use seat belts. Suppose you take a random sample of 40 Kentucky drivers and count the number who wear seat belts. You would expect to get 60% who wear seat belts, but you might get a lot fewer or a lot more. In this activity you will simulate sampling distributions for the proportion of drivers wearing seat belts in samples of size 10, 20, and 40.

GENERATE DATA

*To add cases, choose **Collection | New Cases.***

1. Open a new Fathom document. Make a new collection named Passengers with an attribute named *Buckled*. Add 10 cases to it. This collection represents a sample of 10 drivers.

Consider an "if" statement or randomPick for your formula.

2. Define *Buckled* with a formula that causes the values for the attribute to be either "Yes" or "No" with a 60% chance of being "Yes."

3. Define three measures that compute the size of the sample, the number of successes ("Yes"), and the sample proportion ($\hat{p}$).

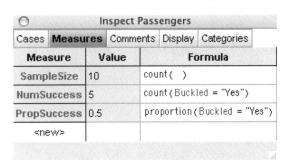

4. Collect measures from the Passengers collection: Select the Passengers collection, then choose **Collection | Collect Measures.**

5. Show the inspector for the measures collection, go to the **Collect Measures** panel, and change to these settings. Click **Collect More Measures.**

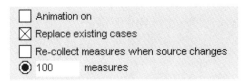

Sampling Distributions of the Sample Proportion—Seat Belts
continued

INVESTIGATE

6. Make a histogram of the 100 sample proportions (*PropSuccess*). On this graph, plot the values of the mean, the mean plus one standard deviation, and the mean minus one standard deviation.

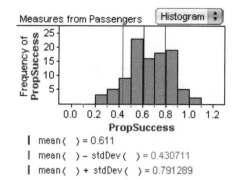

Recall that standard deviation of a sampling distribution is sometimes called the *standard error.*

Q1 Describe your sampling distribution for samples of size 10 in terms of shape, center, and spread.

Q2 If you increase the sample size, how do you think the sampling distribution will be affected?

7. Split the histogram by dropping *SampleSize* on the vertical axis while holding down the Shift key. (The graph won't split yet because you only have samples of size 10.)

Collect *additional* measures by unchecking Replace existing cases in the measures collection's inspector.

8. Change the sample size to 20 by adding 10 cases to the Passengers collection. Collect 100 *additional* measures. Your histogram should now be split into two sample sizes.

Q3 Describe your sampling distribution for samples of size 20 in terms of shape, center, and spread.

Q4 Compare your sampling distributions for samples of size 20 and size 10. Do they have any similarities? Any differences?

9. Change the sample size to 40 and again collect 100 *additional* measures.

Q5 Describe your sampling distribution for samples of size 40 in terms of shape, center, and spread.

Q6 Compare your sampling distributions. Do they have any similarities? Any differences?

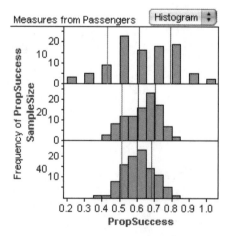

Sampling Distributions of the Sample Proportion—Seat Belts
continued

10. Make a summary table to calculate the mean and standard deviation of your simulated sampling distributions. Remember to hold down the Shift key when dropping *SampleSize*.

Measures from Passengers

		PropSuccess
SampleSize	10	0.18028877
		0.611
	20	0.095879765
		0.607
	40	0.086370414
		0.5985
Column Summary		0.12768422
		0.6055

S1 = stdDev( )
S2 = mean( )

Q7 How does the mean of the sampling distribution compare to the mean of the population? Does the mean of the sampling distribution appear to depend on p? On n?

Q8 How does the standard error of the mean (the SD of the sampling distribution) compare to the standard deviation of the population? Does the standard error appear to depend on p? On n?

Q9 Does the sampling distribution appear to be approximately normal in all cases? Does the shape appear to depend on p? On n?

Q10 For which sample sizes would it be reasonable to use the rule that about 95% of all sample proportions lie within approximately two standard errors of the population proportion? Explain.

Q11 For each sample size, what sample proportions are reasonably likely?

EXPLORE MORE

The stacked box plots provide a hint as to how to proceed.

The ethnicity of about 92% of the population of China is Han Chinese. Explain how to reconstruct the simulation to simulate sampling from this population. Set up this simulation to figure out how large a sample you would need for the sampling distribution of the sample proportion to be approximately normal.

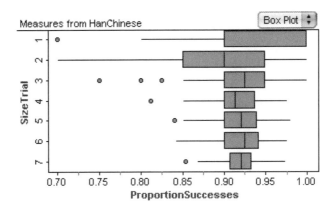

Sampling Distributions of the Sample Proportion— Seat Belts

Activity Notes

Objectives

- Understanding the concept of a sampling distribution of the sample proportion
- Discovering the properties of the shape, mean, and standard deviation of the sampling distribution of the sample proportion
- Recognizing that the mean of the simulated sampling distribution of the sample proportion is approximately the proportion of successes in the population
- Seeing that the standard deviation of the simulated sampling distribution of the sample proportion decreases as the sample size increases
- Reinforcing the Central Limit Theorem: As the sample size increases, the shape of the sampling distribution gets more normal

Activity Time: 20–40 minutes

Setting: Paired/Individual Activity (build simulation) or Whole-Class Presentation (use **SeatBelts.ftm**)

Materials

- *Optional:* A bag of beads, sampling paddle, or other mechanical method of taking random samples from a population with $p = 0.6$

Statistics Prerequisites

- Familiarity with taking a sample
- Familiarity with the distribution of a proportion
- Comparing distributions graphically
- Measures of center and spread

Statistics Skills

- Sampling distributions of the sample proportion
- Properties of the shape, center, and spread of the sampling distribution of the sample proportion
- Introduction to the reason for the conditions that np and $n(1 - p)$ are at least 10
- Reasonably likely sample proportions
- Central Limit Theorem

AP Course Topic Outline: Part III C, D (1, 3, 6)

Fathom Prerequisites: Students should be able to make collections and graphs, plot values, find statistics in a summary table, and define attributes.

Fathom Skills: Students use formulas to create a collection that represents a random sample, define and collect measures, and combine measures for different sample sizes. *Optional:* Students collect cells from a summary table (Extension 1).

General Notes: This activity shows a different way to simulate sampling. Usually, students begin with a collection that represents a population and then is sampled. In this activity the beginning collection represents the sample and a formula is used to randomize it. Fathom reduces the busywork, and helps students concentrate on results.

Procedure: You might begin this activity by taking samples using a mechanical method, such as drawing from a box of beads of two colors. Fill a box with beads of two different colors, say, blue and yellow, such that exactly 60% of the beads are one color (blue) and 40% are the other color (yellow). Then each blue bead will represent a "success" if selected and each yellow bead will represent a "failure." Have a student pick out 1 bead and record whether it's a success (blue) or failure (yellow), then replace the bead, stir the beads, and then repeat the process 9 times to get a sample of size 10. The sample proportion will then be

$$\hat{p} = \frac{\text{number of successes}}{\text{total picked}}$$
$$= \frac{\text{number of blue beads in sample}}{10}$$

Some students must actually "see" the sampling done a few times to catch on to the idea of what is considered a success or failure before they are able to move to tables of random digits, the calculator, or Fathom.

To use this activity as a whole-class presentation, open the document **SeatBelts.ftm**. Select the Passenger collection, then choose **Collection | New Cases**. Add 10 cases. The file is set up to collect 100 measures and plot them each time you add to the original collection.

GENERATE DATA

2. Two ways to write the *Buckled* formula are

$$\text{if(random()} < 0.6) \begin{cases} \text{"Yes"} \\ \text{"No"} \end{cases}$$

randomPick("Yes", "Yes", "Yes", "No", "No")

Sampling Distributions of the Sample Proportion—Seat Belts
continued

Activity Notes

6. Students are asked to make a histogram. A dot plot will also work fine here but will become awkward by step 8 because splitting will prevent the dots from fitting in their allotted space.

INVESTIGATE

Q1 The mean should be close to 0.6 and the standard error close to 0.15. For the histogram shown in the activity, the sampling distribution for samples of size 10 is roughly mound-shaped with a little skew to the left. The center is close to 0.6 and the spread is large (SD 0.18, IQR 0.3), with reasonably likely sample proportions ranging from about 0.3 to 0.9.

Q2 See answers for Q4 and Q6.

Q3 The mean should be close to 0.6 and the standard error close to 0.11. For the histogram shown in the activity, the sampling distribution for samples of size 20 is mound-shaped with only a very slight skew to the left. The center is close to 0.6 and the spread is smaller (SD 0.1, IQR 0.15), with reasonably likely sample proportions ranging from about 0.45 to 0.8.

Q4 The centers are similar and the spreads are different, with the spread for samples of size 10 larger than the spread for samples of size 20.

Q5 The mean should be close to 0.6 and the standard error close to 0.08. For the histogram shown in the activity, the sampling distribution for samples of size 40 is approximately normally distributed with no visible skewness. The center is close to 0.6 and the spread is somewhat smaller (SD 0.086, IQR 0.125), with reasonably likely sample proportions ranging from about 0.45 to 0.75.

Q6 The centers are all around 0.6, and the spreads are different with the larger sample sizes having smaller standard errors.

10. The summary table should clearly show that the mean of each sampling distribution is approximately the population proportion (0.6) and that the standard deviation (standard error) decreases as the sample size increases.

Q7 The means of the sampling distributions definitely depend on p. The sampling distributions have centers close to p regardless of the sample size; the centers do not depend on the sample size.

Q8 The spreads of the sampling distributions decrease as n increases for all values of p. The spreads do depend on the value of p as well (see Extension 2). At this point, however, students will not know whether it does or not.

Q9 The shapes of the sampling distributions become more normal as n increases for all values of p. The shape also depends on p: the farther p is from 0.5, the more skewness (see Extension 2).

Q10 The rule does not work well for samples of size 10, works somewhat better for samples of size 20, and works well for samples of size 40 or more.

Q11 The sampling distribution for samples of size 10 has reasonably likely sample proportions ranging from about 0.3 to 0.9. For samples of size 20, the reasonably likely sample proportions range from about 0.4 to 0.8. For samples of size 40, the reasonably likely sample proportions range from about 0.45 to 0.75.

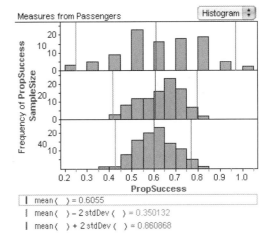

DISCUSSION QUESTIONS

- What does a case in the Passengers collection represent? A case in the Measures from Passengers collection?

- What did you observe about the sampling distribution of the sample proportion as sample size increased?

- In this activity the Passengers collection represented your sample and you randomized the sample with a

Sampling Distributions of the Sample Proportion—Seat Belts
continued

formula. How could you reconstruct the simulation so that you have a population collection from which you collect samples? [One method is to create a population collection that contains six "Yes" and four "No." Then sample this collection *with* replacement.]

EXPLORE MORE

Set up the simulation with an "if" statement. By SizeTrial = 5 ($n = 100$), most of the skew is gone and the distribution is roughly normal. The respective sample sizes that go with the box plot in the activity are 10, 20, 40, 80, 100, 120, and 150.

EXTENSIONS

1. Have students collect measures for at least two more sample sizes, say, 80 and 160. Make a scatter plot of the standard deviation of the sampling distribution versus sample size. Fit a curve through these points. What is the basic relationship between sample size and standard deviation of the sampling distribution of the proportion?

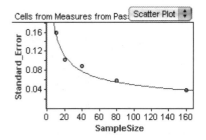

[Students should see an inverse relationship, as defined by $\sigma_{\hat{p}} = \frac{\sigma}{\sqrt{n}}$. One way to do this with Fathom is to make a summary table as shown, then choose **Summary | Create Collection From Cells**. The attributes in the new cells collection will be *SampleSize* and *S1*, where *S1* is the standard error. Students can then make a scatter plot as usual.]

Measures from Passengers

	10	0.15891043
	20	0.10247444
SampleSize	40	0.088221771
	80	0.058226498
	160	0.038281623
Column Summary		0.090974467

S1 = stdDev(PropSuccess)

2. Have students repeat the activity with a probability of success of 0.9 instead of 0.6. Comment on the differences between the new set of distributions and the distributions for $p = 0.6$. [For $\hat{p} = 0.9$, the centers will translate to .9 and the spreads will decrease according to the formula $\sigma_{\hat{p}} = \sqrt{\frac{p(1-p)}{n}}$. The shape will be more skewed for the smaller sample sizes and will still be skewed for samples of size 40.]

Sampling Distributions of the Sample Sum and Difference—Dice Rolls

You are going to use Fathom to simulate rolling two dice. For each roll, you will compute the sum and difference of the two dice. Then you'll examine the sampling distributions of the sum and difference.

GENERATE DATA

1. Create a Fathom collection, named Rolls, with 24 cases. Each case will represent a roll of two dice.

Consider randomInteger or randomPick as functions for generating die rolls.

2. Create attributes as shown. Define formulas that randomly roll the dice (have values 1 to 6), and calculate the sum and difference.

Rolls

	Die1	Die2	Sum	Difference
1	6	6	12	0
2	2	5	7	-3
3	6	1	7	5
4	4	2	6	2

Q1 If each face is equally likely on *Die1*, what should the histogram (or dot plot) look like for 24 rolls?

Q2 Make a histogram of *Die1*. Is the distribution approximately uniform, indicating that each outcome is equally likely, or does the distribution follow some other pattern? If it follows some other pattern, why might that be the case?

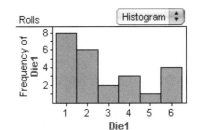

Q3 Predict what the sampling distribution of the sum and the sampling distribution of the difference will look like for the rolls of two dice. Sketch your predictions.

Q4 Make a histogram of *Sum* and a histogram of *Difference*. Compare your histograms with what you predicted in Q3. Can you describe the sampling distributions in terms of shape, center, and spread based on your 24 rolls of two dice? If so, describe the distributions. If not, explain why not.

INVESTIGATE

Usually having more cases helps you see trends.

Q5 If each face is equally likely on *Die1*, what should the histogram look like for 1000 rolls?

3. Add cases so that the collection has a total of 1000 cases.

Sampling Distributions of the Sample Sum and Difference—Dice Rolls
continued

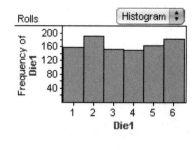

Q6 Does the histogram you plotted for *Die1* look like what you predicted in Q5? Based on this histogram, does it look like each outcome could be equally likely?

Q7 In Q3 you predicted what the sampling distribution of the sum and the sampling distribution of the difference would look like for the rolls of two dice. Compare your predictions with the histogram of *Sum* and the histogram of *Difference*. Are these closer to what you predicted than they were in Q4?

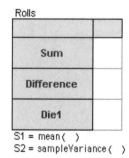

> You can choose **Collection | Rerandomize** to look at new samples of 1000 rolls.

Q8 Describe the sampling distributions of the sum and difference in terms of shape, center, and spread based on your 1000 rolls of two dice.

Q9 How do the sampling distributions of the sum and difference compare with the distribution of a roll of a single die?

Q10 Plot the value of the mean for each distribution. How do the means of the distributions of sums and differences compare with the mean of the distribution of a single roll of a die?

> Fathom has the built-in function variance().

4. Use summary tables to compute the variance of each distribution.

Q11 How do the variances of the distributions of sums and differences compare with the variance of the distribution of a single roll of a die? How do they compare with each other? Can you find formulas that relate them?

EXPLORE MORE

1. Investigate the sampling distribution of the sum of three dice rolls and of four dice rolls. Describe and explain your results.

2. Investigate the sampling distributions of the sum and difference for two dice with a number of sides other than 6. Describe and explain your results.

3. Make a scatter plot of *Difference* versus *Sum*. Explain your results.

Sampling Distributions of the Sample Sum and Difference—Dice Rolls
continued

Consider randomNormal instead of randomInteger to solve this problem.

4. Use what you have learned to answer this question by setting up a similar simulation: Bottle caps are manufactured so that their inside diameters have a distribution that is approximately normal with mean 36 mm and standard deviation 1 mm. The distribution of the outside diameters of the bottles is approximately normal with mean 35 mm and standard deviation 1.2 mm. If a bottle cap and a bottle are selected at random (and independently!), what is the probability that the cap will fit on the bottle?

5. Adjust your simulation in Explore More 4 to answer this question: The distribution of the outside diameters of a set of bottles is approximately normal with mean 35 mm and standard deviation 0.6 mm. The mean of the bottle caps is 36 mm. How small of an error (SD) do you need in your cap-making machine to guarantee that your machine makes a cap that fits 95% of the bottles?

Sampling Distributions of the Sample Sum and Difference—Dice Rolls

Activity Notes

Objectives

- Understanding the concept of sampling distributions of the sum and difference
- Discovering the properties of shape, mean, and *variance* of the sampling distributions of the sum and difference
- Seeing that the mean of the sampling distribution of the sum is approximately the sum of the two population means, and the mean of the sampling distribution of the difference is the difference of the means
- Recognizing that the variance of either sampling distribution is the sum of the population variances

Activity Time: 20–40 minutes

Setting: Paired/Individual Activity (build simulation) or Whole-Class Presentation (use **SumDifference.ftm**)

Optional Document: BottleFit.ftm (Explore More 4 and 5 solutions)

Materials

- *Optional:* One (or two) die for each group of three students

Statistics Prerequisites

- Familiarity with sampling distributions
- Some familiarity with the equally likely outcomes
- Comparing distributions graphically
- Measures of center and spread

Statistics Skills

- Sampling distributions of the sample sum and difference
- Properties of the shape, center, and spread of the sampling distribution of the sample sum and difference
- Central Limit Theorem

AP Course Topic Outline: Part III B, C, D (4, 5, 6)

Fathom Prerequisites: Students should be able to make collections and graphs, plot values, find statistics in a summary table, define attributes, and use randomInteger or randomPick functions.

Fathom Skills: Students use formulas to create a collection that represents a random sample, use Fathom to do simulations, and define measures to create the sampling distributions of the sum and difference.

General Notes: With Fathom, students can accomplish the simulation in a very straightforward manner without having to collate classroom data. Once the simulation has been constructed, it can be rerun instantly by rerandomizing the collection. Furthermore, the sample size can be very large (1000) so that the shape of the distributions is very apparent.

Procedure: Some students must actually "see" the sampling done to get the idea before they can use Fathom. If so, you can begin this activity by having them do steps 1–4 by hand. Divide your class into groups of three. Each group will roll their die twice. One person will record the sum of the numbers on the two rolls. Another partner will record the difference. The third person will record the number on each die. For example, if the first roll is 2 and the second roll is 4, the first partner records 6, the second records −2, and the third records both the 2 and the 4. Have them continue to roll until their group has recorded 24 sums and 24 differences. Have them answer Q1–Q4. After that, they do the same thing using Fathom starting at the beginning of the activity again.

If you'd rather students didn't roll the die, you can use the presentation document **SumDifference.ftm** to do the above as a class, or have them use this file on their own. The file is set up differently than the activity. Here there are two dice in the Rolls collection. When you click the **Rerandomize** button, Fathom "rolls" the two dice. When you click the **Collect More Measures** button, Fathom "rolls" the two dice 10 times. The animation shows each roll and updates each dot plot. The values for the icons in the Measures from Roll collection are the sums of the two dice.

Sampling Distributions of the Sample Sum and Difference—Dice Rolls
continued

Activity Notes

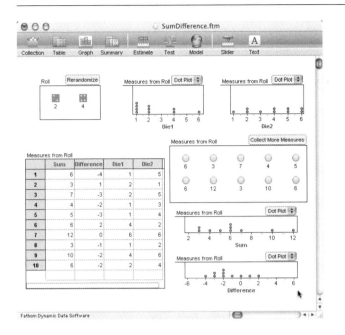

To follow the beginning of the activity, double-click the Measures from Roll collection to show its inspector and change the number of measures to 24.

If your students don't need to "see" the hands-on method, they can start right off with the activity.

GENERATE DATA

2. Either randomPick(1,2,3,4,5,6) or randomInteger(1,6) will generate die rolls.

Q1–Q2 If each face is equally likely, the histogram for *Die1* should be uniform from 1 to 6, with each bin of height 4. More than likely, none of the histograms will look like students' predictions, because of the small sample size.

Q3 Draw attention to the importance of the prediction in this question. It isn't as important that students predict correctly as it is that they think about what is going on before they make it happen.

Q4 The histograms will probably not show any pattern, again because the sample is much too small. Looking at these two distributions, students should see that the difference has a hint of triangularity but the sum definitely does not.

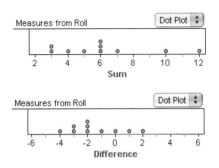

INVESTIGATE

Q5–Q6 If each face is equally likely, the histogram for *Die1* should be uniform from 1 to 6, with each bin height approximately 166.7. More than likely, none of the histograms will look exactly like students' predictions, but they will come fairly close.

Q7–Q8 Both distributions will look approximately triangular. The mean of the sum should be about $3.5 + 3.5 = 7$, or $\mu_1 + \mu_2$. The mean of the difference should be about $3.5 - 3.5 = 0$, or $\mu_1 - \mu_2$. Almost all students will be surprised that the spreads are equal (about 2.4).

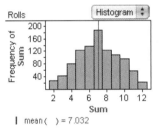

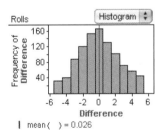

Be sure that students rerandomize the collection at least a few times. This helps them see which features of the histograms are real and which are fluctuations.

Exploring Statistics with Fathom
© 2007 Key Curriculum Press

4: Sampling Distributions | 151

Sampling Distributions of the Sample Sum and Difference—Dice Rolls
continued

(*Note:* The standard deviation of the sums or differences is larger than the standard deviation of an individual roll, but the standard deviations do not add. It is the variances that add.)

Q9 The shape of both distributions is triangular while the shape of the population of outcomes is rectangular.

Q10 The sum of the means of the sum and difference should be approximately equal to the mean of the roll of a single die.

Q11 Both variances should be close to 5.8, or $\sigma_1^2 + \sigma_2^2$. The variance of a single roll should be close to 2.917.

DISCUSSION QUESTIONS

- How is it that the means of the sum and difference distributions are different but the variances are the same?
- How are the means and variances for the sum and difference distributions related to the mean and variance of the distribution of single-die rolls?
- Is there any difference between rolling one die twice and rolling a pair of dice once?

EXPLORE MORE

2. Students could use a slider to redo this simulation for a die with a different number of sides.

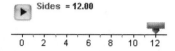

	Die1	Die2	Sum	Difference
=	randomInteger(1, Sides)	randomInteger(1, Sides)	Die1 + Die2	Die1 − Die2

3. The result is a geometric display of the possible sum and difference combinations. By creating a breakdown plot (hold down the Shift key when dropping the attributes to make them categorical), you can get a sense for the probability of getting any particular sum-difference pair.

4.–5. See the document **BottleFit.ftm** for a solution. For Explore More 4, the probability is 0.739. For Explore More 5, the SE must be at most 0.15.

Using Sampling Distributions—By Chance or By Design?

The Westvaco Corporation, which makes paper products, decided to downsize. They laid off several members of their engineering department, and Bob Martin was one of those who lost their jobs. Bob Martin, who was 55, hired a lawyer to sue Westvaco, claiming he had been laid off because of his age. Westvaco's management went through five rounds of planning for a reduction in force. By the time the layoffs ended, after all five rounds, only 22 of the 50 workers had kept their jobs. Here are the ages of the ten hourly workers involved in the second of five rounds of layoffs, arranged from youngest to oldest. The three who were laid off are underlined:

25   33   35   38   48   55   <u>55</u>   <u>55</u>   56   <u>64</u>

The average age of the three who lost their jobs is 58 years. If you pick three of the ten ages at random, is it likely you will get an average age of 58 or more? In this activity you'll use simulation and Fathom to answer this question.

EXAMINE DATA

1. In a new Fathom document, drag a new case table from the shelf. Make a new attribute *Age* and enter the ten workers' ages. Rename the collection Westvaco Workers.

Choose **Graph | Plot Value** and type mean().

2. Make a dot plot of *Age* and plot the mean.

Q1 What is the mean *Age* for the ten hourly workers?

	Westvaco Workers
	Age
1	25
2	33
3	35
4	38
5	48
6	55
7	55
8	55
9	56
10	64

Westvaco Workers

INVESTIGATE

Now, you want to simulate picking three of the ten ages at random for layoff.

3. Click once on the attribute name *Age* to select the column. Choose **Table | Use As Caption.** This will display the age of each worker in the sample collection.

4. Select the collection and choose **Collection | Sample Cases.** By default, Fathom takes a sample of ten cases with replacement and places them in a new collection named Sample of Westvaco Workers. You'll change this to three cases without replacement.

Using Sampling Distributions—By Chance or By Design?
continued

5. Double-click the Sample of Westvaco Workers collection to show its inspector. On the **Sample** panel, change the setting as shown. Click **Sample More Cases.**

   ☒ Animation on   ☐ With replacement
   ☒ Replace existing cases
   ☐ Collect new sample when source changes
   ⦿ 3   cases

6. Resize the Sample of Westvaco Workers collection by dragging its lower-right corner. Make it big enough to see the three sampled cases and their age labels.

7. Click **Sample More Cases** several more times. Observe the ages in your sample each time.

   Q2 From your observations in step 7, do you think it is likely you will get an average age of 58 or more when you choose three workers at random?

Next, you need to compute a summary statistic for the sample.

8. In the Sample of Westvaco Workers inspector, go to the **Measures** panel. Define a new measure as shown. Click **OK.**

Measure	Value	Formula
AverageAge	46	mean(Age)
<new>		

9. Click **Sample More Cases** several times again. Observe the measure *AverageAge* each time you take a new sample. Close the inspector.

   Q3 From your observations in step 9, do you think it is likely you will get an average age of 58 or more when you choose three workers at random?

Now you want Fathom to collect summary statistics and display the distribution.

10. Select the sample collection and choose **Collection | Collect Measures.** By default, Fathom collects five measures and places them in a new collection named Measures from Sample of Westvaco Workers.

11. Make a graph of the distribution for *AverageAge* and plot the mean.

    Q4 Are any of the average ages 58 years or more?

You'll collect more measures in order to better estimate the probability.

If Animation on is checked, this simulation proceeds slowly and you can see a flow of blue balls as samples are collected from the population of ten workers. Then measures from these samples are gathered (green balls), and the dot plot and mean are updated.

Using Sampling Distributions—By Chance or By Design?
continued

12. Double-click the measures collection to show its inspector. On the **Collect Measures** panel, change to these settings. Then click **Collect More Measures.**

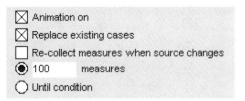

Choose **Graph | Plot Value** and type 58.

13. Plot the value 58 on your plot.

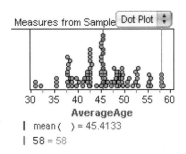

Select the dots in the plot and put your cursor over the measures collection. The number selected will appear in the lower-left corner of the window. Or, make a case table for the measures collection, select *AverageAge,* and choose **Table | Sort Descending.**

Q5 Count the dots that represent an average age of 58 years or more. Calculate the probability by dividing by 100, and interpret the results. Is an average age of 58 years or more reasonably likely?

14. Uncheck Animation on in the inspectors of both the sample and measures collections so that the simulation will proceed more quickly.

15. Make a new summary table and drag the attribute *AverageAge* to either arrow. Choose **Summary | Add Formula** and type proportion(AverageAge≥58). You should get the same value as you did in Q5.

16. Collect 1000 measures, and enlarge your dot plot or change it to a histogram.

Q6 Estimate the probability that if you choose three workers at random, just by chance you will get an average age of 58 years or more. Is an average age of 58 years or more reasonably likely?

EXPLORE MORE

1. Perform the simulation many times with 100 sample means, and observe how the probability fluctuates for a small number of sample means. One way to do this is to define a measure, *Proportion,* for the measures collection using the formula proportion(AverageAge≥58). Then collect measures from the measures collection.

2. Explore what happens when you change the number of measures collected. How does the probability change between 20 sample means, 100 sample means, and 1000 sample means? Does the probability approach a particular value?

Using Sampling Distributions— By Chance or By Design?

Activity Notes

Objectives
- Exploring a set of data related to an actual court case involving alleged age discrimination and practicing relating patterns in data to possible meanings in the applied context
- Learning how to use simulation to decide if an event reasonably can be attributed to chance or whether you should look for some other explanation
- Creating a model of a chance process
- Using the concept of a sampling distribution to find the probability of a given event and to find reasonably likely values

Activity Time: 40–50 minutes

Setting: Paired/Individual Activity (build simulation) or Whole-Class Presentation (use **WestvacoPresent.ftm**)

Materials
- *Optional:* Ten 3 × 5 cards (or small pieces of paper) for each group of students
- *Optional:* Pencil
- *Optional:* Box or other container

Statistics Prerequisites
- Familiarity with sampling distributions
- Some familiarity with reasonably likely outcomes
- Familiarity with the definition of probability
- Measures of center and spread

Statistics Skills
- Sampling distributions of the sample mean
- Simulation of an event to find reasonably likely outcomes
- Creating a model of a chance process
- Using a sampling distribution to make decisions
- Estimating probabilities using plots and summary tables

AP Course Topic Outline: Part III A, D

Fathom Prerequisites: Students should be able to make collections, graphs, and summary tables, plot values, and define attributes.

Fathom Skills: Students sample and collect measures, use Fathom to do simulations, define measures to create the sampling distributions of the mean, and use the **proportion** function to test values. *Optional:* Students define attributes with an "if" or switch statement (Extension 2).

Procedure: It is ideal to do this activity by hand and then use a Fathom simulation. Most students must actually "see" the sampling done before they can use Fathom. If you have time, you can begin with this activity.

Divide students into groups of at least two.

1. Each group should write each of the ten ages on identical pieces of paper or 3 × 5 cards, and put the ten cards in a box. Mix them thoroughly, draw out three (the ones to be laid off), and record the ages.
2. Compute the mean of the sample.
3. Repeat steps 1 and 2 nine times.
4. Pool the class results, and plot the distribution of the summary values.
5. Calculate the number of times the class got a mean age of 58 years or more. Estimate the probability that just by chance the mean age of those chosen would be 58 years or more.
6. Ask students what they conclude from the size of the class's estimate.

Doing the simulation by hand first gives students a better feel for how chance-like behavior works. It is useful, however, to do the activity as written afterward. That way, once students understand how to set up chance models, they will also know how to set up these models using Fathom. If your students have lots of experience at sampling and setting up models, they can easily jump right into Fathom and not use the index cards. If you'd like to do both but don't have the time, you can use the document **WestvacoPresent.ftm** instead. It is set up so that all you need to do is click **Sample More Cases** to sample 3 workers or click **Collect More Measures** to collect 100 average ages.

EXAMINE DATA

Q1 The mean age is 46.4 years.

Using Sampling Distributions—By Chance or By Design?
continued

Activity Notes

INVESTIGATE

Q2–Q4 Students will probably not think that it is very likely to get an average age of 58 or more.

Q5–Q6 The true probability of getting an average age of 58 or more is 6/120 = 0.05. Your class's estimate in Q5 and Q6 won't automatically be close to 0.05, though. With $n = 100$, the margin of error is about 0.04. With $n = 1000$, the margin of error is about 0.014. If your class's estimate is a lot larger than 0.05, talk informally about the relationship between the sample size and the precision of the estimate before going on to obtain a better estimate by increasing the number of repetitions.

Some students may know how to calculate this probability. From ten workers, choose the ways you can pick three workers. To get an average age of 58 or more, we must pick the 64 year old and then any two of the remaining four workers with ages 55, 55, 55, and 56. Thus the probability of getting an average age of 58 or more is

$$\frac{{}_4C_2}{{}_{10}C_3} = \frac{\binom{4}{2}}{\binom{10}{3}} = \frac{6}{120} = 0.05$$

Even if students can do the problem theoretically, it is still important that they learn to design and carry out a simulation and to understand how to use a sampling distribution like the one used in Q5.

EXPLORE MORE

1. The true probability of getting an average age of 58 or more is 6/120 = 0.05. With $n = 100$, the margin of error is about 0.04, so the reasonably likely probabilities should fluctuate between 0.01 and 0.09. Here is a histogram of 100 probabilities when collecting 100 sample means. The mean is 0.0506 and the standard error is 0.0198, which does give a margin of error of about 0.04.

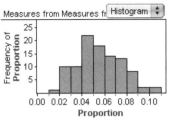

2. When you change the number of measures collected, the probability approaches 0.05. Below are histograms and a summary table for 100 measures with 20 sample means and 1000 measures for 1000 sample means. The distribution for 1000 measures is skewed right with mean 0.0435 and SE 0.045 (the true margin of error is E = 0.0975 and the SE is 0.0487) with reasonably likely probabilities of 0 to 0.15. For 100 sample means, see Explore More 1. The distribution for 1000 measures is roughly mound-shaped with mean 0.04986 and SE 0.00614 (the true margin of error is about 0.014 and the SE about 0.00689).

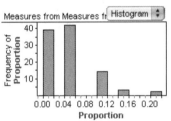

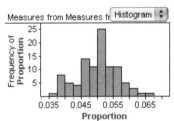

Using Sampling Distributions—By Chance or By Design?
continued

Activity Notes

EXTENSIONS

1. There were ten hourly workers left after Round 1. Their ages were

   25   33   35   38   48   **55**   **55**   **55**   56   **64**

   The ages of the four workers laid off in Rounds 2 and 3 are underlined. They have an average age of 57.25. Have students design a simulation to find the chance of getting an average age of 57.25 or more using the methods of this activity. [Change the number of cases sampled in the sample collection to 4 and then everything is set up after that. The distribution of *AverageAge* should center around 46.5. The theoretical probability of getting an average age of 57.25 or more is only 4/210, or 0.019. That is, getting an average age this high would happen less than 2 times out of every 100, if selections were done randomly. This is strong evidence.]

2. Have students repeat the simulation in this activity, but this time use all 14 hourly workers (ages below) and a different summary statistic. Because the age class protected by law is those 40 or older, use as the summary statistic the number of hourly workers laid off who were 40 or older. Out of ten hourly workers laid off by Westvaco, seven were in the protected class.

   22   25   **33**   **35**   38   48   **53**
   55   **55**   **55**   **55**   56   **59**   **64**

   Design a simulation to pick ten at random for layoff and count how many were in the protected class of 40 or older. What is your estimate of the probability that, just by chance, seven or more of the ten hourly workers who were laid off would be in the protected class? [On the **Measures** panel, define a new measure named *ProtectedClass* by the formula count(AverageAge≥40). Then collect measures as usual and plot *ProtectedClass*. Count the number of dots that are 7 or more. Simulations will vary, but the number of workers in the protected class should center around 6.4, which is the average number of workers aged 40 or more that you would get when drawing 10 workers to lay off from the 14 hourly workers. The theoretical probability of drawing 10 ages from the 14 and getting 7 or more who are aged 40 or more is 455/1001, or 0.45.]

5
Probability

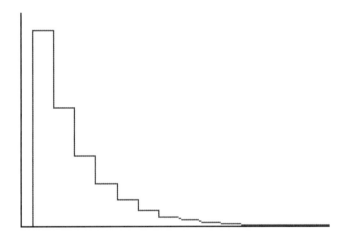

Constructing a Probability Model—Spinning Pennies

You will need
- one penny per student

For flipping a penny, heads and tails have the same probability. You might think that the same probability model is true for spinning a penny.

COLLECT DATA

1. Use one finger to hold your penny on edge on a flat surface, with the head right side up, facing you. Flick the penny with the index finger of your other hand so that it spins around many times on its edge. When it falls over, record whether it lands heads up or tails up.

2. Repeat until you have a total of 40 spins. Count the number of heads for your 40 spins.

3. Plot your value for the number of heads in 40 spins on a dot plot with the values of the other members in your class.

Q1 Are the data consistent with the model that heads and tails are equally likely outcomes, or do you think that the model can safely be rejected?

4. Pool your data for the number of heads with the rest of your class to get the total number of heads for your class. Record that value and the total number of spins for your class. Compute $\hat{p}$ for your class. Save this for later.

INVESTIGATE

Now you'll use Fathom to find out whether the "equally likely" probability model could generate results similar to the ones you observed.

To enter a formula, choose **Table | Show Formulas** and double-click in the formula cell.

5. In a new Fathom document, make a case table with the attribute *Face*. Add 40 new cases and name the collection Spins. Define *Face* with the formula randomPick("H","T"), which randomly selects heads or tails for each spin.

6. Show the inspector. On the **Measures** panel, define a measure as shown.

7. Select the Spins collection and choose **Collection | Collect Measures.** By default, Fathom collects five measures in a collection called Measures from Spins. Each measure is the result of spinning a penny 40 times.

Constructing a Probability Model—Spinning Pennies
continued

To speed things up, uncheck Animation on.

8. Make a dot plot of *NumberOfHeads*.

9. Show the Measures from Spins inspector. On the **Collect Measures** panel, change the number of measures collected to 100 measures and check Replace existing cases. Click **Collect More Measures.**

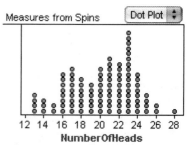

Q2 How often did the number of heads from your spins appear in the measures? How many values were greater than your value?

Q3 Do you think the "equally likely" probability model applies to spinning pennies? Explain.

Now you'll look at possible values under the "equally likely" probability model for your whole class and see if that model holds.

You'll need to add the class total of spins – 40.

10. Add cases to the Spins collection so that the total number of cases in the collection is the same as the total number of spins your class collected.

Each measure will now be the result of spinning a penny the same number of times as your class did as a whole.

11. Select the Measures from Spins collection and choose **Collection | Collect More Measures.** Your dot plot will update. Here is a dot plot for a class with 16 students (640 spins).

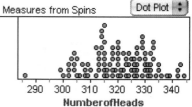

Q4 How often did the number of heads from your class's spins appear in the measures? How many values were greater than your class's value?

Q5 Do you think the "equally likely" probability model applies to spinning pennies? Explain.

EXPLORE MORE

1. If getting heads or tails is equally likely, what is the probability of getting *exactly* 250 heads out of 500 spins? Use your simulation to estimate this probability.

2. Suppose you spin a penny and get 11 heads and 19 tails. Perform a simulation that tells you how often 11 (or fewer) heads could occur, assuming heads and tails are equally likely.

3. Suppose the actual probability of spinning heads is 0.4. Make a simulation to determine about how many spins are required before you can detect the difference between the 0.4 and 0.5 probability models.

Constructing a Probability Model— Spinning Pennies

Activity Notes

Objectives
- Comparing actual results to a model to evaluate whether the observed results are consistent with the model
- Using simulation to estimate the probability of obtaining the observed results under the assumed model

Activity Time: 40–50 minutes

Setting: Paired/Individual Activity or Whole-Class Presentation (collect data, combine data as a class, then build simulation individually or as a class)

Optional Document: SpinSimulator.ftm (Explore More 3 solution)

Materials
- One penny per student (for U.S. pennies, those from the 1960s are better than newer ones, if you can get them)

Statistics Prerequisites
- Familiarity with sampling distributions
- Some familiarity with equally likely outcomes
- Definition of probability

Statistics Skills
- Probability simulation
- Working with the definition of probability and equally likely outcomes
- Comparing actual data to a hypothesized model
- Detecting differences between models

AP Course Topic Outline: Part III A, C, D

Fathom Prerequisites: Students should be able to make collections and graphs, add cases, and define attributes and measures.

Fathom Skills: Students use random generators to create a collection that represents a random sample, use Fathom to do simulations, and collect measures to compare models. *Optional:* Students collect measures from various size samples and use two parameters to test various models (Explore More 3).

General Notes: This activity demonstrates to students the need for data when the hypothetical model (in this case, the "equally likely" principle) lets them down. The activity uses Fathom to repeatedly sample from a population in which spinning a coin has an equal probability of heads or tails. This allows students to focus on the underlying idea that testing a hypothesis involves comparing a particular result with a hypothetical sampling distribution.

Procedure: When U.S. pennies are spun rather than flipped, the data often support a model that has something other than 0.5 for the probability of getting heads. The probability of getting heads by spinning seems to be related to the year in which the pennies were minted. For example, 1990 pennies have a probability of around 0.4 for heads, whereas 1961 pennies have a probability of only about 0.1 for heads. So, in step 2 and especially in step 4, the proportion of heads will largely depend on the ages of your class's collection of pennies. However, most likely in step 4, the proportion of heads will be less than 0.5, which is not what students expect. Whether your students reject the model that spinning a penny is fair will largely depend on how far the proportion is from 0.5. If their answers change from Q3 to Q5, you have another opportunity to talk about the difference between small samples and large ones.

Steps 5–11 help students build a Fathom simulation to analyze the results, assuming that spinning a penny is fair. For steps 9 and 11, students should arrange their Fathom screen so that they can see the case table and dot plot change. It is recommended that students leave animation on in step 9 while collecting measures so that they can see the dot plot grow. They might need to turn animation off in step 11, depending on the speed of their computer.

COLLECT DATA AND INVESTIGATE

Q1–Q4 If students are using newer pennies ($p = 0.4$), they probably won't reject the equally likely model because with only 40 spins, it is quite plausible to get around 20 heads. For example, using the same type of

Constructing a Probability Model—Spinning Pennies
Activity Notes
continued

simulation setup as the activity except using a model with $p = 0.4$:

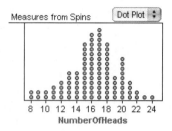

Have students find the number of measures that are greater than or equal to the class's number of heads. They can do this using a summary table for the Measures from Spins collection.

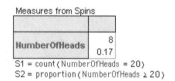

In this simulation, for 17 cases out of 100, the number of heads was 20 or more when a penny was spun 40 times.

Q5 It is likely that students will be able to reject the model. For example, in the sample data there were 640 spins. The equally likely model would predict 320 heads. However, out of 100 simulations of 640 spins (with $p = 0.4$), *none* of the 100 simulations approached 320.

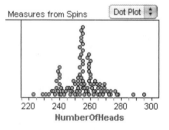

DISCUSSION QUESTIONS

- What could explain the difference between the observed results and the equally likely probability model? Are you really convinced the model is wrong?

- Suppose you spin a coin four times and get no heads. What would you think about the coin at that stage? How many spins without any heads would you need before you became convinced something was "wrong"?

EXPLORE MORE

1. 0.0357. Keep 500 cases in the Spins collection and collect measures. Count how many are exactly 250. Values between 0.01 and 0.06 are acceptable.

2. 0.1002. Keep only 30 cases in the Spins collection and collect measures. Count how many are less than or equal to 11. Values between 0.04 and 0.15 are acceptable.

3. There are many ways to tackle this problem, and there is no exact answer. One approach is to replace the formula given in step 5 with randomPick("H","H","T","T","T") and run the simulation for different sample sizes. Compare the results of each sample size with the results for randomPick("H","T") of the same sample size. Look for the sample size at which the results are distinctly different.

For another solution, see the document **SpinSimulator.ftm**. Here sliders control the sample size and the probability. With p set at 0.4, 100 measures were taken of 20 spins, then 40, then 60, and so on, until the number of measures greater than or equal to 0.5 was less than 5%. This happened between 60 and 80 spins in the simulation shown. To start this simulator from the beginning, select the measures collection and choose **Edit | Select All Cases,** then **Edit | Delete Cases.** Your measures collection will empty and you can start your own simulation by changing n or p and clicking **Collect More Measures** in the measures collection.

The Law of Large Numbers

You will need
- a bucket of beads or a random digit table
- BucketOfBeads.ftm
- BucketOfBeads 2.ftm

The first commercial Internet providers appeared in the former Yugoslavia in 1996. A study in 2002 found that out of the total population of Internet users, about 40% started using the Internet in the last year, 30% started one to two years ago, 20% started three to four years ago, and 10% started more than four years ago. (Source: http://soemz.euv-frankfurt-o.de/media-see/newmedia/main/articles/l_bacevic.htm)

Suppose you wanted to take a sample from this population to check these proportions. How would you design the sample? How big of a sample would you need to take to get the proportions you expect? Would the sample sizes need to be different for the different percentages? In this activity you'll address these questions.

COLLECT DATA

1. Describe how you would use a random digit table or a bucket of beads to simulate selecting one person from this population of Internet users in Yugoslavia to see if they started using the Internet in the last year.

Your sample proportion will be either 0/1 or 1/1.

2. For your first pick, use your method in step 1 to select one person. Record the number of selections you have made so far (1) and whether or not your selection started using the Internet in the last year. Compute your sample proportion for the number of people in your sample who started using the Internet in the last year.

Consider a "success" to be picking someone who started using the Internet in the last year.

3. Continue picking until you have a total of 25 picks. For each pick, record the number of picks you have made so far, the *cumulative* number of successes, and the cumulative proportion of successes.

4. Plot the cumulative proportion of successes versus the number of picks.

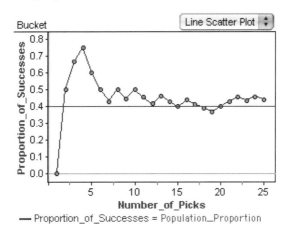

Q1 Compare your plot with those of others in your class. Does it look like the sample proportion of successes will converge to the population proportion of successes (40%)?

Exploring Statistics with Fathom   5: Probability

The Law of Large Numbers
continued

GENERATE DATA

Now you'll use a Fathom simulation to investigate further whether the sample proportion of successes will converge to the population proportion of successes. First you need to build the simulation. It will be exactly like the by-hand simulation.

5. Open the Fathom document **BucketOfBeads.ftm.** You'll see a collection of 10 beads where each color represents one of the groups from Yugoslavia. There is also a slider set to the population of interest, which in this case is blue.

6. Select the Bucket collection and choose **Collection | Sample Cases.** By default, Fathom takes a sample of ten cases with replacement and places them in a new collection named Sample of Bucket. Drag the lower-right corner of the sample collection to see your sample.

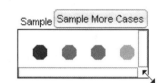

7. Double-click the sample collection to show its inspector. On the **Sample** panel, change the number of cases to 1 and uncheck Replace existing cases.

8. Go to the **Measures** panel and define three measures as shown.

Color_of_Interest is under Global Values in the functions list of the formula editor.

Proportion_of_Successes	0.1	proportion(Ball = Color_of_Interest)
Number_of_Successes	1	count(ball = Color_of_Interest)
Number_of_Picks	10	count( )

Select the sample collection. Choose **Edit | Select All Cases,** then **Edit | Delete Cases.**

9. Delete all the beads in the sample collection.

10. With the sample collection selected, choose **Collection | Collect Measures.** You should see Fathom take five samples from the Bucket collection and place a measure in a new collection named Measures from Sample of Bucket.

11. Show the inspector for the new collection. On the **Collect Measures** panel, change the number of measures collected to 1.

12. Make a plot of *Proportion_of_Successes* versus *Number_of_Picks* from the measures collection. Change the graph to a line scatter plot. Also, make a case table for the measures collection.

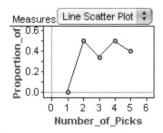

The Law of Large Numbers
continued

INVESTIGATE

Taking a Sample

Now you have everything set up to take a sample one selection at a time.

You may need to resize the sample collection to see the newest bead.

13. Click **Collect More Measures.** One more bead is picked and the measure is stored in the collection. Click **Collect More Measures** one more time.

Q2 Explain how this process of sampling in Fathom is identical to what you did in steps 1 and 2.

14. Click **Collect More Measures** until you have a total of 25 measures.

15. Choose **Graph | Plot Function** and type 0.4.

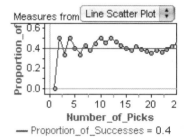

Q3 After 25 picks, what was your final proportion of successes? How close is that to 0.4?

Q4 What was the range of values you got for your proportion of successes between 1 and 25 picks?

The law of large numbers says that the difference between a sample proportion and a population proportion must be small (except in rare instances) when the sample size is large. So, you need more measures.

16. Go to the **Collect Measures** panel in the measure's inspector and change the number of measures collected to 25. Click **Collect More Measures.** Now you have 50 measures in the measures collection.

Q5 After 50 picks, what was your final proportion of successes? How close is that to 0.4?

Q6 What was the range of values you got for your proportion of successes between 25 and 50 picks?

17. Repeat step 16 for 75 picks and for 100 picks. Record your final proportion of successes each time.

Q7 What was the range of values you got for your proportion of successes between 50 and 75 picks? Between 75 and 100 picks?

The Law of Large Numbers
continued

Use your answers to Q4, Q5, and Q7 to answer this question.

Q8 Does it look like the sample proportion of successes will converge to the population proportion of successes of 40%? Explain.

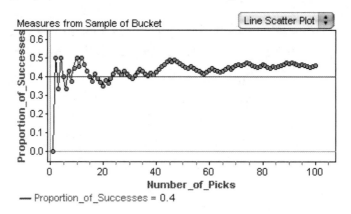

Let's see another run of 100 picks.

18. Empty the sample and measures collections.

You might want to turn Animation off.

19. Change the number of measures collected to 100. Click **Collect More Measures**. Observe the fluctuations and record the final sample proportion.

Q9 Repeat steps 18 and 19 a few times. What was the smallest sample proportion you got? The largest? Did your series of 100 picks always end up around 0.4, or did any of them surprise you?

On the slider, click on Blue and type Yellow.

20. Change the *Color_of_Interest* to yellow. Double-click the equation below the graph and change it to the appropriate proportion for yellow.

Don't forget that each time, you'll need to empty the sample and measures collections.

21. Run the simulation for yellow a few times. Observe the fluctuations and record the final sample proportion.

Q10 What was the smallest sample proportion you got? The largest? Did your series of 100 picks always end up around the expected population proportion, or did any of them surprise you?

Q11 Compare the simulations for blue with the simulations for yellow. How were they different? How were they the similar? Did one fluctuate more than the other?

The Law of Large Numbers
continued

Changing the Sample Size

Now we'd like to see bigger samples.

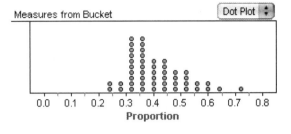

22. Open the Fathom document **BucketOfBeads2.ftm.** In this document you'll see a sample of 25 beads and the corresponding line scatter plot. The measures collection in this document collects 50 samples of size 25, and for each sample it records the final sample proportion and plots that proportion on a dot plot.

23. Click **Collect More Measures** a few times to see how this simulation works.

Q12 What is the range of proportions you get with samples of size 25? Describe the distribution of the sample proportion for samples of size 25 when the *Color_of_Interest* is blue.

Q13 Are the values you got in Q1 and Q3 within the range of proportions you got in Q12?

To add new cases, choose **Collection | New Cases.**

24. Add 75 cases to the Bucket collection so that the sample size is 100. Collect more measures.

Q14 What is the range of proportions you get with samples of size 100? Describe the distribution of the sample proportion for samples of size 100 when the *Color_of_Interest* is blue.

Q15 Are the values you got in Q10 within the range of proportions you got in Q13?

25. Add 900 cases to the Bucket collection for a sample of size 1000. Collect more measures.

Q16 What is the range of proportions you get with samples of size 1000? Describe the distribution of the sample proportion for samples of size 1000 when the *Color_of_Interest* is blue. Describe how the distribution of sample proportions changes as the sample size gets larger.

Q17 Does it look like the sample proportion of successes will converge to the population proportion of successes of 40%? Explain.

The Law of Large Numbers

Activity Notes

Objectives

- Designing a simulation to take a sample from a population to check given proportions
- Understanding that because of random behavior, in small samples the range of likely sample proportions will fluctuate more than in larger samples and that even in large samples there is fluctuation
- Seeing that the difference between a sample proportion and a population proportion must be small (except in rare instances) when the sample size is large or that the sample proportion of successes will converge to the population proportion of successes

Activity Time: 40–80 minutes for activity or 30–40 minutes for presentation

Setting: Paired/Individual Activity (use **BucketOf Beads.ftm** and **BucketOfBeads2.ftm**) or Whole-Class Presentation (use **BucketOfBeadsPresent.ftm**). See the Procedure section for details.

Optional Documents: Cards.ftm, Cards2.ftm (Whole-Class Presentations or Extensions)

Materials

- One random digit table per student or a bucket of beads with 4 colors (40% blue, 30% red, 20% green, and 10% yellow)

Statistics Prerequisites

- Familiarity with sampling distributions
- Some familiarity with probability distributions
- Definition of probability
- Familiarity with designing simulations

Statistics Skills

- Probability simulation
- Working with the definition of probability and probability distributions
- Counting successes versus failures
- Comparing actual data to a hypothesized model
- The law of large numbers

AP Course Topic Outline: Part III A (1, 2)

Fathom Prerequisites: Students should be able to make graphs, define attributes and measures, and work with different collections: the original, sample, and measures collections.

Fathom Skills: Students use sample and measures collections to compute cumulative proportions and values, use Fathom to create a simulation of a probability distribution, collect measures to compare models, collect measures from various size samples, delete cases, and use a slider for a non-numerical variable.

General Notes: This activity demonstrates the variability in sampling due to randomness in small and large samples. Students see that if they want to estimate a proportion, it is better to take a larger sample than a smaller one. The activity uses Fathom to repeatedly sample from a population in which outcomes are not equally likely. Students compare the effect of sample size on the sampling distribution of the sample proportion, and they also compare how these distributions change depending on the population proportion.

The data used in this activity come from "The Development of Internet in Yugoslavia" by Ljiljana J. Bacevic, (in *New media in Southeast Europe,* ed. O. Spassov and Ch. Todorov (Sofia: Southeast European Media Centre, 2003)).

Procedure: There are many ways to organize this activity. You can have your students do the whole activity or just parts of it.

For the hands-on activity, students are asked to design a simulation for sampling from the population of Internet users in Yugoslavia. They then use their design to simulate taking a sample of size 25, at each stage calculating the cumulative proportion of successes. You can have each student do his or her own simulation, or you can do the simulation as a class with one plot on the board or overhead. As a class, the bucket of beads method works best. Have a bucket with 10 beads: 4 blue, 3 red, 2 green, and 1 yellow. Sample with replacement, mixing the beads before the next student picks. As a class, keep track of the cumulative number of picks, the cumulative number of successes, and the cumulative proportion of successes. You can have each student plot his or her proportion while the next person picks, or you could wait until all selections have been made and then make the plot.

If you don't have time to use the bucket of beads idea, use the random digit table method, which goes fairly quickly.

The Law of Large Numbers
continued

Activity Notes

Depending on your time, you can either have the students start with the Generate Data section and build the simulation themselves or have them start at the Investigate section (step 13) with the simulation already built. If you choose the latter, have the students start their activity with the Fathom document **BucketOfBeadsPresent.ftm** instead of the document **BucketOfBeads.ftm.** Then they can proceed with the rest of the activity as written.

The last option is that you can use the document **BucketOfBeadsPresent.ftm** or **Cards.ftm** as a presentation for the whole class. Proceed through the steps as described in the activity with either file. (Note that in **Cards.ftm** the population proportion is 0.25.)

COLLECT DATA AND INVESTIGATE

Q1–Q4 Some plots will look like the sample proportion and will converge to 0.4, but other plots might fluctuate wildly. Reasonably likely cumulative sample proportions range from 0.2 to 0.6, but as shown below, others do pop up (0.72). The range of values between 1 and 25 picks will vary greatly, depending on students' first picks. For example, for the plot on the student worksheet, the range is from 0.333 to 1; other plots could range from 0 to 0.7.

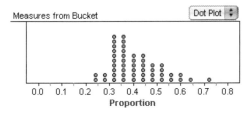

Q2 Fathom selects one bead at a time without emptying the collection first, so each measure collected is a cumulative measure. *Proportion_of_Successes* is the cumulative proportion of blue beads picked so far, *Number_of_Successes* is the cumulative number of blue beads picked so far, and *Number_of_Picks* is the total number of beads that have been picked so far.

Q5–Q6 The reasonably likely range of values is 0.26 to 0.54, although smaller and larger values are possible. The range of possibilities from 25 to 50 picks will be smaller than in Q4. Typically the range will be somewhere within the band from 0.2 to 0.6.

Q7 The reasonably likely range of values for samples of size 75 is 0.28 to 0.52, although smaller and larger values are possible. The range of possibilities from 50 to 75 picks will be smaller than in Q6. Typically the range will be somewhere within the band from 0.25 to 0.55.

For samples of size 100, the reasonably likely range of values is 0.3 to 0.5, although smaller and larger values are possible. The range of possibilities from 75 to 100 picks will be somewhere within the band from 0.3 to 0.5.

Q8 As the sample size grows, between successive 25 marks, the range gets smaller as if it is converging. So, yes, it looks reasonable.

Q9 The answer is the same as in Q7. Here is a dot plot of 100 sample proportions. Notice there is one dot at 0.52, a rare event.

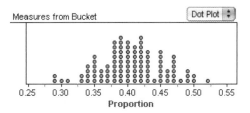

Q10–Q11 Here, a sample of size 100 will often give surprising values. Here is the distribution of sample proportions for samples of size 100 with $p = 0.1$. The median is 0.09 and the distribution is skewed right. Possible values range from 0.03 to 0.17. Typically, both will fluctuate a great deal, but the plots for $p = 0.4$ converge a little quicker.

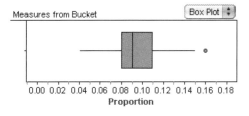

22. In this section, students generate sampling distributions of 50 sample proportions. Because 50 is a relatively small number, some of these distributions can show unexpected or rare cases, even with as large a sample size as 1000 (see plot for Q16–17).

The Law of Large Numbers
continued

Q12–Q13 As shown in the plot in the student activity, the distribution of the sample proportion for samples of size 25 can still be skewed. Typically, their center is around 0.4, although that too can vary.

Q14–Q15 For samples of size 100, the distribution is closer to normal, with a mean closer to 0.4 and a smaller spread (see Q7).

Q16–Q17 For samples of size 1000, the distribution should be relatively close to normal with a mean close to 0.4 and a smaller spread. Reasonably likely values range from 0.37 to 0.43. The distribution shown here has mean 0.4016 and standard deviation 0.014 and is somewhat skewed right.

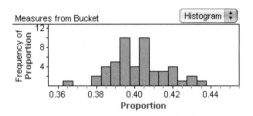

EXTENSIONS

1. Have students delete the cases in the Bucket collection, then add 25 cases to the Bucket collection and change the *Color_of_Interest* to yellow. Explore how the distribution of sample proportions changes as the sample size gets larger. How is it different from the sampling distribution of sample proportions for blue?

2. Students could explore one of the other colors in the Bucket collection. Explore how the distribution of sample proportions changes as the sample size gets larger. How is it different from the sampling distribution of sample proportions for blue or for yellow?

3. Use either **Cards.ftm** or **Cards2.ftm** as a whole-class presentation. What population proportion is being investigated in these documents?

Addition Rule—Scottish Children

You will need
- Scottish Children.ftm

In this activity you'll use sample data from a survey of Scottish children to explore when you can find $P(A \text{ or } B)$ if you know $P(A)$ and $P(B)$.

EXAMINE DATA

1. Open the Fathom document **ScottishChildren.ftm**. You will see a collection of 500 children from Scotland who were randomly selected to participate in a survey.

The last attribute, Subject, is the number this child was in the study. This is a random sample of the original sample of 5387 children.

2. Double-click the collection to show the inspector. Make a ribbon chart for *Hair_Color*, then drag *Eye_Color* into the *interior* of the graph. (*Eye_Color* is now the legend.) The entire bar represents all the children. Each slice represents a different hair color. The four sections of each slice represent the four different eye colors.

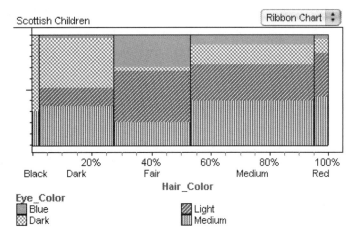

3. Click on one section of the ribbon chart. Hold down the Shift key and click on additional sections. Now move your cursor over the *collection* (not the graph) and notice that in the lower-left corner of the Fathom window you can read the number of selected cases.

Use this method to answer these questions.

Q1 How many children have *either* medium *or* red as their hair color?

Q2 How many children have *either* blue eyes *or* medium hair color?

INVESTIGATE

Addition Rule

4. Define a new attribute, *IsMedium*, with the formula Hair_Color="Medium". This attribute will be true if the child's hair color is medium. Similarly, define

Addition Rule—Scottish Children
continued

an attribute *IsRed* that will be true if the child's hair color is red, and define an attribute *IsBlueEyed* that will be true if the child has blue eyes.

> Formulas such as count(IsBlueEyed) will be useful.

5. Go to the **Measures** panel of the inspector and define the measures *Medium* (the number of children whose hair color is medium), *Red* (the number of children whose hair color is red), and *BlueEyed* (the number of children who have blue eyes).

Q3 From the values of these three measures alone, can you determine the number of children whose hair color is medium or red? If so, explain how to do it. If not, what additional information do you need?

Q4 From the values of these three measures alone, can you determine the number of children who have blue eyes or who have medium as their hair color? If so, explain how to do it. If not, what additional information do you need?

> In the formula editor, use the **or** button on the keypad for the logical expression "or."

6. Define the measures *MediumOrRed* (the number of children whose hair color is medium or red) and *BlueEyedOrMedium* (the number of children who have blue eyes or who have medium as their hair color). If possible, use the measures that already exist as part of your formulas.

Q5 Write your formulas for the measures you defined in step 6 and record their values.

Q6 True or false: $P(Medium \text{ or } Red) = P(Medium) + P(Red)$. Explain.

Q7 True or false: $P(BlueEyed \text{ or } Medium) = P(BlueEyed) + P(Medium)$. Explain.

Q8 What conclusions can you draw so far from doing this activity?

Extending the Rule

You should have found that with the measures *BlueEyed* and *Medium* alone, you can't determine the number of children who are blue-eyed or who have medium as their hair color.

> In the formula editor, use the **and** button on the keypad for the logical expression "and."

7. Define a new measure, *BlueEyedAndMedium*, which counts the number of children who are blue-eyed and have medium as their hair color.

Q9 Write your formula for the measure you defined in step 7 and record its value.

Q10 From the values of *BlueEyed*, *Medium*, and *BlueEyedAndMedium* alone, can you determine the number of children who are blue-eyed or who have medium as their hair color? If so, explain how to do it. If not, what additional information do you need?

Addition Rule—Scottish Children
continued

8. Make a summary table with *Eye_Color* for the rows and *Hair_Color* for the columns. The summary table is equivalent to a two-way table.

Scottish Children

		Hair_Color					Row Summary
		Black	Dark	Fair	Medium	Red	
Eye_Color	Blue	0	5	38	21	1	65
	Dark	9	55	5	36	3	108
	Light	0	21	59	69	9	158
	Medium	4	44	26	85	10	169
Column Summary		13	125	128	211	23	500

S1 = count( )

Q11 If you select a child at random from your collection, what is the probability that the child has blue eyes or has medium as a hair color?

Q12 What is the probability that a randomly selected child has black hair?

EXPLORE MORE

1. Make a second ribbon chart, with *Eye_Color* on the horizontal axis and *Hair_Color* as the legend attribute. Select sections of this graph and predict which sections will highlight in the first ribbon chart.

2. Make a bar chart of *Hair_Color* and drag *Eye_Color* into the interior of the graph, or make a bar chart of *Eye_Color* and drag *Hair_Color* into the interior of the graph. Explore how each section of the bar chart corresponds to each section of the ribbon chart. Compare and contrast the two types of graphs.

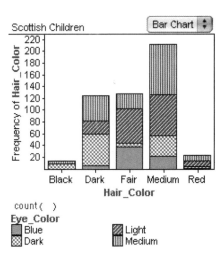

You can change the order of the categories by dragging a label.

3. Make a breakdown plot with *Hair_Color* on the horizontal axis and *Eye_Color* on the vertical axis. Determine which sections of the breakdown plot correspond to which cells of the summary table.

Addition Rule—Scottish Children

Activity Notes

Objectives

- Recognizing that $P(A \text{ or } B) = P(A) + P(B)$ only for disjoint categories
- Extending the addition rule to $P(A \text{ or } B) = P(A) + P(B) - P(A \text{ and } B)$ when A and B are not disjoint

Activity Time: 30–50 minutes

Setting: Paired/Individual Activity or Whole-Class Presentation (use **ScottishChildren.ftm** for either)

Statistics Prerequisites

- Definition of probability
- Familiarity with the concepts of "or" and "and" (not the formulas)

Statistics Skills

- Addition rule for disjoint events
- Working with the definition of probability
- General addition rule
- Calculating probabilities with graphs and two-way tables
- *Optional:* Stacked bar charts (Explore More 2)

AP Course Topic Outline: Part III A

Fathom Prerequisites: Students should be able to make bar and ribbon charts and to define attributes and measures.

Fathom Skills: Students define formulas using the logical expressions "or" and "and," use a summary table as a two-way table, work with ribbon charts to calculate probabilities, and work with legends in ribbon charts. *Optional:* Students make stacked bar charts (Explore More 2) and breakdown plots (Explore More 3).

General Notes: This important activity helps students see why you can't always add $P(A)$ and $P(B)$ to get $P(A \text{ or } B)$. It also shows practical ways to extend the addition rule to compute $P(A \text{ or } B)$ when A and B are not disjoint. Fathom eases the computational burden of this activity, allowing students to focus on the concepts. The collection of 500 children is actually a random sample of a much larger sample of 5387 children. (Source: D. J. Hand et al., *A Handbook of Small Data Sets* (London: Chapman and Hall, 1994), 146. Their source: L. A. Goodman, "Association models and canonical correlation in the analysis of cross-classifications having ordered categories," *Journal of the American Statistical Association* 76 (1981): 320–334.))

Students write logical expressions to count the number of children with various values for attributes. This activity also reviews the ribbon chart, an important tool for reasoning about sets.

EXAMINE DATA

3. A common mistake students make here is dropping *Eye_Color* on the vertical axis instead of in the graph's interior. If they make this mistake, they can choose either **Edit | Undo** or **Graph | Remove Y Attribute**.

4. Because answering questions based on a ribbon chart can appear simple but become confusing, encourage students to spend time gaining understanding of ribbon charts.

Q1 234 children

Q2 255 children

INVESTIGATE

5. Here are the formulas for the complete set of attributes for the Scottish Children collection.

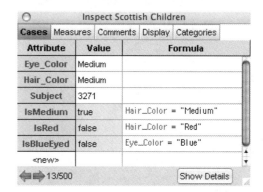

Q3 The categories *Medium* and *Red* are disjoint, so the answer can be found with the sum count(IsMedium)+count(IsRed). This value is equivalent to count(IsMedium or IsRed).

Q4 The categories overlap. You need to know how many have blue eyes and select *Medium* as their hair color. Hence, count(IsBlueEyed)+count(IsMedium) is not equivalent to count(IsBlueEyed or IsMedium).

Addition Rule—Scottish Children
continued

Activity Notes

6.–7. Here is the complete set of measures. The last, *BlueEyedNotMedium*, is not explicitly requested in the activity but is useful in understanding that if you know this quantity, then you can add it to *Medium* to get *BlueEyedOrMedium*. You may want to challenge students to use the alternative "not" statement (*MediumNotBlueEyed*) and explain how to use it to find *BlueEyedOrMedium*.

Measure	Value	Formula
Medium	211	count(IsMedium)
Red	23	count(IsRed)
BlueEyed	65	count(IsBlueEyed)
MediumOrRed	234	count(IsMedium or IsRed)
BlueEyedOrMedium	255	count(IsBlueEyed or IsMedium)
BlueEyedAndMedium	21	count(IsBlueEyed and IsMedium)
BlueEyedNotMedium	44	count(IsBlueEyed and ¬IsMedium)

Q5–Q7 The formulas are given in the inspector above, as are the values.

Q6 True; they are disjoint (see explanation in Q3).

Q7 False; they overlap (see Q4).

Q8 The conclusion to be drawn so far is that only for disjoint categories does $count(A \text{ or } B) = count(A) + count(B)$.

Q9 See the inspector above for the formula. There are 21 children who are blue-eyed and have medium as their hair color.

Q10 Students should conclude that *BlueEyedOrMedium*, previously defined by the formula count(IsBlueEyed or IsMedium), is equivalent to *BlueEyed* + *Medium* − *BlueEyedAndMedium*.

8. The summary table that students make here is exactly analogous to the ribbon chart they made in step 2.

Because Q11 and Q12 ask for probabilities and not numbers, you might want to show students how to change the summary table's default formula from counts to proportions. Double-click count() and enter count()/GrandTotal. This will give the proportion of cases in each cell.

Q11 The probability can be calculated two ways. Using the formula:

$$P(BlueEyed \text{ or } Medium) = P(BlueEyed) + P(Medium) - P(BlueEyed \text{ and } Medium) =$$
$$\frac{65}{500} + \frac{211}{500} - \frac{21}{500} = \frac{255}{500} = \frac{51}{100}$$

Using the summary table, count just the shaded cells, counting 21 once. So

$$P(BlueEyed \text{ or } Medium) =$$
$$\frac{5 + 38 + 21 + 1 + 36 + 69 + 85}{500} = \frac{255}{500} = \frac{51}{100}$$

You could also use the total of the column "Medium" and add to that the shaded cells in the "Blue" row that are not in the "Medium" column:

$$P(BlueEyed \text{ or } Medium) =$$
$$\frac{211 + 5 + 38 + 1}{500} = \frac{51}{100}$$

Scottish Children

Eye_Color	Hair_Color					Row Summary
	Black	Dark	Fair	Medium	Red	
Blue	0	5	38	21	1	65
Dark	9	55	5	36	3	108
Light	0	21	59	69	9	158
Medium	4	44	26	85	10	169
Column Summary	13	125	128	211	23	500

S1 = count( )

Q12 13/500

DISCUSSION QUESTIONS

If possible, ask the following questions while projecting the Fathom document in front of the class.

- How could you use this ribbon chart to find the number of children with blue eyes?
- Which sections of the ribbon chart do you need to select to find the number of children whose hair color is *Red* or *Medium*?
- Which sections of the ribbon chart do you need to select to find the number of children who have light eyes or who have red as their hair color?
- What formulas did you use for measures *MediumOrRed* and *BlueEyedOrMedium*? How many different yet equivalent formulas can you write?

Addition Rule—Scottish Children
continued

- What did you conclude in Q10?
- If there are two not necessarily disjoint events A and B, how can you compute $P(A \text{ or } B)$?

EXPLORE MORE

2. The fundamental difference is that the ribbon chart shows percentages whereas the bar chart shows actual quantities.

3. This breakdown plot is in the same order as the summary table.

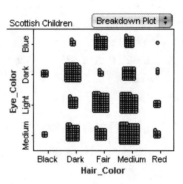

Establishing Independence with Data

Data often serve as a basis for establishing a probability model or for checking whether an assumed model is reasonable. In using data to check for independence, however, you have to be careful. You might ask, "Why do you have to be careful?" In this activity you'll answer that question.

COLLECT DATA

1. Record whether you are right-handed or left-handed.

2. Determine (and record) if you are right-eyed or left-eyed. Hold your hands together in front of you at arm's length. Make a space between your hands that you can see through. Through the space, look at an object at least 15 feet away. Now close your right eye. Can you still see the object? If so, you are left-eyed. Now close your left eye. Can you still see the object? If so, you are right-eyed.

Q1 Would you expect being right-handed and being right-eyed to be independent? That is, if you know a person is right-handed, does that change the probability that he or she is right-eyed?

INVESTIGATE

Hands and Eyes

3. Enter the data for your class into a Fathom document. Make a two-way table (that is, a summary table) with *Handedness* along one table dimension and *Eyedness* along the other.

> Click on a section of the ribbon chart. (Hold down the Shift key to click additional sections.) Move your cursor over the *collection*. You can read the number of selected cases in the lower-left corner of the Fathom window.

4. Make a bar chart for *Handedness*. Change it into a ribbon chart, then drag *Eyedness* into the *interior* of the graph. (*Eyedness* is now the legend.) The entire bar represents your whole class. Each slice represents a different handedness. The two sections of each slice represent left-eyed and right-eyed.

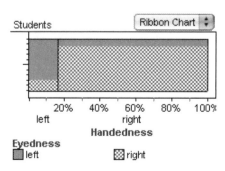

Q2 What area on your ribbon chart represents the proportion $P(\text{right-eyed} \mid \text{right-handed})$?

Establishing Independence with Data
continued

Q3 Based on your ribbon chart, compare these two proportions: $P(\text{right-eyed} \mid \text{right-handed})$ and $P(\text{right-eyed})$. Which proportion is larger, or are they equal?

Q4 After examining the two-way table and the ribbon chart, would you say that being right-handed and being right-eyed are independent?

Q5 Use the definition of independence to check your answer to Q4.

Coin Flips

Now you'll simulate two coin flips and study the results.

Consider randomInteger or randomPick as functions for generating coin flips.

5. Create a collection with 100 cases and two attributes, *FirstFlip* and *SecondFlip*. Use formulas to randomly generate heads or tails for each attribute.

Q6 Would you expect the results of the first flip and the second flip to be independent? That is, if you know the outcome of the first flip, does that change the probabilities for the second flip?

6. Make a two-way table and a ribbon chart to analyze your data.

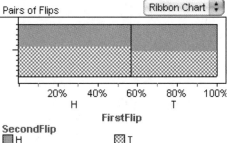

Q7 What area on your ribbon chart represents the proportion $P(\text{SecondFlip is heads} \mid \text{FirstFlip is heads})$?

Q8 Based on your ribbon chart, compare these two proportions: $P(\text{SecondFlip is heads} \mid \text{FirstFlip is heads})$ and $P(\text{SecondFlip is heads})$. Which proportion is larger, or are they equal?

Q9 After examining the two-way table and the ribbon chart, would you say that the results of the first flip and the second flip are independent?

Q10 Use the definition of independence to check your answer to Q9.

Q11 Did you get the results you expected in Q10? Explain.

Q12 What does a ribbon chart look like when two attributes are independent? When they are not independent?

Establishing Independence with Data
continued

EXPLORE MORE

1. Change your ribbon chart to a bar chart (stacked bar graph). What does a stacked bar graph look like when two attributes are independent? When they are not independent?

2. Open the Fathom document **Scottish Children.ftm.** Make a ribbon chart or stacked bar graph and determine if the attributes *Hair_Color* and *Eye_Color* are independent. Explain your reasoning.

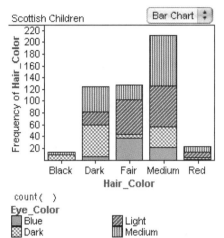

3. Make a simulation of eyedness and handedness in which the attributes aren't independent.

4. Invent a measure of how the results in a two-way table differ from what you would expect if the two attributes are independent. It might be, for example, that if you get exactly what you expect, the value of the measure is zero and the measure increases the farther the results are from what you expect. Apply your measure to your collections from the activity.

5. Using your measure from Explore More 4, repeatedly collect measures for your coin-flip collection and build a distribution of measures for many sets of 100 flips.

Establishing Independence with Data

Activity Notes

Objectives

- Recognizing the difficulty of establishing independence with real-world data
- Using the definition of independent events, $P(A \mid B) = P(A)$, and the multiplication rule for independent events, $P(A \text{ and } B) = P(A) \cdot P(B)$, to check for independence
- Understanding that because the two proportions in any experiment are probably not equal, $P(A \mid B) \neq P(A)$ does not necessarily mean that the events are not independent
- Seeing that data do not often show independence clearly but sometimes data can demonstrate that independence is not a reasonable assumption

Activity Time: 30–50 minutes (25 minutes as a presentation)

Setting: Paired/Individual Activity (collect data or use **Independence.ftm**) or Whole-Class Presentation (use **IndependencePresent.ftm**)

Optional Document: ScottishChildren.ftm (Explore More 2)

Statistics Prerequisites

- Definition of probability
- Familiarity with the definition of independent events, $P(A \mid B) = P(A)$
- Familiarity with the concept of independent events
- Familiarity with the conditional $P(A \mid B)$

Statistics Skills

- Multiplication rule
- Working with the definition of conditional probability
- Working with the definition of independence
- Comparing probabilities with graphs and two-way tables
- Recognizing independence or dependence with graphs and two-way tables
- *Optional:* Stacked bar charts (Explore More 1 and 2)

AP Course Topic Outline: Part III A (1, 3, 5), B (1)

Fathom Prerequisites: Students should be able to make case tables, summary tables, and ribbon charts and define attributes based on a random function.

Fathom Skills: Students use summary tables to numerically check for independence, use ribbon charts to visually check for independence, use a summary table as a two-way table, work with ribbon charts to compare probabilities, and work with legends in ribbon charts.

General Notes: This activity demonstrates that samples often do not meet the mathematical definition of independence even when we have reason to believe that the events are independent. Sometimes, however, data can demonstrate that independence is not a reasonable assumption.

Procedure: Eye dominance and hand dominance are associated, so a random sample should produce data that look dependent: $P(\text{right-eyed} \mid \text{right-handed}) > P(\text{right-eyed})$. One note of caution: The percentage of left-handers is rather small, so you may need to have a fairly large group of students to see any left-handed students.

The coin flips are independent, and the data should support this even though students are not likely to find that $P(H \text{ and } H)$ is exactly equal to $P(H \text{ on first flip}) \cdot P(H \text{ on second flip})$ in any set of sample data. If you don't have time for your students to collect the data, you can have them use the Fathom document **Independence.ftm** instead. It has a collection of data for 30 students that your students can use to work through the activity. Alternatively, **IndependencePresent.ftm** has everything made. With this document projected, you could do the questions as a class activity.

INVESTIGATE

Q2 The large checkered area on this ribbon chart represents the proportion $P(\text{right-eyed} \mid \text{right-handed})$. Go to the section that is right-handed, then look at just the right-eyed group in that section.

Establishing Independence with Data
continued

Activity Notes

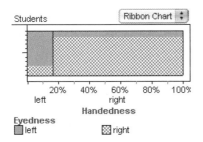

Q3 The proportion of the shaded region above that represents *P*(*right-eyed* | *right-handed*) compared to the whole section of right-handedness is larger than the proportion of *right-eyed* to the whole (both sections). (You might want to remind students that they are looking at proportions, not at counts). Using the table: There are 21 right-eyed students in the right-handed row. That proportion is 21/25. The proportion for right-eyed is 22/30 ≈ 0.73.

Q4–Q5 The table and ribbon chart suggest dependence, not independence. Using the definition: *P*(*right-eyed* | *right-handed*) = 21/25 = 0.84 > 22/30 = *P*(*right-eyed*). Using the multiplication rule: *P*(*right-eyed* and *right-handed*) = 21/30 = 0.7, but *P*(*right-eyed*) · *P*(*right-handed*) = 22/30 · 25/30 ≈ 0.6.

Q7 The top-right area below is the proportion *P*(*SecondFlip is heads* | *FirstFlip is heads*).

Q8 Based on the ribbon chart below, the two proportions *P*(*SecondFlip is heads* | *FirstFlip is heads*) and *P*(*SecondFlip is heads*) look like they could be equal. They aren't exactly equal—which is the problem.

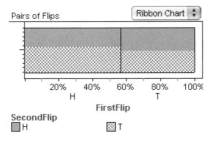

Q9–Q10 The coin flips are independent but they will fail the test. Using the definition: *P*(*SecondFlip is heads* | *FirstFlip is heads*) = 28/54 ≈ 0.52 and *P*(*SecondFlip*) ≈ 56/100 = 0.56. Using the multiplication rule: *P*(*SecondFlip is heads* and *FirstFlip is heads*) = 28/100 = 0.28, but *P*(*SecondFlip is heads*) · *P*(*FirstFlip is heads*) = 56/100 · 54/100 = 0.3024. Not exactly equal but close.

Q12 A ribbon chart will have roughly the same *proportion* of each category in each section when the attributes are independent.

DISCUSSION QUESTIONS

- What does a ribbon chart look like when two attributes are independent? When they are not independent?
- What did you find about the independence of the first flip and the second flip? How much variation from "expected" did you find?

EXPLORE MORE

1. A stacked bar chart will have roughly the same *proportion* of each category in each section when the attributes are independent. This is harder to see in a stacked bar chart because the bar chart works with counts and the ribbon chart works with proportions.

2. There is an association between hair color and eye color, so they are not independent. As can be seen in the ribbon chart, the proportions of the legend attribute are not the same across all sections. For example, the fair-haired children have a greater proportion of light eyes than do, say, the dark-haired

Establishing Independence with Data
continued

children. So, just looking at the light-eyed rectangle in each section, it is clearly proportionally not the same across the different hair colors.

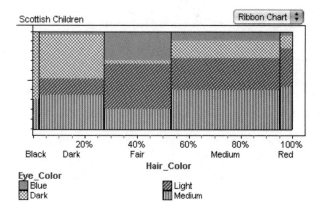

4. This question emphasizes the need for a statistical test of independence. Chi-square, χ^2, is one such measure. For now, leave this as an open-ended extension and allow students to create their own measures—there is no right or wrong answer.

Exploring Sampling—With or Without Replacement

You will need
- WithWithout.ftm

In this activity you'll use Fathom's sampling capabilities to see if there is a difference between sampling with replacement and sampling without replacement.

GENERATE DATA

1. Open the Fathom document **WithWithout.ftm.** In this document there is a collection called Bag, with six cases. Each case (colored square) represents an equal-sized slip of paper.

2. Add a new attribute, *Name,* to the case table. Enter six different names. Make yours one of them.

3. Select the Bag collection and choose **Collection | Sample Cases.** A new collection appears named Sample of Bag. Make a case table for that collection.

   Q1 How many names are in the case table for the sample collection? Does Fathom appear to be sampling with or without replacement? Explain.

INVESTIGATE

Bag Simulation

4. Double-click the sample collection to show its inspector. Go to the **Sample** panel and change the number of cases to 5.

*You can also click **Sample More Cases** in the inspector.*

5. Drag the lower-right corner of the sample collection until you can see the **Sample More Cases** button. Click **Sample More Cases** repeatedly.

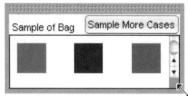

   Q2 Describe what happens as you sample. Does Fathom appear to be sampling with or without replacement? Explain.

6. In the inspector, change the number of cases to 3. Click **Sample More Cases** again. Count the number of times you have to click **Sample More Cases** until you see the same person appear twice in the same sample of 3.

   Q3 How many times did you have to sample? Try the experiment a few more times. Were your results about the same?

   Now you will sample *without replacement.*

Exploring Statistics with Fathom
© 2007 Key Curriculum Press

Exploring Sampling—With or Without Replacement
continued

7. In the inspector, uncheck With replacement. Click **Sample More Cases** repeatedly.

Q4 Will the same name ever appear twice in the same sample of 3? Explain. What do you think would happen if you changed the sample size to 6?

8. Check your prediction by changing the number of cases to 6—the number of names in your collection. Click **Sample More Cases** repeatedly.

Q5 What happens in the case table as you sample?

9. Now, increase the number of cases to 10. Watch the inspector closely when you click **Sample More Cases.**

Q6 Describe what happens as you sample.

Deck Simulation

Now you'll use what you've learned about sampling to simulate picking cards from a deck of cards.

10. Scroll down in the document until you see the Deck collection. Double-click this collection to show the inspector.

Q7 How many cases are in the collection? What are the attributes?

11. Select the collection and choose **Collection | Sample Cases.** A new collection appears, called Sample of Deck. Open the collection by dragging the lower-right corner.

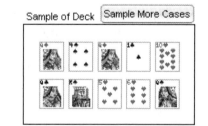

Q8 Click **Sample More Cases** several times. Describe what happens.

Now you'll make a simulation to help you determine the chance of getting a pair when you draw two cards. A pair occurs when the *Number* values of the two cases are the same.

12. Change the number sampled from 10 (the default) to 2. Uncheck Animation on. When you click **Sample More Cases** (whether in the inspector or in the collection), you now get two cards.

Exploring Sampling—With or Without Replacement
continued

Three functions that might help are first(), last(), and uniqueValues(). They're all in the Statistical section in the Functions list.

13. To keep track of the number of pairs, you can record the results in a *measure*. Go to the **Measures** panel in the inspector. Create a new measure, *Pair*, with a formula that will be true if the two cards in the collection are a pair and false if they are not.

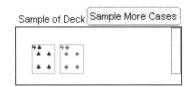

Q9 What is your formula?

Now you'll collect some different samples—different sets of two cards.

14. With the Sample of Deck collection selected, choose **Collection | Collect Measures.** A new collection appears, called Measures from Sample of Deck. Show this collection's inspector and go to the **Cases** panel.

Q10 What attributes does this collection have? How many cases does it have?

15. Go to the **Collect Measures** panel and enter these settings. Click **Collect More Measures.** Be patient. You are collecting 200 samples of two cards.

Q11 Make a bar chart of *Pair*. How many of your 200 samples were pairs? How do you know?

Q12 In the entire class, how many pairs were there? Out of how many samples?

16. Show the inspector for the Sample of Deck collection—not the measures collection. On the **Sample** panel, uncheck With replacement. You will now sample *without replacement*.

17. Show the inspector for the Measures from Sample of Deck collection, and go to the **Collect Measures** panel. Make sure Replace existing cases is checked so that you'll get 200 new samples. Click **Collect Measures.** You'll (slowly) get the data on 200 samples.

Q13 How many of those samples were pairs? In the entire class, how many pairs were there? Out of how many samples?

Q14 What's the empirical probability of getting a pair when drawing from a 52-card deck *with replacement*? What's the probability *without replacement*? Explain, in words, why one probability is greater than the other.

Exploring Sampling—With or Without Replacement
continued

EXPLORE MORE

1. Find the theoretical probabilities for drawing a pair from a 52-card deck with replacement and without replacement.

2. Modify the Deck simulation to compute the probability of drawing two cards and having one be an ace and the other a king, queen, jack, or ten (getting a blackjack).

   The uniqueValues function will help.

3. Modify the Deck simulation to compute the probability of getting only one pair in a five-card sample (a pair in poker).

   The uniqueValues function will help. The histogram also shows a hint.

4. Design a Bag simulation that counts the number of samples it takes to get the same name to appear twice in a sample of 3. Then generate a distribution of the number of samples it takes to get the same name to appear twice in a sample of 3 and find its center and spread.

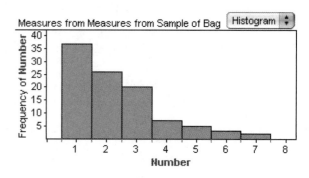

5. Open the document **Westvaco.ftm**. The Westvaco Corporation, which makes paper products, decided to downsize. They laid off several members of their engineering department and Bob Martin was one of those who lost their job. He claimed he had been laid off because of his age. Westvaco's management went through five rounds of planning for a reduction in workforce. In the first two rounds, 20 people were fired, 16 of whom were 50 or older.

   Using what you have learned in this activity, design a simulation to calculate the probability of firing 20 people and getting 16 or more people who are 50 or older.

Exploring Sampling—
With or Without Replacement

Activity Notes

Objectives
- Using simulation to calculate empirical probabilities
- Creating a model of a chance process
- Using a sampling distribution to find the probability of a given compound event by sampling with replacement and without it

Activity Time: 30–45 minutes

Setting: Paired/Individual Activity or Whole-Class Presentation (use **WithWithout.ftm** for either)

Optional Documents: Westvaco.ftm (Explore More 5) and **Westvaco2.ftm** (Explore More 5 solution)

Statistics Prerequisites
- Definition of probability
- Familiarity with the definition of independent events, $P(A \mid B) = P(A)$
- Familiarity with the conditional $P(A \mid B)$
- Familiarity with sampling

Statistics Skills
- Multiplication rule
- Working with the definition of conditional probability
- Working with the definition of independence
- Calculating probabilities with graphs and two-way tables
- Sampling with and without replacement and how that affects probabilities and independence
- *Optional:* Collecting measures of measures (Explore More 4)

AP Course Topic Outline: Part III A (1–3, 5), B (1), D (6)

Fathom Prerequisites: Students should be able to make case tables and bar charts, create attributes, and use the formula editor.

Fathom Skills: Students sample with and without replacement, change the properties of a sample, use an inspector to create a measure, and create a collection of measures.

General Notes: Sampling is a central idea in statistics. This activity explores the differences when sampling with or without replacement and then uses sampling to investigate probability. Using Fathom allows students to quickly take many samples of different types and easily build sampling distributions.

Procedure: The Bag Simulation section simulates putting 6 names on equal-sized slips of paper—one name on each slip—then putting the slips of paper in a bag, shaking it up, then drawing out a sample from the bag, one name at a time. In steps 3–6, the slip of paper would be returned to the bag and the slips well mixed before choosing another name, whereas in steps 7–9, the slips are not returned to the bag. You might want to demonstrate each of these with a real bag and six pieces of paper—just to reinforce the method. You can also suggest that students resize the sample collection so that they can see the whole sample. Some students will find it easier and faster to spot repeats by looking for the repeating colors rather than names.

In the Deck Simulation section, students need to find out how many pairs were drawn in all the samples from the entire class. You will need to facilitate this data collection in any way that is practical. This is done to increase the number of samples and to therefore decrease the sampling error of the probability.

GENERATE DATA

Q1 There are 10 names by default, so Fathom is sampling with replacement. This is evident because there are only 6 names in the bag, but 10 in the sample. Some must be repeating.

INVESTIGATE

Q2 Each time **Sample More Cases** is clicked, a new set of 5 names appears in the case table. With only 6 names to choose from, most likely some will repeat.

Q3 Here is one set of results for 100 trials where the sampling was continued until the first sample was drawn that had a repeated name. The mean for these 100 trials was 2.33 but as you can see below, 2 trials took 14 samples before a name repeated in a sample of 3. (See Explore More 4 for how this simulation was set up.)

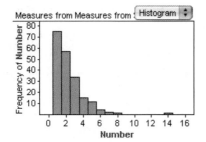

Q4–Q5 The same name will never appear twice in the sample. If the sample size is changed to 6, all 6 names will come up every time (but the order will change).

Q6 When the **Sample More Cases** button is clicked, the inspector changes the number of cases to 6. This is the maximum number of cases for sampling without replacement.

Q7 There are 52 cases in the collection. The attributes are *Suit*, *Number*, *Name*, and *CardID*.

Q8 Each time **Sample More Cases** is clicked, a new set of 10 cards is drawn from the deck. Cards may repeat.

Q9 The activity doesn't tell students how to make the formula for *Pair*—the Boolean formula that assesses whether the two cards in the sample are a pair. Any formula that accomplishes the task is acceptable. Here are two possibilities:

first(Number)=last(Number)
uniqueValues(Number)=1

Students are more likely to come up with the first formula, which works nicely. If they go on to the Explore More section, you should introduce them to the second formula. For the question on getting a pair in a five-card sample, the formula is uniqueValues(Number)=4. (The equals sign—as opposed to ≤—is correct. If uniqueValues(Number)=3, for example, that means the sample has two pair or three of a kind.)

Q10 There should be 5 cases if the sample size is set to its default of 5. This collection has attributes *CardSize* and *Pair*.

Q11 The bar chart in the activity has 18 pairs. You can put the cursor over the bar and look in the lower-left corner of the Fathom window.

Q12 Make sure that you have recorded class values before moving on. The next part of the activity will wipe out the previous data.

Q14 Make sure students know that empirical probability is the same as experimental probability. Use the class results to calculate these answers. Sample answer: The probability would be 18/200, or 0.09, for sampling with replacement and 13/200, or 0.065, for sampling without replacement.

Here are 50 runs of 200 samples of two cards. The distribution of proportions from sampling with replacement (top) is centered a bit more to the right (at 0.75) than the distribution of the proportions from sampling without replacement (centered at 0.6), as it should be.

With Replacement

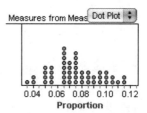

Without Replacement

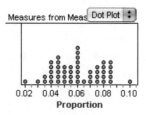

Exploring Sampling—With or Without Replacement
continued

Activity Notes

EXPLORE MORE

1. With replacement: 4/52; without replacement: 3/51.

2. Note that the value of *Number* for an ace is 1. You need to test the first and last (second) cards in the sample to see whether they are aces or ten cards. (Face cards have a value of 10.) One way to do this is to create new measures in the original collection with these formulas:

   HasAce=(first(Number)=1) or (last(Number)=1)
   HasTen=(first(Number)>9) or (last(Number)>9)
   Blackjack=(HasAce) and (HasTen)

   The theoretical probability for getting blackjack without replacement is

   $$\frac{4}{52} \cdot \frac{16}{51} + \frac{16}{52} \cdot \frac{4}{51} \approx 0.048$$

   With replacement, the probability is

   $$\frac{4}{52} \cdot \frac{16}{52} + \frac{16}{52} \cdot \frac{4}{52} \approx 0.047$$

3. Change the number of cases for the sample collection to 5. To get a pair, use the formula uniqueValues(Number)=4. This will return the cases where there are only four unique numbers, indicating that two cards must match.

4. In the sample collection, define the measure *SamePerson* with the formula uniqueValues(Name)<3. In the measures collection, create a measure *Number* defined by the formula count(). On the **Collect Measures** panel, change the panel to sample until SamePerson=true as shown. Then select the measures collection, and choose **Collection | Collect Measures.** The new collection, Measures from Measures from Sample of Bag, counts the number of samples that were taken before a repeating name popped up. See the histogram and mean in the answer to Q3 for possible means.

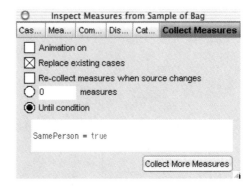

5. See the document **Westvaco2.ftm** for one solution.

Binomial Distributions—Can People Identify the Tap Water?

You will need
- 40 identical small paper or plastic cups
- tap water
- bottled water
- about 20 volunteer subjects

In this activity you'll design an experiment to see if people can tell bottled water from tap water. Then you'll simulate the situation where none of them can tell bottled water from tap water, so they guess which is which. You'll compare your results to this simulation.

GENERATE DATA

Q1 Design a study with 20 subjects to see if people can identify which of two cups of water contains the tap water. Make sure to follow the principles of good design. Describe your design.

Before you perform your study, suppose that none of your 20 subjects actually can tell the difference and so their choice is equivalent to selecting one of the cups at random. You want to make a probability distribution of the number who will choose the tap water. First you'll make a simulation.

*To add cases, choose **Collection | New Cases**.*

1. Open a new Fathom document and make a collection named Subjects with an attribute named *Guess*. Add 20 cases to it. This collection represents your volunteer subjects.

Consider an if statement or randomPick for your formula.

2. Define *Guess* with a formula that causes the values for the attribute to be either "Incorrect" or "Correct" according to the model that they cannot tell the difference and are just randomly guessing.

Subjects	
	Guess
1	Incorrect
2	Correct
3	Incorrect

3. Define a measure, *Number_Correct*, that counts the total number of subjects who correctly select the tap water.

4. Collect 200 measures from the Subjects collection. Choose **Collection | Collect Measures**. Show the inspector for the measures collection and change it to collect 200 measures. Click **Collect More Measures.**

*To plot the mean, choose **Graph | Plot Value**. To rerandomize, select the collection and choose **Collection | Rerandomize**.*

5. Make a histogram of the 200 measures, *Number_Correct*. On this graph, plot the mean. Rerandomize the Subjects collection a few times to see several distributions.

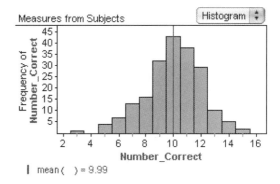

Q2 Describe the distributions of *Number_Correct* in terms of shape, center, and spread.

Exploring Statistics with Fathom

Binomial Distributions—Can People Identify the Tap Water?
continued

Q3 If only 10 people correctly select the tap water, you don't have any reason to conclude that people actually can tell the difference. Why not?

Q4 Decide how many people will have to make the correct selection before you are reasonably convinced that people can tell the difference. Explain your choice.

INVESTIGATE

Now you'll plot the true binomial distribution for this situation and see how well the simulation compares. First you'll plot it on its own graph.

6. Make a new graph and choose **Function Plot** from the pop-up menu.

The easiest way to adjust the axes is to double-click the graph and use the inspector.

7. Choose **Graph | Plot Function** and enter binomialProbability(round(x),20). Adjust the axes to see the curve.

8. Trace the function plot using the cursor.

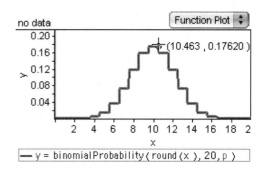

Use tracing and your function plot to answer these questions.

Q5 Make a probability distribution of the number of subjects who correctly choose the tap water when they cannot tell a difference between tap water and bottled water, so they randomly guess.

Q6 If only 10 people correctly select the tap water, you don't have any reason to conclude that people actually can tell the difference. Why not?

Q7 Decide how many people will have to make the correct selection before you are reasonably convinced that people can tell the difference. Explain your choice. Did your choice change from what you decided in Q4?

Q8 Verify that correctly identifying the tap water is a binomial situation.

EXPERIMENT

Q9 Conduct the study you designed in Q1 and write a summary of your findings. In your summary, don't forget to compare your results to both your simulation and the function you plotted in step 7.

Binomial Distributions—Can People Identify the Tap Water?
continued

EXPLORE MORE

1. Plot the binomial probability function on your histogram. (You will need to multiply by a particular constant or change scales to see them both on the same plot.) How does the simulation compare to the true distribution curve? How many measures would you need to take to improve the accuracy of your simulation?

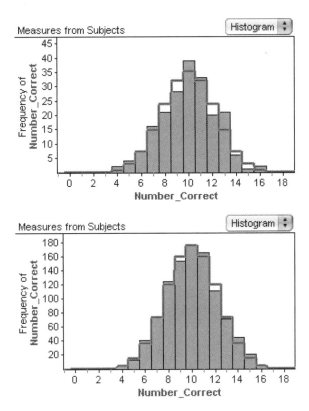

2. Learn about the binomialQuantile function in Fathom Help. Use it to answer Q7.

3. Change your simulation so that the subjects guess correctly only 10% of the time. Run the simulation and report your findings.

Binomial Distributions— Can People Identify the Tap Water?

Activity Notes

Objectives

- Using simulation to calculate empirical probabilities
- Creating a model of a chance process
- Understanding and using the binomial probability model as a theoretical model for a real-life situation
- Seeing that the binomial probability model can be used to check a real-life situation

Activity Time: 30–50 minutes

Setting: Paired/Individual Activity (build simulation) or Whole-Class Presentation (use **TapWaterPresent.ftm**). All students should collect data in Q9.

Materials

- About 40 identical paper or plastic cups
- Tap water
- Bottled water
- 20 volunteer subjects

Statistics Prerequisites

- Definition of probability
- Familiarity with the definition of independent events
- Familiarity with the principles of designing an experiment
- Familiarity with sampling

Statistics Skills

- Binomial distribution and probabilities
- Setting up a simulation of a binomial situation
- Doing a simulation to compare experimental results to a theoretical model to see if the results are reasonable
- Calculating probabilities using simulation and the binomial distribution
- Setting up and doing an experiment using randomization
- Experimental protocol

AP Course Topic Outline: Part II A (3), C; Part III A (1, 4, 5), C (3)

Fathom Prerequisites: Students should be able to make a new collection based on a random function, define attributes and measures, make histograms and plot values, and collect measures.

Fathom Skills: Students work with the binomial distribution function, set up a binomial population to sample from, compare results using plots and a distribution function, and create a collection of measures.

General Notes: This activity is essential as an introduction to the use of the binomial distribution in significance testing. The information is presented within an easily understandable context—a taste test. Using Fathom allows students to quickly build a sampling distribution of their null hypothesis (that people cannot tell tap water from bottled water.)

Procedure: Students will probably want to use a "designer" brand of bottled water. They may wish to use another comparison such as brands of soda or diet soda vs. regular soda of the same brand. Their subjects could be another classroom of students or the students in this class. All students should answer Q1 (design their study). At this point, you can have your students use Fathom to do the activity or use the presentation document. This file is set up to begin taking 200 measures for $p = 0.5$. You can change the value for p and do another binomial distribution as well because the collection is based on the function if(random()<p). The binomial distribution function is plotted, and you can use the summary table to find different cumulative probabilities. All students should collect data in Q9.

GENERATE DATA

Q1 Characteristics of a good design would include having the two types of water at the same temperature and poured at about the same time, randomizing which type of water is given first to each subject, and using identical cups for each type of water except for perhaps a small mark on the bottom for identification purposes.

Q2 For $p = 0.5$, the binomial distribution is roughly symmetric and mound-shaped for any sample size. The center should be approximately at np, or 10. The spread for $n = 20$ should be around 2.236. The sample histogram in the student activity has mean 9.99 and standard deviation 2.130.

Binomial Distributions—Can People Identify the Tap Water?
Activity Notes continued

Q3 On average, 10 people will select the tap water even if no one can tell the difference. According to the sample histogram, when all subjects were guessing, roughly 10 chose correctly and 10 chose incorrectly.

Q4 This distribution represents what would happen if people cannot tell the difference. So the results from the study would have to be in the upper portion to convince us people weren't just guessing. The standard is that the result from the sample has to be in the upper 5% of the distribution. Fourteen or more people correctly selecting the tap water occurs roughly 6% of the time according to the sample histogram. Fifteen or more occurs about 2% of the time according to the histogram. To be in the top 5%, you would have to have 15 or more people select the tap water.

INVESTIGATE

Q5 For this distribution, $n = 20$ and $p = 0.5$.

x	0	1	2	3	4	5	6	7
p	0	0	0.0002	0.0011	0.0046	0.0148	0.037	0.0739

x	8	9	10	11	12	13	14
p	0.1201	0.1602	0.1762	0.1602	0.1201	0.0739	0.037

x	15	16	17	18	19	20
p	0.0148	0.0046	0.0011	0.0002	0	0

Q6 The answer is the same as in Q3.

Q7 The standard is that the result from the sample has to be in the upper 5% of the distribution. Fourteen or more people correctly selecting the tap water occurs roughly 0.037 + 0.0148 + 0.0011 + 0.002, or 5.77% of the time, and 15 or more occurs about 2.07% of the time. To be in the top 5%, you would have to have 15 or more people select the tap water. Depending on students' histograms, it is quite possible their answers will change.

Q8 **B:** They are binomial—each trial can be classified as a "success" (correct) or as a "failure" (incorrect).

I: Each trial is independent of the others. That is, the probability that one subject correctly identifies the tap water does not change the probability that another subject correctly identifies the tap water.

N: There are a fixed number of trials, $n = 20$.

S: It is given that all subjects are guessing, so there is the same probability of a success on each trial, $p = 0.5$.

EXPERIMENT

Q9 It is common to find no convincing evidence that people can tell the bottled water from tap water.

EXPLORE MORE

1. A probability distribution can serve as a theoretical model for a real-life situation, or a probability distribution can serve as a model to check a real-life situation. In either situation, the probability distribution will not exactly match the chance behavior that you observe. Increasing the number of measures will help. The second histogram in the student activity was made with 1000 measures. Here is one using 500 measures.

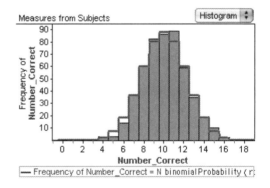

2. Students could use the **binomialQuantile** function to calculate the number of correct guesses that occur some specified percentage of the time. As shown here, if subjects cannot tell the difference between the two kinds of water, then about 14 (or fewer) out of 20 correct choices would occur 95% of the time (the actual cumulative percentage is 97.9%). Because the binomial distribution is a discrete distribution, that would mean 15 (or more) out

of 20 correct choices would occur less than 5% of the time.

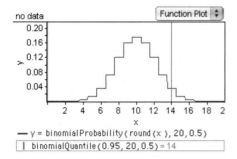

3. Shape: For small values of p (close to 0), the binomial distributions are highly skewed toward the larger values of x. As p increases to 0.5, the distributions get more symmetric, with perfect symmetry achieved at $p = 0.5$. As p increases from 0.5 to 1, the distributions get increasingly skewed toward the smaller values of x. As n increases, the skewness decreases for values of p different from 0.5, and the distributions tend to look more and more normal.

Center: Because the mean is located at np, it increases with both n and p.

Spread: Again, the standard deviation increases with n. For a fixed n, the standard deviation is largest around $p = 0.5$ and gets smaller as p gets closer to 0 or 1. Here is the distribution of 200 measures for $p = 1$.

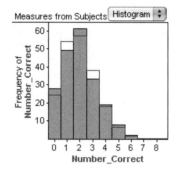

Geometric Distributions—Waiting-Time Problems

You will need
- one die
- RollingDice.ftm

In the general population, about 40% of people have type A blood. Suppose that a worker in a blood bank needs type A blood today and wants to know about how many blood donations he might have to process before he finds the first donation that is type A. Or suppose you want to know how many times, on the average, you have to roll a die to get a six. Is the sample below unusual?

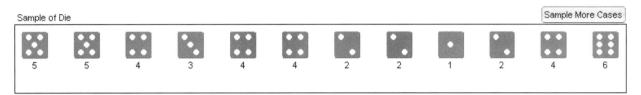

Both of these situations are called *waiting-time problems* because the variable in question is the number of trials you have to wait before the event of interest happens. The number of trials isn't fixed—you simply count the number of trials until you get the first success. In this activity you'll design simulations for these waiting-time situations and use them to construct approximate probability distributions.

COLLECT DATA

1. Roll a die. Is it a six? In other words, was your waiting time to success (a six) just one trial?

2. If you were successful on your first roll, then stop this run of the simulation. If you did not roll a six on the first trial, then continue rolling your die until you get your first six. Record the value of x, the number of the trial (roll) on 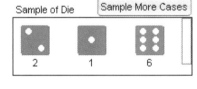 which the first six occurred. For example, in this set of rolls, it took 3 trials (rolls) to get a six, so you would record a 3 and then start all over again.

3. Repeat steps 1 and 2 at least ten times, recording the value of x each time.

4. Combine your values of x with others from the class and construct a plot to represent the simulated distribution of X. Describe the shape of this distribution and find its mean.

INVESTIGATE

Rolling Dice

Now you'll design a simulation in Fathom.

5. Open the document **RollingDice.ftm**. It has two collections: Die and Sample of Die. Make sure you can see the case table or the collection for Sample of Die.

Geometric Distributions—Waiting-Time Problems
continued

6. Open the sample collection's inspector. On the **Sample** panel, you want to set up the simulation to roll the die until it shows a six. Click the **Until condition** button. Enter the expression at left. This expression tells Fathom when to stop collecting samples.

Q1 Click **Sample More Cases** repeatedly. Describe what happens to the case table. What was the greatest number of rolls required to get a six? About how many times did you have to roll to get a six?

Now you will collect the numbers of rolls that it takes to get a six. Counting the number of rolls will be the same as counting the number of cases in the Sample of Die collection.

7. On the **Measures** panel in the inspector, define a new measure, *Number_of_Rolls*, with the formula count().

Select the Sample of Die collection and choose **Collection | Collect Measures.**

8. Collect 5 measures from the sample collection.

9. Make a dot plot of *Number_of_Rolls* and plot the mean.

10. Show the measures collection's inspector and go to the **Collect Measures** panel. Change the number of measures to 45. Click **Collect More Measures.** The simulation runs 45 times, collecting the numbers of rolls it took to get a six and recording them in the measures collection.

Q2 Describe your distribution of *Number_of_Rolls* in terms of shape, center, and spread. Which number of rolls is the most likely? What is the maximum number of rolls your simulation took to get a six?

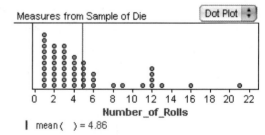

If you decide to collect a large number of measures, it will speed things up to uncheck Animation on. Also, you might want to change your dot plot to a histogram.

Q3 Collect measures again, watching the graph. Continue to collect measures to answer the original question: What is the mean number of rolls you need to get a six?

Q4 Based on the simulation, what do you think is the *most likely* number of rolls you need to get a six (that is, the number that happens most frequently)? Explain. Sketch or refer to a graph if you need to.

Waiting for Type A Blood

Now you'll set up the situation about waiting for type A blood. This time, the basic idea is to use a collection with one randomly generated donation, then sample from the collection until you get a donation of type A blood.

Geometric Distributions—Waiting-Time Problems
continued

*You can either scroll down to an empty space in the document or create a new document by choosing **File | New**.*

11. Create a collection named Donor with one case. Define an attribute named *BloodType* using the formula here.

You can also rerandomize by pressing Ctrl+Y (Win) ⌘+Y (Mac).

12. Repeatedly choose **Collection | Rerandomize**. Observe the effect on *BloodType*. How many times do you have to rerandomize to get a type A donation?

13. Select the Donor collection and choose **Collection | Sample Cases**. Show the sample collection's inspector. On the **Sample** panel, use what you learned in the first part of this activity to write the formula that tells Fathom to sample cases *until* the first "A."

*To collect measures, select the sample collection and choose **Collection | Collect Measures**.*

14. As you did in step 7, define a new measure, named *How_Many*, that counts the number of samples until there is a donation of type A blood. Collect 5 measures.

15. Make a dot plot or histogram of *How_Many* and plot the mean.

16. Use the measures collection's inspector to collect at least 100 measures.

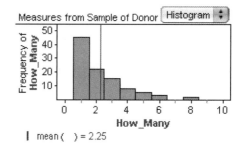

Q5 Describe the shape, center, and spread of the distribution of measures.

Q6 Estimate the probability that the first donation with type A blood is the second one checked.

Q7 What is your estimate of the probability that at most five donations are checked to get one that is type A? At least two donations? At most four donations?

Q8 Which donation is the most likely to be the first one that is type A?

EXPLORE MORE

1. Repeat the simulation for the 7% of donors having type O− blood. What is the new mean? Repeat the simulation with other probabilities. Make a conjecture about the relationship between the probability and the mean of the distribution.

2. Fathom has a function for generating random numbers from a geometric distribution. Investigate the distribution of such numbers and see if there is a connection to the blood-donation simulation.

3. On the average, how many rolls does it take to get a five *and* a six? A five *or* a six? Make predictions, then simulate both situations.

4. On the average, how many rolls does it take to get all six values: 1, 2, 3, 4, 5, and 6?

Geometric Distributions—Waiting-Time Problems

Activity Notes

Objectives

- Understanding the nature of the geometric distribution and distinguishing the geometric distribution from other distributions (particularly being skewed right)
- Understanding that even though the results (measures) vary from one iteration to the next when a random process is repeated over and over, knowing the distribution of the measures enables you to decide whether any *single* observation is probable or improbable
- Using simulation to calculate empirical probabilities
- Using the geometric probability model as a theoretical model for a real-life situation

Activity Time: 35–50 minutes

Setting: Paired/Individual Activity (collect data, use **RollingDice.ftm**, then build simulation or use **TypeADonor.ftm**)

Optional Document: RollingDiceExp.ftm (Explore More 3 and 4 solutions)

Materials

- One six-sided die for each student

Statistics Prerequisites

- Definition of probability
- Familiarity with the definition of independent events
- Describing distributions in terms of shape, center, and spread
- Familiarity with sampling

Statistics Skills

- Geometric distribution and its probabilities
- Setting up a simulation of a geometric situation
- Doing a simulation to compare experimental results to a theoretical model to see if the results are reasonable
- Calculating probabilities using the plot of a simulation of the geometric situation
- Setting up and doing a simulation using randomization
- Sampling and collecting measures

AP Course Topic Outline: Part II A (4), B; Part III A, B (1), D (6)

Fathom Prerequisites: Students should be able to make a new collection, define attributes and measures, make histograms, and collect measures.

Fathom Skills: Students set up a geometric situation to sample from, compare results using plots and a distribution function, sample until a criterion is met, and use an "if" statement to generate a population. *Optional:* Students work with the geometric distribution function (Explore More 2).

General Notes: This activity is essential so that students can understand the nature of the geometric distribution. It helps students understand what distinguishes a geometric distribution from other distributions. Using Fathom, students can easily simulate waiting-time problems and build the sampling distributions.

Procedure: The first part of this activity has students roll a die until they roll a six, recording the number of trials it took to roll a six. They then repeat this process ten times. You can have your class plot the distribution on the board or on an overhead. Usually, this part of the activity goes quickly and they then have a better understanding of what they do with Fathom.

The Rolling Dice section uses Fathom to explore how many trials it will take to roll a six on a six-sided die. If the probability of an event is p, the expected number of trials you need to get a success is $1/p$. In this case the probability of rolling a six is 1/6, so the expected number of trials to roll a six is 6.

Geometric Distributions—Waiting-Time Problems
continued

Activity Notes

When the students open the document **RollingDice.ftm,** their screen will look like this.

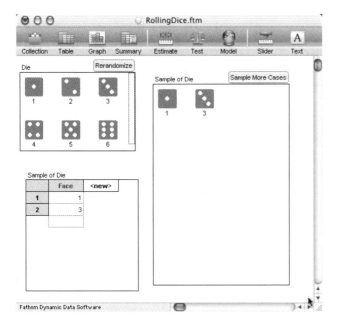

The collection Die is the population of possible rolls when rolling a six-sided die. The sample collection at this point is set up to roll a die twice. The sample collection window is large so that when the students do sample until they get a six, they can see most of the rolls in their samples if the samples end up being large.

In the next part of the activity, students use Fathom to simulate how many blood donations a technician might have to process before she or he finds the first donation that is type A, given that in the general population, about 40% of people have type A blood (Source: www.redcross.org/services/biomed/blood/supply/usagefacts.html).

You can change this to any waiting-time simulation because Fathom makes it easy to change the probability of an event, thereby allowing students to explore the effects of different probabilities on the distribution.

If you do not have time to do this part of the activity in full, your students can use the document **TypeADonor.ftm** instead. They can then resume the activity at Q5.

INVESTIGATE

Q1 As the sample is collected, the number of cases in the case table increases or decreases to accommodate the number of rolls. The last number in the case table is always 6. The greatest number of rolls required will vary; some students have gotten values as large as 36.

Q2 This situation gives rise to a skewed distribution with the skew toward the larger values. With this kind of distribution, mean > median > mode. The first bin will probably be the tallest, although with only 50 measures any of the bins from 1 to 6 could be the tallest. The bins probably (but not necessarily) decrease in size moving to the right (it is a geometric distribution). Again, the maximum will vary. The sample dot plot in the student activity has mean 4.86 and standard deviation 4.45. The median is 3, with an IQR of 4 (6 − 2). The maximum is 21.

Q3 The mean is 6, although depending on the number of measures students collect, their answers will vary. Here is a histogram of 500 measures with mean 6.68.

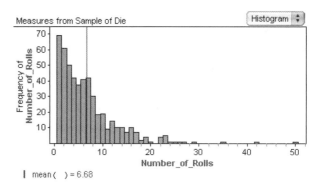

Q4 The most likely number of rolls to get a six is the mode, which is 1.

Q5 The histogram in the activity shows a sample distribution of 100 measures. The mean of the 100 waiting times is 2.25, slightly smaller than the expected waiting time of 2.5. The estimated mean typically will be too small because a small simulation usually won't produce any of the large possible values for X. The distribution is skewed right toward the larger values, and the standard deviation is about 1.6. (For $p = 0.4$, the standard deviation is about 1.94.) The median is 2 and the IQR is 3 − 1, or 2.

Note: If your students would like their captions to show "A" or "Not A" as in **TypeADonor.ftm,** show the inspector for the Donor collection and go to the **Display** panel. Double-click in the formula cell for the attribute *caption* and enter BloodType.

Geometric Distributions—Waiting-Time Problems
continued

Q6–Q7 Simulated answers will vary. Theoretical answers: $P(X = 2) = 0.24$, $P(X \le 5) = 0.92224$, $P(X \ge 2) = 1 - P(X = 1) = 0.6$, and $P(X \le 4) = 0.8704$. For the above histogram, these would be 0.22, 0.95, 0.55, and 0.9, respectively. You might need to remind students that the easy way to find these is to select the bars in the histogram that they want to include (hold down the Shift key to select more than one). Move the cursor over the measures collection to see the total in the lower-left corner of the document window.

Q8 The first checked donation is the donation most likely to be the first one that is type A. That probability is 0.4. Each subsequent checked donation has the same chance of being type A itself, but a number of failures have to occur before it even has the chance of being the first one that is type A. You can use the discussion questions to help students make sense of this.

DISCUSSION QUESTIONS

- Explain how the "if" statement chooses a blood type for a single donation.
- What was the most frequent number of donations to wait for? Explain why that makes sense.
- Generally, the frequency of each higher number of donations to wait for is less than the previous number. Why does this make sense?
- If the probability that a person has type A blood is much less than 0.4, what do you think would happen to the distribution of measures? Why? (Use Explore More 1 to test your conjecture.)

EXPLORE MORE

1. Students may conjecture that the mean (expected value) of the distribution is 1 divided by the probability of the event, or $1/p$.

2. Students should use Fathom Help to learn about the randomGeometric function. Using this random number generator to define an attribute for a collection of many cases enables students to redo the activity without sampling or collecting measures. Encourage students to change the probability value, possibly by using a slider.

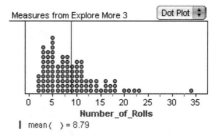

3. To sample until you get a five *and* a six, use the formula (count(Face=6)>0) and (count(Face=5)>0), then collect measures. For 100 such measures, the mean number of trials to get a five and a six was 8.79, as shown below. The average number of trials to get a five or a six was 2.4 (the theoretical answer is 1/(1/3), or 3). See the document **RollingDiceExp.ftm** for these setups.

4. In one experiment, the average number of trials to get all six values was 14.81. Students can use the until condition uniqueValues(Face)=6. See the document **RollingDiceExp.ftm**.

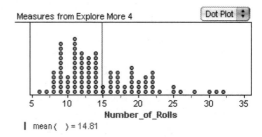

Geometric Distributions: Expected Value

In this activity you will discover the formula for the expected value (mean) of a geometric distribution. You will answer the question "If the probability of success is *p*, how many trials, on average, will you need (waiting time) to get a success?"

GENERATE DATA

To change the slider's range, double-click below the axis to show the inspector.

1. In a new Fathom document, make a slider named *p* whose range goes from 0 to 1. Set *p* to 0.5 for now.

2. Make a collection named Trials with 100 cases in it. Define an attribute, *k*, with the formula caseIndex. This attribute will represent the *k*th trial for the random variable *X*.

3. Create an attribute named *Probability_of_Success* and define it with the formula that calculates the probability of success on the *k*th trial, given that the probability on any given trial is *p*.

INVESTIGATE

Now you want to create and collect some measures for this distribution.

Hint: Use the formula for the expected value of a probability distribution for the random variable X.

4. Create a measure, *Expected_Wait_Time*, and define it with a formula that computes an approximation of the expected value of this distribution.

Q1 Will the value of *Expected_Wait_Time* be a bit larger or a bit smaller than the theoretical expected value? Explain.

5. Create a measure, *SD_of_Wait*, and define it with a formula that computes an approximation of the standard deviation of this distribution.

6. Create another measure, *Probability*, whose formula is just *p*. This measure will keep track of the slider value.

7. Set the value of the slider by double-clicking the current value and typing 0.1.

8. Collect measures from the Trials collection.

Checking Re-collect measures when source changes means that you collect one new measure whenever the slider changes.

9. Open the inspector for the collection Measures from Trials and set it as shown here. Click **Collect More Measures.**

Geometric Distributions: Expected Value
continued

10. Uncheck Replace existing cases. This allows you to collect a measure for several different values of *p* in the same collection.

You may want to rescale your axes. Double-click the graph to show the inspector.

11. Make a scatter plot with *Expected_Wait_Time* versus *Probability*. The graph will show only one point at first.

12. Double-click the slider to show its inspector. Type 0.1 for *Restrict_to_multiples_of*. This restricts the slider value to a precision of one-tenth. You can now drag the slider to change the slider's value to 0.2. Then continue changing the value of the slider by tenths until $p = 1$.

Q2 What is the shape of the scatter plot? Notice that you get another point in the scatter plot each time you change the value of the slider.

Q3 State your hypothesis for the formula for the expected value of a geometric distribution.

Select the graph and choose **Graph | Plot Function.**

13. To check your hypothesis, plot your function.

Q4 Does your function fit the scatter plot? If so, state your function. If not, try another one until you find a function that does fit the scatter plot.

Now you want to find a formula for the standard deviation. This is a bit harder.

14. Make a scatter plot of *SD_of_Wait* versus *Probability*.

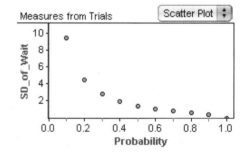

Hint: The shape is one clue. Also, note what happens at p = 1.

Q5 What is the shape of the scatter plot? State your hypothesis for the formula for the standard deviation of a geometric distribution, and check it by plotting your hypothesis.

EXPLORE MORE

1. Q1 asked whether the estimate of expected wait time would be an overestimate or an underestimate. Use Fathom to explore which it is. Add or delete cases from the Trials collection and investigate how far off the estimate is with more or fewer trials.

2. Explore values of *p* between 0 and 0.1. Why does the relationship between *Expected_Wait_Time* and *Probability* appear to break down in this region?

3. Simulate how many trials, on average, it takes to get two successes.

Geometric Distributions: Expected Value

Activity Notes

Objectives
- Learning that the expected value (mean) of a geometric distribution is inversely proportional to the probability of success
- Learning that the standard deviation of a geometric distribution is proportional to the square root of $1 - p$ and inversely proportional to the probability of success

Activity Time: 20–35 minutes

Setting: Paired/Individual Activity (build simulation) or Whole-Class Presentation (use **ExpectedValue.ftm**)

Statistics Prerequisites
- Definition of probability
- Familiarity with the geometric distribution
- Formula for calculating the probability of success on the kth trial for a given p
- Definitions and formulas for the expected value and standard deviation of a probability distribution
- Familiarity with the function $y = 1/x$ and its shape

Statistics Skills
- Geometric distribution and its mean and standard deviation
- Fitting a function to a scatter plot

AP Course Topic Outline: Part III A (4, 6)

Fathom Prerequisites: Students should be able to make a new collection, define attributes and measures, make scatter plots, collect measures, and use a slider.

Fathom Skills: Students use a slider to dynamically collect measures and plot a function to fit a scatter plot.

General Notes: In this activity, students discover that the expected value (mean) of a geometric distribution is $1/p$. Fathom eases the computational burden and allows students to easily accomplish the simulation.

Procedure: This activity can be exploratory, or you can use the document **ExpectedValue.ftm** as a whole-class presentation. In this document the collections are set up, as is the scatter plot, so you can start the activity at step 12 by dragging the slider.

GENERATE DATA

3. There are a number of formulas students could use for *Probability_of_Success*. One is $p(1 - p)^{k-1}$.

4. Students should come up with *Expected_Wait_Time* = $\mu_X = \Sigma(x_i \cdot p_i)$ = sum(k • Probability_of_Success).

INVESTIGATE

Q1 The students' values will be slightly smaller because they used only the first 100 values in the distribution.

5. This formula is equivalent to
   $SD = \sqrt{\Sigma(x_i - \mu_X)^2 \cdot p_i}$.

Q2–Q4

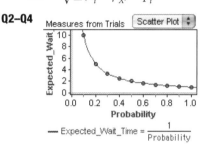

Q5 At $p = 1$, the SD is zero, suggesting a zero in the numerator of the function. Trial and error suggests the square root.

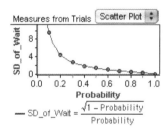

DISCUSSION QUESTIONS

- How did you define *Probability_of_Success*? Why does your formula work?
- What is the meaning of the measure *Expected_Wait_Time*? How did you define it?

EXPLORE MORE

3. In general, the expected number of trials needed until the nth success is $n \cdot \dfrac{1}{p}$.

Transformations and Simulation Explorations

Use what you have learned so far about simulations to answer these questions.

STRIPED BASS

The Maryland Department of Natural Resources Striped Bass Stock Assessment Project is responsible for assessing the status of striped bass spawning stock. This research generates data that are critical for management of the entire East Coast striped bass population. Their research has shown that the lengths of the male striped bass are approximately normally distributed with mean 25.33 inches and standard deviation 6.71 inches. The lengths of the female striped bass are approximately normally distributed with mean 36.96 inches and standard deviation 2.88 inches. (Source: www.dnr.state.md.us/fisheries/spring_survey/index.shtml.)

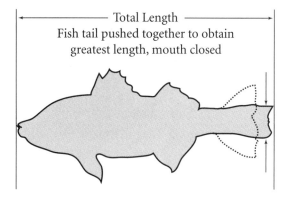

Q1 Let X be the random variable for the length of female striped bass and Y be the random variable for the length of male striped bass. Will the distribution for $X + Y$ be approximately normal with mean 62.29 inches and standard deviation 7.30 inches? Design a simulation in Fathom to check your answer.

Q2 The New Jersey Department of Environmental Protection Division of Fish and Wildlife works closely with the Maryland Department of Natural Resources Striped Bass Stock Assessment Project. During their fishing season, March 1–December 31, the possession limit for striped bass is two fish. One fish must be greater than or equal to 24 inches and less than 28 inches in length (called a *slot fish*), and the other fish must be 34 inches or greater. All other lengths must be released back into the water. It does not matter which fish is caught first. (Source: www.state.nj.us/dep/fgw/marfhome.htm.)

For example, here is one person's catch. She had to catch 10 fish to reach the limit. The throw backs were either too small or between 28 and 34 inches.

Transformations and Simulation Explorations
continued

Legally, she can take only one of the big fish (34 inches or longer), so two of these big fish would have to be thrown back as well.

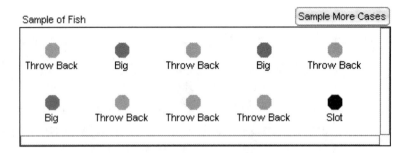

Design a simulation to see how many fish on average must be caught before a person catches the limit of two appropriately sized fish. For this simulation, you can assume that females and males are equally likely to be caught. (*Hint:* You will first need to create an attribute that determines whether the fish caught is female or male before you can define the length of the caught fish.)

Q3 Actually, females and males are *not* equally likely to be caught. Females are caught 44.4% of the time and males are caught 55.6% of the time. Redo your simulation from Q2, taking this fact into consideration.

FILIPINO CHILDREN

The Laguna Province in the Philippines is located south of Manila and covers 1759 km^2. There is mandatory schooling from the first academic year after a child reaches age seven until completion of elementary education; that is until the child is approximately fifteen years old. A study was conducted to determine how educational attainment varied within the community and within individual families. The sampled households are located in 20 different villages or communities, also known as *barangays*. Educational attainment is measured in years of schooling. The study started in 1975 and ended in 1998, so "children" means that the subjects were students when the study began.

The final size of the sample is 790 children from 126 households. The boys have an average length of schooling of 8.5 years (with standard deviation 2.7 years), while the girls' average length of schooling is 9.7 years (with standard deviation 2.8 years). Both distributions are approximately normally distributed. (Source: www.eco.rug.nl/~esp2002/portner.pdf.)

Q4 Suppose that one boy and one girl are selected at random. Design a simulation in Fathom to find the approximate probability that the boy's length of schooling is longer than the girl's length of schooling.

Transformations and Simulation Explorations
continued

Q5 The average number of children per family was 6.6. You randomly pick one family with seven children: 4 boys and 3 girls.

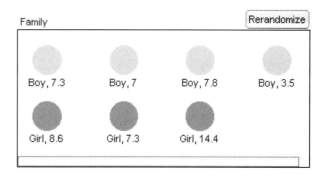

a. Use a simulation to explore the distribution of the total length of schooling for a family with 4 boys and 3 girls. What are the reasonably likely values for the total length of schooling for a family with 4 boys and 3 girls? What are the mean and standard deviation?

b. Use the formulas to explore the distribution of the total length of schooling for a family with 4 boys and 3 girls. Assuming independence, what are the reasonably likely values for the total length of schooling for a family with 4 boys and 3 girls? What are the mean and standard deviation?

c. The actual within-family standard deviation of length of schooling for boys is 1.41 years, while the within-family standard deviation length of schooling for the girls is 1.54. Using these values, rerun your simulation to find the reasonably likely values for the total length of schooling for a family with 4 boys and 3 girls. What are the mean and standard deviation?

d. Explain how your answers to parts a–c show that the total length of schooling within a family is not independent.

CHEST AND SHIRT SIZES

Chest sizes for men are approximately distributed with mean 40.8 inches and standard deviation 4.05 inches. A merchant sells shirts with the shirt and chest sizes given in the table. On average, he has 200 customers per month.

Shirt Size	Chest Size (in.)
XS	$32 \leq$ chest size < 35
S	$35 \leq$ chest size < 38
M	$38 \leq$ chest size < 42
L	$42 \leq$ chest size < 46
XL	$46 \leq$ chest size < 50
XXL	$50 \leq$ chest size < 54
XXXL	$54 \leq$ chest size < 58

Hint: Go to Fathom Help to learn about the switch() function.

Q6 Assuming that the 200 customers are a random sample of the population, design and run a simulation to determine how many of each size shirt the merchant should have in stock.

Transformations and Simulation Explorations

Activity Notes

Objectives

- Using simulation to calculate empirical probabilities
- Creating a model of a chance process
- Using a simulated sampling distribution to find the probability of a given compound event
- Using the geometric probability model as a theoretical model for a real-life situation
- Working with transformations of random variables

Activity Time: 30–50 minutes (depending on which problems are chosen)

Setting: Paired/Individual Activity (build simulations) or Whole-Class Presentation (use **MalePlusFemale.ftm, FishSepEquallyLikely.ftm, FishActualPercentages.ftm, FilipinoChildren.ftm, Family.ftm, FamilyActual.ftm,** or **ChestSizes.ftm**)

Statistics Prerequisites

- Probability concepts
- Familiarity with the definition of independent events
- Designing simulations and samples
- Familiarity with the rules for combining random variables

Statistics Skills

- Linear transformations of a random variable
- Practice in setting up simulations to generate distributions
- Working with the definition of independence
- Calculating probabilities with graphs and two-way tables
- Using transformation formulas to verify results found by sampling

AP Course Topic Outline: Part III A, B, D (5, 6)

Fathom Prerequisites: Students should be able to make graphs, create collections using random functions, define attributes and measures, sample and collect measures, and use "if" statements and Boolean expressions.

Fathom Skills: Students design simulations, change the properties of a sample, and generate distributions based on given information.

General Notes: This activity is exploratory in nature. These questions present problems that students can address using what they have learned so far in their statistics course and in Fathom. Each question involves designing a simulation to model the situation given. You can have your students work individually or in groups, or you could choose to do the questions together as a class. Some hints and methods are given below, and each question or part of a question has an accompanying solution document. Some of these problems are similar to various AP Free Response questions from over the years—with some little twists.

STRIPED BASS

Further information about striped bass and the Maryland Department of Natural Resources Striped Bass Stock Assessment Project can be found at www.dnr.state.md.us/fisheries/spring_survey/index.shtml. Further information about the New Jersey Department of Environmental Protection Division of Fish and Wildlife and their regulations concerning striped bass can be found at www.state.nj.us/dep/fgw/marfhome.htm and www.scottsbt.com/fishids/regsrecs/regsNJ.htm.

Q1 This simulation has students exploring the distribution of $X + Y$, where X is the random variable for the length of female striped bass and Y is the random variable for the length of male striped bass. In this simulation the means do add, as do the variances. So the distribution should be approximately normal with mean $25.33 + 36.96$, or 62.29 inches, and variance $6.71^2 + 2.88^2$, or 53.31. Then collect measures from this collection to see if you can get even closer to the hypothesized mean and standard deviation of the sum. The presentation or solution document is **MalePlusFemale.ftm**.

Q2 This simulation is a little more realistic. To do this one, make a collection with one case. Randomly choose whether the fish is male or female. Based on the gender, assign a length. Then define attributes to test what kind of fish was caught: slot, throw back, or big (34 inches or greater). Select the collection and choose **Collection | Sample Cases.** Define a measure

Transformations and Simulation Explorations
continued

Activity Notes

in the sample collection named *Number_Caught* and defined by count(). Change the **Sample** panel as shown.

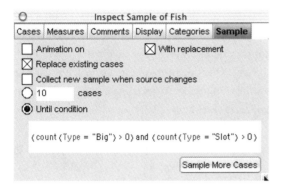

You want to sample until one big fish is caught *and* one slot fish is caught. The formula above counts the number of big fishes caught and slot fishes caught. As soon as those counts are both greater than 0, the sampling process stops and records the number of fishes caught. Next, select the sample collection and choose **Collection | Collect Measures.** Collect 200 measures. As expected, the distribution is geometric. The plot below shows 50 sample means (collected from 100 measures). The 50 means are roughly mound-shaped with mean 8.9. Reasonably likely means for 100 trials then are from about 7.9 to 10.1. The presentation or solution document is **FishSepEquallyLikely.ftm**.

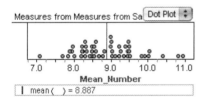

Q3 The only change for this problem is how *Gender* is determined. Because catching a male is different from catching a female, an if statement is used instead of randomPick. The rest of the setup is identical to Q2. As expected, the distribution is geometric. Below is the plot of 50 sample means (collected from 100 measures). The 50 means are skewed left with mean 8.2. Reasonably likely means for 100 trials then are from about 7.5 to 9. The solution document is **FishActualPercentages.ftm**.

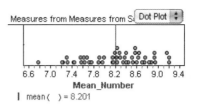

FILIPINO CHILDREN

This information is part of a larger project, Birth Order and the Intrahousehold Allocation of Time and Education, by Mette Ejrnæs, a professor from the Institute of Economics at the University of Copenhagen, Denmark, and Claus Chr. Pörtner from the World Bank. The complete write-up of this project can be found at www.eco.rug.nl/~espe2002/portner.pdf.

Q4 This simulation has students exploring the distribution of $X - Y$, where X is the random variable for the length of schooling for girls and Y is the length of schooling for boys. In this simulation the means do subtract and the variances add. So the distribution should be approximately normal with mean $9.7 - 8.5$, or 1.2 years, and variance $2.7^2 + 2.8^2$, or 15.13. The results from 400 samples in one simulation had mean 1.16 and SD 3.7. The proportion that the boy's schooling is longer than the girl's schooling was 0.3725 in this simulation. Define measures for the mean, SD, and proportion. Then collect measures from this collection to see if you can get even closer to the hypothesized mean and standard deviation of the sum. The presentation or solution document is **FilipinoChildren.ftm**.

Q5 a. There are a variety of ways to set this up. The easiest way is to make seven cases in a collection and assign three to be girls and four to be boys, then use randomNormal to define their average length of schooling. Define a measure named *TotalSchooling* with sum(Length_of_Schooling). One simulation of 500 sums collected had mean 63.3 years and SD 7.52 years. Using a summary table, the reasonably likely totals were from about 48.6 to 77.2.

Then in the measures collection, define the measures, mean and SD. You may want to set up

Transformations and Simulation Explorations
continued

Activity Notes

your original sample to be smaller, say, 50 or 100. That would speed collection up considerably. Then collect measures from the measures collection. The solution document is **Family.ftm**.

b. Using the formulas, the mean total length of schooling for a family of 4 boys and 3 girls would be $4(8.5) + 3(9.7)$, or 63.1 years. The variance would be $4 \cdot 2.7^2 + 3 \cdot 2.8^2$, or 52.68. The standard deviation would be 7.26 years. The reasonably likely values would be $63.1 \pm 1.96(7.26)$, or 48.87 to 77.33 years.

c. The setup for part c is the same as for part a except that the standard deviations change in the randomNormal function. One simulation of 500 samples of 7 children had mean about 63 years with standard deviation 3.73. The reasonably likely family totals for this histogram were from about 55.7 years to 70.3 years. Collecting 100 measures of 50 "families" gave an SD of 3.82. The solution document is **FamilyActual.ftm**.

d. If total length of schooling was independent regardless of family, the variance should be 52.68 and the standard deviation should be 7.26 years. The actual standard deviation for the within-family total is around 3.82 years.

CHEST AND SHIRT SIZES

Q6 There are a variety of ways to set this up. The easiest way is to make 200 cases in a collection with attributes and measures as defined in **ChestSizes.ftm**. Then collect measures and examine the counts of each size to find the mean count for each size. To do that all at once, select the measures collection and choose **Collection | Stack Attributes.** This makes a new collection with two attributes: *Group* and *Value*, where *Group* is the measure attribute and *Value* is the actual counts for each collection of 200 customers. Then plot or make a summary table to find the means:

Stacked Measures from Chest Sizes

	Number_Large	46.15
	Number_Medium	76.205
	Number_Small	41.34
Shirt_Size	Number_XLarge	11.945
	Number_XSmall	23.165
	Number_XX	1.14
	Number_XXX	0
	Column Summary	28.563571

S1 = mean(Number_of_Shirts)

From this simulation, the merchant should keep on hand about 23 XSmall, 41 Small, 76 Medium, 47 Large, 12 XLarge, 1 XXLarge, and no XXXLarge. The solution document is **ChestSizes.ftm**.

Inference for Means and Proportions

Reasonably Likely Values and Confidence Intervals

In this activity you will collect some data about the proportion of students in your class who are right-eye dominant. Then you will make a chart that allows you to see the reasonably likely outcomes for all population proportions p when the sample size is 40.

COLLECT DATA

1. Determine which eye is your dominant eye. Hold your hands together in front of you at arm's length. Make a space between your hands that you can see through. Through the space, look at an object at least 15 feet away. Now close your right eye. Can you still see the object? If so, your left eye is your dominant eye. If not, your right eye is your dominant eye.

2. Gather the results about eye dominance from exactly 40 students.

Q1 What proportion of students in your sample are right-eye dominant?

Q2 Suppose you can reasonably consider your sample of 40 students a random sample of all students. From the information you've gathered, is it plausible that 10% of all students are right-eye dominant? Is it plausible that the true percentage is 60%? What percentages do you find plausible?

INVESTIGATE

Reasonably Likely Outcomes by Simulation

Suppose you take repeated samples of size 40 from a population with 60% success. What proportions of successes would be reasonably likely in your sample? Steps 3–8 help you build a simulation to answer this question.

To change the slider's range easily, double-click below the axis to show the inspector.

3. Open a new Fathom document and make a slider named p with range 0 to 1. Set the slider to 0.6 to represent 60% success. (This slider will help you change the population proportion later on.)

4. Create a collection named Population Sample, with 40 cases and one attribute, *Success*. Define *Success* to randomly determine whether a case is a success ("true") based on the population proportion, as shown.

Exploring Statistics with Fathom
© 2007 Key Curriculum Press

Reasonably Likely Values and Confidence Intervals
continued

The formula for *SampProp* finds the proportion of cases for which *Success* is "true." You do not need to include true in the formula, though.

5. Go to the **Measures** panel and define two measures for the collection, *PopProp* and *SampProp*, as shown. *PopProp* records the population proportion, and *SampProp* calculates the sample proportion.

Measure	Value	Formula
PopProp	0.6	p
SampProp	0.5	proportion(Success)
<new>		

6. Select the collection and choose **Collection | Collect Measures.** Show the inspector and change the settings on the **Collect Measures** panel as shown, then collect more measures.

- [ ] Animation on
- [x] Replace existing cases
- [ ] Re-collect measures when source changes
- (•) 100 measures

To rescale, double-click the graph to show the inspector, and change *Lower* and *Upper*.

7. Make a scatter plot of *PopProp* versus *SampProp*. Scale both axes from 0 to 1. The scatter plot should be a horizontal "segment" of points.

Although the "segment" is made of discrete points, imagine that it is a solid segment.

Q3 Based on the simulation, what proportions of successes would be reasonably likely in your sample?

Q4 Compare the proportion of students you found to be right-eye dominant in Q1 to your answer in Q3. Is it reasonable to assume that 60% of all students are right-eye dominant? Explain your reasoning.

Adding Other Population Proportions

You now know whether 60% is a reasonably likely population proportion for right-eye dominance. You'll use your slider to collect measures and consider the likelihood of other population proportions.

8. To build a chart of reasonably likely outcomes, you'll want to add measures for other population proportions *p* to those you collected for 60%. Go to the **Collect Measures** panel of the measures collection's inspector and uncheck Replace existing cases. Drag the lower-right corner of the measures collection so that you can see the **Collect More Measures** button.

The next time you drag the slider, it will jump to the next number by 0.05.

9. Double-click the slider to show its inspector and enter the values shown. Click **Collect More Measures** to collect 100 measures for $p = 0.05$.

Inspect Slider — Properties

Property	Value
p	0.05
Max_updates_per_second	
Lower_	0
Upper_	1
Restrict_to_multiples_of	0.05
Reverse_scale	false

218 6: Inference for Means and Proportions

Reasonably Likely Values and Confidence Intervals
continued

10. Drag the population proportion (slider *p*) to 0.1. Click **Collect More Measures** to collect 100 measures for *p* = 0.1. Notice that the scatter plot updates to include a "segment" of points for 10%.

11. Repeat step 10 for 15%, 20%, 25%, and so on, up to 95%. (Skip 60%— you already did that in step 6.)

Choose **Graph | Plot Value**.

12. Plot a vertical line, through the *SampProp* axis, that represents your sample proportion. Here is the scatter plot for one class whose sample proportion was 0.75.

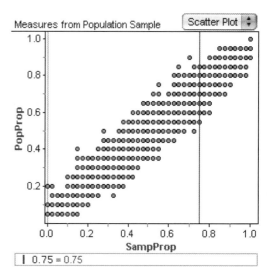

Q5 Refer to your results from Q1 and Q2 about eye dominance and look at your scatter plot. For which population proportions is your sample proportion reasonably likely?

Reasonably Likely Outcomes by Theoretical Values

The chart you made in steps 3–12 was based on simulation. Because of variability, your chart may look rather ragged. Next you'll make a smoother chart by graphing theoretical lower and upper bounds.

Q6 Recall that, in theory, about 95% of all sample populations will fall within about 2 standard errors of the population proportion, provided *np* and *n*(1 − *p*) are at least 10. So, for any such population proportion *p* and sample size *n*, what is the lower bound of the reasonably likely confidence interval? The upper bound?

13. Open a new Fathom document. Make a new graph and change it to a function plot. Adjust the bounds of the *x*- and *y*-axes to go from 0 to 1. For this graph, the *x*-axis represents the sample proportion and the *y*-axis represents the population proportion.

Select the graph and choose **Graph | Plot Function**.

14. Plot a function for the theoretical lower bound of all the population proportions, x. For now, assume the sample size is 40.

15. Repeat step 14 for the theoretical upper bound.

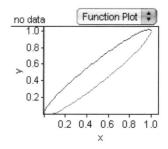

Reasonably Likely Values and Confidence Intervals
continued

Q7 Explain how this chart relates to the one you created in steps 3–12.

16. Create two new sliders, *p* and *pHat*, whose ranges are both 0 to 1. These allow you to set the value of your population proportion (*p*) and the sample proportion (*pHat*).

17. Plot the value *pHat*, which will give you a vertical line. Then plot the function *p*, which will give you a horizontal line.

Q8 What does this horizontal line relate to in the chart you created in steps 3–12?

18. Set the slider *pHat* to the value of the sample proportion for eye dominance that you got in Q1. Drag the slider *p* until the horizontal and vertical lines intersect at the lower-bound curve.

Q9 What is the approximate value of the theoretical lower bound (the value of slider *p* at the intersection)?

Q10 What is the approximate value of the theoretical upper bound?

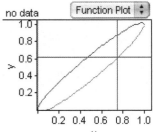

Q11 For which population proportions is your sample proportion reasonably likely? How do these theoretical values compare with your results by simulation?

EXPLORE MORE

Repeat steps 14 and 15, but do them for the scatter plot you created in step 11. (You'll need to use one of your attributes instead of *x* for the dependent variable.) You should find that some results from your simulation are not appropriately confined to by the lower and upper bounds. Which population proportions seem poorly confined? Explain why your function might be invalid for these population proportions. Investigate other functions that might remedy this problem.

A Complete Chart of Reasonably Likely Outcomes for *n* = 40

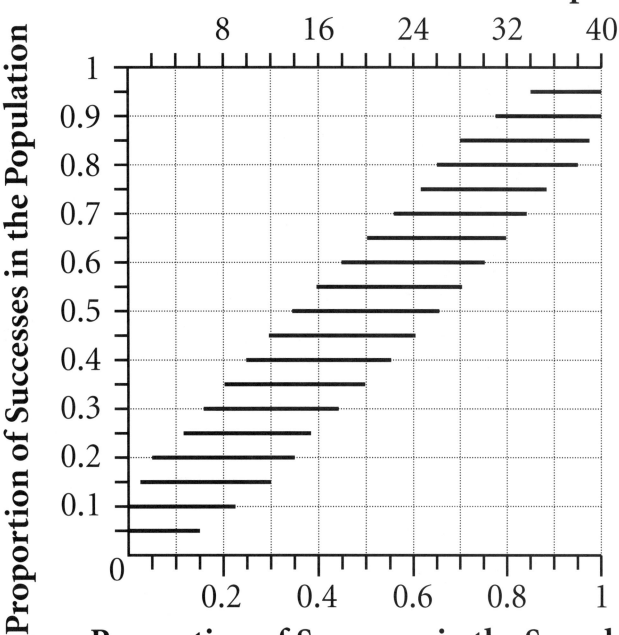

Reasonably Likely Values and Confidence Intervals

Activity Notes

Objectives
- Considering which possible population proportions could likely produce an observed sample proportion
- Developing an intuitive understanding of confidence intervals
- Finding a confidence interval graphically
- Understanding a confidence interval as consisting of those population proportions for which the result from the sample is reasonably likely

Activity Time: 40–90 minutes (depending on data collection and if all sections are done—see breakdowns of time in General Notes)

Setting: Paired/Individual Activity (collect data and build simulations) or Whole-Class Presentation (collect data and use **ChartPresent.ftm** and **TheoryPresent.ftm**)

Materials
- *Optional:* One copy per student of "A Complete Chart of Reasonably Likely Outcomes for $n = 40$"

Statistics Prerequisites
- Definition of reasonably likely events
- The binomial distribution
- Definition of probability
- Familiarity with sampling

Statistics Skills
- Constructing confidence intervals
- Working with the definition of reasonably likely and rare events
- Sampling distribution of the sample proportion and binomial distributions
- Finding plausible population proportions given a sample proportion
- Beginning inference by comparing actual results with the theoretical results

AP Course Topic Outline: Part III D (1, 6); Part IV A (1–4)

Fathom Prerequisites: Students should be able to make case tables and bar charts, create attributes, and use the formula editor.

Fathom Skills: Students create a population based on a probability distribution, update a collection when the population changes, build a collection of measures from different populations (that is, a population collection that changes before each new set of measures is collected), and create formulas using Boolean expressions.

General Notes: This activity is essential for students to get an intuitive understanding of confidence intervals. The activity basically has three parts: a hands-on activity in Collect Data (10 minutes), a simulation in the first two Investigate sections (35–40 minutes), and using formulas in the last Investigate section (35–40 minutes).

Procedure: In Collect Data, students determine which eye is their dominant eye. They then calculate the sample proportion for a group of 40 students. You will need the results from exactly 40 students. Most classes will not have 40 students, so you may want to gather data beforehand from one of your other classes. If your class has more than 40 students, eliminate some students at random to get the total down to 40.

The first two Investigate sections use simulation to create a line-segment chart of reasonably likely outcomes for samples of size 40. (See the chart following the activity.) The simulation is fairly complex but it allows each individual student or group of students to make the entire chart themselves, thereby building a deeper understanding of how the chart is created.

If you would prefer to have your students skip making this simulation and start with the chart all ready to go, then use the document **ChartPresent.ftm**. Open the document and click **Collect More Measures** on the measures collection and have your students answer Q3 and Q4. Then change the slider p to 0.05 and click **Collect More Measures** again. Then follow steps 10 and 11 to make the rest of the chart. For step 12, plot the value of any of your students' sample proportions to answer Q5.

If you choose to have your students build the simulation, it starts with a collection that represents a randomly generated sample and then collects measures rather than first creating a collection for the population, sampling

222 | 6: Inference for Means and Proportions

Reasonably Likely Values and Confidence Intervals
continued

from the population, then collecting measures from the sample. Doing it this way makes the simulation less complicated.

In the third Investigate section, students graph the theoretical lower and upper bounds of the chart, using the formula

$$p \pm 1.96 \cdot \sqrt{\frac{p(1-p)}{n}}$$

If you choose to have your students skip building this simulation, the document **TheoryPresent.ftm** is set up to start at step 18. Drag the sliders *Lower_p* and *Upper_p* until they intersect your sample *pHat*.

COLLECT DATA

1. Make sure students understand that if they cannot still see the object, the closed eye is the dominant eye.

   An alternative method for determining eye dominance is to cut out a 1″ by 1″ square from the middle of an 8 1/2″ by 11″ piece of paper. Hold the paper at arm's length with both hands. Look through the square at a relatively small object across the room. Without changing position, close your right eye. If you can still see the object, your left eye is your dominant eye. If not, your right eye is your dominant eye.

Q1 For a typical group of 40 students, about 25 or 30 will be right-eye dominant.

Q2 Answers will vary according to the results from Q1. Probably, students won't think that 10% is plausible, but that 60% is. If, for example, 90% of the 40 students are right-eye dominant, students might be willing to make a statement such as "We are pretty sure that more than half of all students are right-eye dominant." At this stage, students' responses probably won't be precise, although it would be great if they realize that with a sample size of 40, the value of $\hat{p}$ shouldn't be farther than about 15% from p:

$$1.96 \cdot \sqrt{\frac{p(1-p)}{n}} = \sqrt{\frac{0.5(1-0.5)}{40}} \approx 0.15$$

Note that it may or may not be a reasonable assumption that your students are representative of all students with respect to eye dominance.

INVESTIGATE

Q3 The middle 95% of the sampling distribution for $p = 0.6$ is theoretically cut off by the two points

$$p \pm 1.96 \cdot \sqrt{\frac{p(1-p)}{n}} = 0.6 \pm \sqrt{\frac{0.6(1-0.6)}{40}} \approx$$
$$0.6 \pm 0.1518$$

So the horizontal "segment" of points should go from approximately 0.4482 to 0.7518. For the plot given in the student activity, here is the dot plot of just the sample proportions. Eliminating the top and bottom 2.5% gives an interval from 0.475 to 0.725.

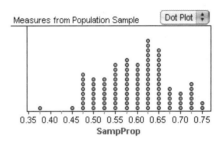

Q4 If your class's sample proportion from that activity was between approximately 0.448 and 0.752 (or the values they give in Q3), then it is reasonable to assume that the true value of p might be 0.6. The reason is that if 0.6 is the true proportion of students who are right-eye dominant, then it is perfectly reasonable to get the result in a sample of size 40 that your class did.

11.–12. A sample scatter plot appears on the student worksheet. If you wish, distribute copies of the completed, theoretical chart. This chart, invented by Jim Swift, a high school teacher in British Columbia, is the key to understanding the confidence interval. Be sure students understand that the confidence intervals are read vertically and the reasonably likely outcomes are the horizontal line segments.

Q5 For the sample plot, the value of the sample proportion (0.75) is plotted as a vertical line intersecting the populations for which a sample proportion of 0.75 is reasonably likely, so in this case, 0.55 to 0.85.

Reasonably Likely Values and Confidence Intervals
continued

14.–15. Students should adapt

$$p \pm 1.96 \cdot \sqrt{\frac{p(1-p)}{n}} \text{ into } x - 1.96 \cdot \sqrt{\frac{x(1-x)}{40}}$$

$$\text{and } x + 1.96 \cdot \sqrt{\frac{x(1-x)}{40}}$$

Please note that these functions are technically invalid for population proportions less than 0.25 and greater than 0.75 because you shouldn't use the normal approximation when either np or $n(1-p)$ is less than 10.

Q6 $p \pm 1.96 \cdot \sqrt{\frac{p(1-p)}{n}}$

Q7 Simulated values were done in steps 3–12; these are the theoretical values.

Q8 The horizontal line $y = p$ corresponds to the population proportions or the slider.

Q9–Q11 See the plot below. With a sample proportion of $\hat{p} = 0.75$ (*pHat*), for example, you can adjust the slider p to approximate the bounds of the confidence interval at 0.616 and 0.884. (An extra slider was added here to show the results of both the lower and upper values at the same time.)

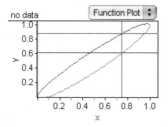

EXPLORE MORE

Students will need to plot the functions below the graph. For population proportions less than 0.25, many points fall outside the left bound. For proportions greater than 0.75, many points fall outside the right bound. The functions are invalid for these values because either np or $n(1-p)$ is less than 10. Students can use the **binomialQuantile** function to remedy this problem.

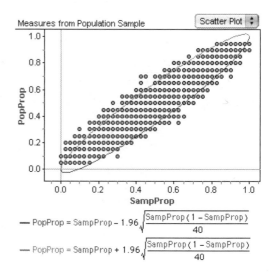

EXTENSION

The functions that students graphed in steps 14 and 15 probably contain 1.96, which determines the interval that contains 95% of sample proportions. Ask, "How would you change the functions to use the interval that contains 90% or 99% of sample proportions?" Explore what happens to the graph when you make this change. (You could try a slider that changes this value.) This function also contained 40 for the sample size. Have students create a slider, n, and use it in the functions to dynamically change the sample size. Explore what happens to the graph when you change the sample size. [Students could make sliders, *Level* and n, for the confidence level and sample size, respectively. Then make another slider, z, defined with the formula $\text{normalQuantile}\left(\frac{\text{level}}{100} + \frac{(100 - \text{level})}{200}\right)$. Finally, make the plot using the slider's values.]

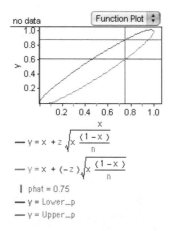

Confidence Interval for a Proportion—Capture Rate

You will need
- calculator or table of random digits
- **CaptureRate.ftm**

If you construct one hundred 95% confidence intervals, you expect that the population proportion, p, will be in 95 of them. This statement is not as obvious as it may seem at first. You will test this statement in this activity.

COLLECT DATA

1. If your instructor does not assign you a group of 40 random digits, generate 40 random digits on your calculator.

2. Count the number of even digits in your sample of 40 and calculate the sample proportion.

Q1 Use the formula to construct a 95% confidence interval for the proportion of random digits that are even.

Q2 What is the true proportion of all random digits that are even?

Q3 Did the confidence interval you constructed in Q1 capture the true proportion you gave in Q2? Is this what you expected? Explain.

INVESTIGATE

Either your confidence interval captured the true population proportion or it didn't. Now you want to find the percentage of such confidence intervals that capture the true proportion.

3. Open the Fathom document **CaptureRate.ftm.** The collections in this document have been set up for you to get a nice display, so don't delete them or their attributes.

4. Choose **Collection | Rerandomize.** Notice that you get a new set of 40 digits in the case table. A certain proportion of your 40 digits are even. You're interested in whether the confidence interval for that proportion includes what you know is the *true* proportion, namely, 0.5.

Digits

	Digit	IsEven
=	randomInteger(0, 9)	even(Digit)
1	9	false
2	3	false
3	6	true
4	8	true
5	7	false

Confidence Interval for a Proportion—Capture Rate
continued

Consider using an "and" statement to test whether the true proportion, 0.5, has been captured.

Q4 Show the inspector for the Digits collection and go to the **Measures** panel. There are four measures, none of which have formulas. This table describes the four measures. Define each measure with an appropriate formula and record your formulas.

Measure	Description
pHat	Proportion of even digits in the sample
Upper	Upper bound of the 95% confidence interval for the proportion
Lower	Lower bound of the 95% confidence interval for the proportion
Capture	True if 0.5 is captured in the confidence interval; otherwise, false

5. The measures collection is set up to collect measures from the Digits collection and display the results graphically. Click the **Collect More Measures** button on the measures collection. The Digits collection rerandomizes 100 times, and the results appear as gray or red bars in the measures collection. Gray means that the confidence interval captured the true proportion of 0.5, and red means that it did not.

Q5 How many of the 100 bars are red? What percentage of the confidence intervals captured the true proportion? Is this what you expected? Explain.

To split the graph, drag Capture to the vertical axis.

6. Make a histogram or dot plot of *pHat*, split by *Capture*. On the plot, choose **Graph | Plot Value** and plot your sample proportion from step 2.

Q6 Describe the values of *pHat* that are false. What must be true about the value of *pHat* so that its confidence interval captures the population proportion?

Q7 Where does the sample proportion that you got in step 2 fall in your plot, with the *pHat*s that are true or the ones that are false? Does your answer here agree with the answer you gave in Q3?

EXPLORE MORE

1. Either pool the percentage you got in Q5 with other members of your class and find the mean percentage for your class or define a new measure in the measures collection that counts the percentage of captures. Then collect these measures from the measures collection. Describe your results.

2. Modify the simulation to use 80 digits instead of 40. Compare your results with the original simulation.

3. Modify the simulation to use a 90% or 99% confidence interval. Are the results what you expect?

Confidence Interval for a Proportion— Capture Rate

Activity Notes

Objectives

- Testing whether 95% of 95% confidence intervals do indeed capture the population proportion
- Understanding that 95% confidence means that we expect that the true value of p will be in 95 out of every 100 of the 95% confidence intervals we construct
- Developing a better understanding of confidence intervals and their interpretation
- Finding a confidence interval graphically

Activity Time: 30 minutes

Setting: Paired/Individual Activity (use **CaptureRate.ftm** or **CaptureRatePresent.ftm**) or Whole-Class Presentation (use **CaptureRatePresent.ftm**)

Materials

- Calculator or table of random digits

Statistics Prerequisites

- Definition of reasonably likely events
- Definition and formula for the confidence interval for a proportion
- Probability concepts
- Familiarity with sampling on the calculator or with a random digit table

Statistics Skills

- Constructing confidence intervals for the sample proportion
- Working with the definition of reasonably likely and rare events
- Sampling distribution of the sample proportion and binomial distributions
- Finding capture rates
- Testing the advertised rate of 95% with actual results

AP Course Topic Outline: Part III A (5), D (1); Part IV A (1–5)

Fathom Prerequisites: Students should be able to make graphs and plot values, create measures, and use the formula editor.

Fathom Skills: Students work with true/false formulas (Boolean expressions) and with "and" statements. *Optional:* Students update a collection when the population changes (Explore More 2), build a collection of measures from different populations (Explore More 1), and collect measures of measures (Explore More 1).

General Notes: This activity deals with the fundamental idea that in estimating a parameter of a population, we construct a confidence interval for that parameter using a method that "captures" the true value of the parameter a certain percentage of the time. In the Fathom simulation, students will be able to see this capturing happen.

Procedure: Both the hands-on activity in Collect Data and the Fathom part in Investigate use random digits and pose the question "What proportion of random digits are even?" One potential *dis*advantage of the Fathom version is that it lacks human error. When students count even digits by hand, they are likely to make errors in the counts. Then the capture rate might not be 95%, which leads to a discussion and investigation into what went wrong—a good thing!

A variation on this activity is to give each student a bag of plain M&M's and have them count out 40 and get a confidence interval for the percentage of red M&M's. The true percentage of plain M&M's that are red is 20%.

Students start the Investigate section with a partially constructed Fathom document, **CaptureRate.ftm.** This relieves them from having to construct the rather complex display of confidence intervals. If you do not have time for your students to define the measures in Q4, you or your students can use the document **CaptureRate Present.ftm** instead. Follow the steps of the activity but instead of defining measures in Q4, open the inspector to the **Measures** panel and discuss the formulas that are there.

COLLECT DATA

1.–2. With a TI-83, students can get 40 random digits from 0 to 9 and place them in list L_1 using the command randInt(0,9,40)STO→L_1. Students then can count the number of even digits in list L_1 by using the command INT(L_1/2) = L_1/2 STO→L_2. This tests if the

Confidence Interval for a Proportion—Capture Rate
continued

Activity Notes

digit is even and, if so, stores a 1 in its position in list L_2. If the digit is odd, the command stores a 0 in its position in list L_2. Then students can count how many nonzero entries are in list L_2, or they can sum up the number of even digits (represented now by 1's) by using the command sum(L_2).

They could also just plot a histogram with bin width 1 and trace the histogram tallying up the even bars.

Be sure students have enough time to do this carefully. If they don't count accurately, you have another lesson: the confidence interval doesn't have a 95% chance of capturing the true value of p if there are errors in the survey such as miscounting the number of successes in the sample.

Q1 Intervals will vary depending on the number of even digits students found. One student's number of successes in step 2 was 21, or $pHat = 21/40 = 0.525$. Then the student's confidence interval for the proportion of random digits that are even is

$$0.525 \pm 1.96 \sqrt{\frac{0.525(1 - 0.525)}{40}} \approx 0.525 \pm 0.155,$$

or 0.37 to 0.68

Q2 50%

Q3 For the student example in Q1, the true proportion was captured because 0.5 is between 0.37 and 0.68.

INVESTIGATE

Q4 Students have the opportunity to decide what formulas are appropriate and the chance to work with Boolean expressions. One way to define the formulas is shown here. For *Upper* and *Lower,* students could use 40 as the denominator, but using count() makes it easy to modify the simulation for larger samples (as in Explore More 1). Based on prior experience with proportion, students might alternatively define *pHat* with the formula proportion(IsEven=true). (*Note:* true is a special value in Fathom and does not require quotation marks.) There are alternative Boolean expressions to define *Capture*, too. If students want to use ≤ or ≥, show them that holding down Ctrl (Win) Option (Mac) changes the keypad within the formula editor.

Measure	Value	Formula
pHat	0.575	proportion(isEven)
Upper	0.728198	pHat + 1.96·√(pHat(1 − pHat)/count( ))
Lower	0.421802	pHat − 1.96·√(pHat(1 − pHat)/count( ))
Capture	true	(0.5 < Upper) and (0.5 > Lower)

Q5 The percentage should be close to 95%. The display confidence intervals in the student activity (Measures from Digits) show six segments that didn't capture 0.5, so this example is 94%.

A confidence interval comes with a specification of a *level.* The interpretation of this level by students often comes in the mistaken form "We are 95% confident that the population proportion is between 0.37 and 0.68." They need to come to understand that our confidence is in the process, not the estimate. What we know is that following the procedure will result in a capture of the population proportion a certain percentage of the time given by the chosen level. It's a subtle difference, the need for which is often lost on students and instructors alike.

Another subtlety centers on the fact that it is correct to say that we expect the true value of p will be in 95 out of every 100 of the 95% confidence intervals we construct. However, it is not correct to say, after we have found a confidence interval, that there is a 95% chance that p is in that confidence interval.

Q6 The dot plot shows one student's results. The values of *pHat* that are false are on the tail ends of a mound-shaped distribution. The value of *pHat* must be a reasonably likely sample proportion for the population proportion 0.5—which, according to this dot plot, goes from about 0.375 to 0.625.

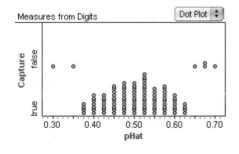

228 | 6: Inference for Means and Proportions

Confidence Interval for a Proportion—Capture Rate
continued

Q7 For the student example, the sample proportion of 0.525 falls in with the captured *pHats*.

DISCUSSION QUESTIONS

- What formulas did you use for the measures? If possible, discuss different yet equivalent formulas.
- How many "misses" did you get when the confidence level was 95%? Why didn't everyone get the same number of misses? What would you think if someone got 20 misses out of 100?
- What range of sample proportions results in confidence intervals that include the population proportion? Did everyone get the same range? If so, why is there no variation?
- Are the segments in the display of confidence intervals the same length? Why or why not? Can you calculate the minimum and maximum lengths?

EXPLORE MORE

1. The mean percentage should be close to 95%, and it may be 100% if you have a small class. So, if your class is small, you may wish to have each student construct several confidence intervals. If you do a large number of confidence intervals, because the formula for a confidence interval uses several approximations, the capture rate you get may be quite a bit less than 95%. The formula is an approximation in several ways: The normal distribution is used as an approximation of the binomial distribution; the formula is symmetric around $\hat{p}$, but it shouldn't be because the binomial distribution is slightly skewed for all $p \neq 0.5$; and the sample proportion $\hat{p}$ is used in the formula for the standard error as an approximation of p.

   As a result, the formula for a confidence interval gives a capture rate that tends to be less than the 95% advertised. (You can demonstrate that to your students with a computer simulation. Use, for example, $n = 40$ and $p = 0.25$ with several thousand trials.) Imposing the condition that both $n\hat{p}$ and $n(1 - \hat{p})$ are at least 10 eliminates the more highly skewed distributions where the capture rate could be as small as 80% if the formula were used.

   Here are 55 capture percentages from this activity. The mean is 0.925, lower than the 95% expected.

   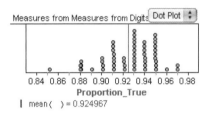

2. Confidence intervals for samples of size 80 will be shorter because the horizontal line segments will be shorter. From their work with the sampling distribution of the sample proportion, students should understand that the spread of a binomial distribution of the proportion of successes decreases as the sample size n increases.

3. For 90%, the intervals will be shorter because 1.645 is used instead of 1.95. For 99%, the confidence intervals will be longer because the endpoints of the horizontal line segments for 99% confidence are farther apart than the endpoints for 95% confidence.

Significance Test for a Mean—Body Temperature

You will need
• Temperatures.ftm

What is the average body temperature of men under normal conditions? Medical researchers interested in this question collected data from a large number of men. A random sample from these data of size 10 is shown. These temperatures seem a little low, but is that "lowness" the kind of chance variation you would expect from an approximately normal situation where the mean temperature is 98.6 degrees?

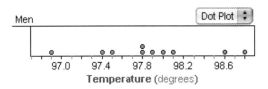

In this activity you'll use Fathom to explore hypothesis tests and confidence intervals and the relationship between them.

EXAMINE DATA

You'll start with a hypothesis test and connect it to confidence intervals.

1. Open the document **Temperatures.ftm.** Note slider *TemperatureTarget*—it controls the mean of the population against which *Temperature* will be tested. It's set to 98.6 degrees.

You might want to convert the dot plot to a box plot to see if there is reason to suspect that the population is not normal.

Q1 Check the conditions required to use a significance test for a mean. Are the conditions satisfied? Discuss any concerns you might have about these data.

Working with the *t*-statistic relies on an assumption about the population from which the measurements were drawn, namely, that the values in the population are normally distributed. Is this a reasonable assumption for these data? Fathom can help you determine qualitatively whether this sample is unusual.

2. Make a new attribute in the collection. Call it *SimTemp* for simulated temperature. Define *SimTemp* with the formula

    randomNormal(mean(Temperature),stdDev(Temperature))

Select *SimTemp* and choose **Edit | Edit Formula.**

This formula tells Fathom to generate random numbers from a normal distribution whose mean and standard deviation are the same as in the original data.

3. To compare the distribution of these simulated temperatures with the distribution of the original data, drop *SimTemp* on the plus sign to add it to the horizontal axis of the dot plot.

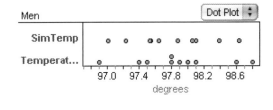

One set of simulated data doesn't tell the whole story. You need to look at many sets.

Significance Test for a Mean—Body Temperature
continued

4. Choose **Collection | Rerandomize** several times. Each time you rerandomize, you get a new set of 10 values from a population with the same mean and standard deviation as for the original data.

Q2 Does it appear that the original distribution is very unusual, or does it fit in with the simulated distributions? In other words, does it seem reasonable to assume that the original data came from a population that is approximately normally distributed?

Now that you have checked the conditions, the next step is to state your hypotheses.

Choose **Edit | Show Text Palette** to format text and create mathematical expressions.

Q3 State the null and alternative hypotheses for this situation. You can write out these two hypotheses in Fathom in a text object to be stored with your document. From the shelf, drag a text object into the document and type in your hypotheses.

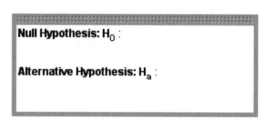

Null Hypothesis: H_0 :

Alternative Hypothesis: H_a :

INVESTIGATE

Now it is time to compute the test statistic.

5. Drag a test object from the shelf. From the pop-up menu, choose **Test Mean.**

6. Drag *Temperature* from the case table to the top pane of the test where it says "Attribute (numeric): unassigned."

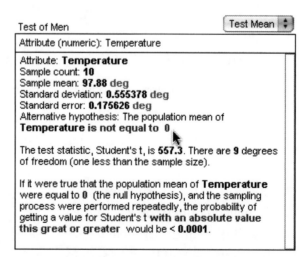

7. Highlight the **0** of **is not equal to 0** in the Alternative hypothesis and choose **Edit | Edit Formula.** Enter *Temperature_Target*. The value of *Temperature_Target* should replace the 0.

Significance Test for a Mean—Body Temperature
continued

Q4 The last paragraph of the test describes the results and shows the *P*-value for the sample, which is 0.0027. This is very small. Write a conclusion in context, linked to the computations.

It is helpful to be able to visualize the *P*-value as an area under a distribution.

8. With the test selected, choose **Test | Show Test Statistic Distribution.** The curve shows the probability density for the *t*-statistic with 9 degrees of freedom. The shaded area corresponds to the *P*-value. However, because the *P*-value is small, you can't see the shading unless you rescale your axes.

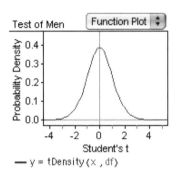 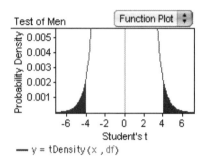

Let's explore how the *P*-value depends on the test mean, which is currently set to 98.6 degrees.

9. Select the dot plot and choose **Graph | Remove X Attribute: SimTemp.**

If you move the points far enough to the right, you will be able to see the shading without zooming in.

10. Draw a selection rectangle around the three lowest points. Drag these points around the dot plot. This changes their value. Observe what happens to the values in the test and the shading in the distribution plot.

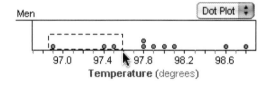

Q5 Many people use $\alpha = 0.05$ as the significance level—the maximum *P*-value for which they will reject the null hypothesis. Move the three lowest points until the *P*-value is as close to 0.05 as you can get it. Now look in the analysis: What's the value of Student's *t* in this situation?

df stands for degrees of freedom.

Q6 This is a critical value for *t* at the 5% level of significance. Verify it by looking it up in a statistics book or by using a calculator. Look under $df = 9$ and in the section for two-tailed tests. What do you find? How well does it match?

*You could also choose **File | Revert Collection.***

11. Put the data back where they were by repeatedly choosing **Edit | Undo.**

Significance Test for a Mean—Body Temperature
continued

You can drag the pointer to change the value, or edit the number itself.

Q7 What happens to the *P*-value if you set *Temperature_Target* to 98.25 degrees? What's the new value? Explain why it's larger than the value for 98.6 degrees.

Q8 If a person's temperature was supposed to be 98.25 degrees (instead of 98.6 degrees), what would you conclude from these data?

Now let's find the range of values for which your data are plausible.

12. Drag the slider to change the values for *Temperature_Target*. Look for the smallest and largest values on the slider for which $P \geq 0.05$. (Anything outside the range—such as 98.6—would be an unreasonable estimate for the temperature of the male population.)

Q9 What is the range that you found for *Temperature_Target* for which your data are plausible?

You can use Fathom's interval estimate tool to find this range directly.

13. Drag a new estimate from the shelf. Choose **Estimate Mean** from the pop-up menu and drag *Temperature* into the top of the box, as you did with the test.

14. If everything went well, the range of values in this confidence interval is the same as the range you found in Q9. Compare them. If they are not the same, figure out what to do to get these numbers as the limits when you adjust *Temperature_Target*, as you did in Q9.

Q10 Interpret this confidence interval in terms of the context of the situation.

EXPLORE MORE

Scroll down in your document window until you see the collection named Women. Make a plot to check conditions, then test whether the mean may differ from the advertised standard of 98.6 degrees for women. Use a 95% confidence interval to estimate the mean body temperature of women. Could you say that the average body temperature of women is different from the average body temperature of men? Explain, and be careful.

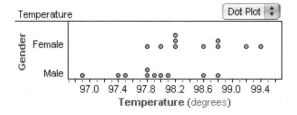

Significance Test for a Mean—Body Temperature
Activity Notes

Objectives

- Performing a test of significance for deciding whether it is reasonable to reject a claim that a sample was drawn from a population with a specified mean
- Understanding the concept of a confidence interval for estimating a population mean, computing a confidence interval, and interpreting the results
- Discovering that critical *t*-values for hypothesis tests are special "edge" values of a statistic
- Exploring how hypothesis tests and confidence intervals for the mean are closely related
- Deepening the understanding of these terms in the context of inference for means: *confidence interval*, *null hypothesis* and *alternative hypothesis*, *test statistic*, *level of significance*, and *P-values*
- Computing and interpreting in context a *P*-value in testing a mean

Activity Time: 40–50 minutes

Setting: Paired/Individual Activity or Whole-Class Presentation (use **Temperatures.ftm** for either)

Statistics Prerequisites

- Familiarity with the language for testing a hypothesis
- Familiarity with the structure and procedure for testing a hypothesis
- The *t*-test statistic
- Familiarity with the structure and procedure for constructing confidence intervals

Statistics Skills

- Constructing confidence intervals for the sample mean
- Checking conditions for testing the mean using simulation and comparison
- Performing a hypothesis test for the sample mean with unknown SD
- Finding *P*-values and relating them to the plot of the distribution of the test statistic
- Determining confidence levels, critical values, and significance levels.

AP Course Topic Outline: Part III D (2, 6, 7); Part IV A (1–3, 6), B (1, 4)

Fathom Prerequisites: Students should be able to make graphs and drag points, create attributes, and use the formula editor.

Fathom Skills: Students use Fathom's hypothesis test and interval estimate tools, split a dot plot without using a categorical variable, make a graph of the test statistic's distribution, and use a slider to change the null hypothesis.

General Notes: This activity explores essential concepts in hypothesis testing. You could use it as an introduction to traditional tests, but it may be more effective to wait until students have solved some straightforward problems using whatever technology is most common in your class. These activities do not, for example, explain what a *t*-statistic is, so students who already know that—even shakily—may get more out of the experience.

This activity introduces Fathom's analyses—both hypothesis tests and confidence intervals (estimates). Students explore *P*-values, critical *t*-values, and the correspondence between hypothesis testing and estimation. This activity is designed to be suitable as an introductory activity; that is, you don't have to know much about Fathom.

EXAMINE DATA

Q1 This is a random sample from a group of men examined by the researchers, so the final result generalizes only to this select population. As to the shape of the distribution, body temperatures cannot stray too far from the mean (or else the person would not have been available to have his temperature taken!). So the population cannot be very skewed. The dot plot gives no reason to suspect that the population is not normal. Here is the box plot.

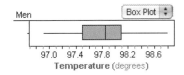

Q2 The simulation shows further that there is no reason to suspect that the population is not normal. In fact, the simulations look pretty close to the data.

Significance Test for a Mean—Body Temperature
continued

Q3 The research claim here is that the mean may differ from the advertised standard of 98.6 degrees. This claim forms the alternative (or research) hypothesis. The standard against which the sample mean is compared forms the null hypothesis. The hypotheses are null hypothesis: H_0: $\mu = 98.6$, where 98.6 is the mean body temperature of all persons in the population under study; and alternative hypothesis: H_a: $\mu \neq 98.6$.

INVESTIGATE

Q4 The small *P*-value implies that if the null hypothesis were true, the chance of seeing a sample mean this unusual would be extremely small. Thus, you should reject 98.6 as a plausible value for the mean body temperature of all males in this population.

Q5–Q6 The value they should get is 2.262. Students may not be able to land exactly on a *P*-value of 0.05, but they should be able to get within 0.002 of 0.05. *t*-values then range between 2.243 and 2.283.

Q7–Q8 The *P*-value jumps to 0.064 and $t = -2.107$. It is larger because the hypothesized mean is now closer to the sample mean for these data. The conclusion then becomes: With a *P*-value of 0.064, this sample result is reasonably likely to occur under the null hypothesis, so you cannot reject the null hypothesis; the claim that the mean body temperature for males is 98.25 is plausible. Note that the test has been switched to non-verbose mode (choose **Test | Verbose**).

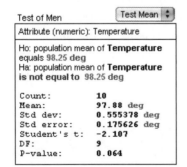

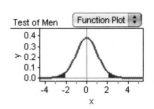

Q9–Q10 The confidence interval is 97.4827 to 98.2773. Students may need to rescale their slider axis to get exactly 0.05. You are 95% confident that the mean body temperature of the men in this population is in the interval (97.48, 98.28). Any population of body temperatures with a mean in this interval could have produced the sample results seen here as a reasonably likely outcome.

DISCUSSION QUESTIONS

- What is the possible range of values for *P*? What about for *t*?
- How do you compute the *t*-statistic? Verify it using numbers from one of your tests.
- What does it really mean for a *P*-value to be small? What if it's large?
- Why do you suppose Fathom's statistical tests have so many words in them? (Fathom also has a non-verbose mode: select the test and choose **Test | Verbose**.)

EXPLORE MORE

The dot plot (on the student worksheet) gives no reason to suspect that the population is not normal. The sample was randomly taken from a larger population. The hypotheses are the same as those for the men. For the women, the sample mean is 98.52 and the sample standard deviation is 0.527. You are 95% confident that the mean body temperature of the women in this population is in the interval (98.14, 98.90). Any population of body temperatures with a mean in this interval could have produced the sample results seen here as a reasonably likely outcome.

Significance Test for a Mean—Body Temperature
continued

To test the same hypothesis for the women, everything is the same, except that the observed *t*-statistic is −0.48 and the resulting *P*-value is 0.6426. This sample mean is quite likely to occur under the null hypothesis, so you cannot reject the null hypothesis; the claim that the mean body temperature of women is 98.6 is plausible.

From the sample means, it may appear that the average body temperature of women is higher than that of men, but beware! There is some overlap of the confidence intervals, so that conclusion may not be valid.

EXTENSIONS

1. Drag the slider *Temperature_Target* slowly and observe the changes that take place. For what value of the slider is all of the area under the curve shaded? Explain why it should be this particular value. [The value of the slider that makes all of the area under the curve shaded is at the mean of the sample, or 97.88.]

2. Change the test for male temperatures to test the claim that the mean body temperature for men is lower than 98.6 degrees. What is the *P*-value for this test and what is your conclusion? [For the alternative hypothesis, $H_a: \mu < 98.6$, the *P*-value is 0.0013. So, again, there is sufficient evidence to reject 98.6 as a plausible value for the mean body temperature of all males in this population, and there is support for the claim that it is less than 98.6 degrees.]

Significance Test for a Mean: Types of Errors

You will need
• TypesOfErrors.ftm

A hypothesis test helps you infer things about the population based on a sample. In this activity you'll begin with a document that is already set up with a sample drawn from a known population—a normal distribution. You'll see what happens when you try to use that sample to learn things about the population. Will it tell you the truth?

EXAMINE DATA

1. Open the Fathom document **TypesOfErrors.ftm.** You should see a collection named Sample, a dot plot of the attribute *Value*, sliders that determine the population mean *mu* and its standard deviation *SD*, and a hypothesis test, testing whether *mu* = 0.

   Q1 Move the *mu* slider. What happens to the dot plot and the values in the test? Explain why this happens.

   Q2 Move the *SD* slider. What happens to the dot plot and the values in the test? Explain why this happens.

Or press Ctrl+Y (Win) ⌘+Y (Mac) to rerandomize.

2. With *mu* set to 0, select the sample collection and choose **Collection | Rerandomize.**

   Q3 Write down the new *P*-value from the test. Collect five *P*-values in this way and record them.

Drag the slider to 2.00 or type the value.

   Q4 Set the *mu* slider to 2.00 and repeat step 2. Collect five *P*-values and record them.

   Q5 Your values for *mu* = 2 should be small and close together; those for *mu* = 0 should range widely between 0 and 1. Explain why.

INVESTIGATE

Now you'll perform this experiment many times by collecting measures.

3. Reset *mu* to 0. Then, select the test and choose **Test | Collect Results As Measures.** A new collection named Measures from Test of Sample appears.

4. Show the inspector for the new collection. Look at the **Cases** panel to see the various attributes you have collected. They include *tValue* and *pValue*, the values for the *t*-statistic and the *P*-value, respectively.

Significance Test for a Mean: Types of Errors
continued

5. Go to the **Collect Measures** panel and change the settings to match those below. Click **Collect More Measures**. Fathom rerandomizes your sample 100 times, tests the mean of the sample, and collects the information for each *t*-test in the measures collection.

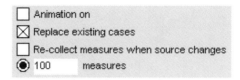

Q6 Make histograms of *tValue* and *pValue*. Sketch your histograms.

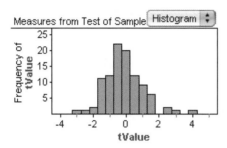 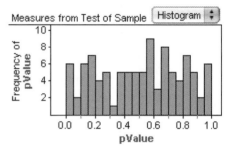

To review, each histogram shows 100 cases. Each case was a statistical test on a different sample—a test to see if it is plausible that a normal population with mean 0 and SD 1 could have produced the observed sample mean. Each test produced a *t*-statistic—it was zero if the sample mean was zero and became larger (in absolute value) the more the sample mean differed from zero. Each test also produced a *P*-value—the probability of seeing a result from a random sample of size 10 that is as extreme as or more extreme than the one you got from your random sample if the null hypothesis is true.

> Select the first bin, then select the second bin while pressing the Shift key.

Q7 Select the two tallest bins in the middle of the histogram for *tValue*. What part of the histogram for *pValue* is selected? Explain why this is so.

Q8 Select the bottom bin of the histogram for *pValue* ($pValue < 0.05$). What part of the histogram for *tValue* is selected? Explain why this is so.

Q9 The cases in this bottom bin are where $pValue < 0.05$. If your significance level was 5%, you would reject the null hypothesis for these cases. Yet in this simulation, the null hypothesis is *true*, because $mu = 0$. What type of error have you made if you reject the null hypothesis when it's true?

Q10 What percentage of cases are highlighted in the histogram of *tValue* that would have led to this type of error? Is that about what you would expect? Explain.

Q11 What is one way you can decrease this type of error? Try it with your collection and see if your idea truly decreases this error.

Significance Test for a Mean: Types of Errors
continued

Now you'll see what happens if you set *mu* = 1 but still test the null hypothesis that the mean is 0.

6. Set *mu* to 1. Select the measures collection and choose **Collection | Collect More Measures.**

Q12 Describe the distributions of *tValue* and *pValue*. How do they compare to the distributions that you graphed in Q6?

Q13 Use a selection rectangle to highlight the *pValues* that are 0.05 or greater. The cases selected are where *pValue* ≥ 0.05. If your significance level was 5%, you would fail to reject the null hypothesis (*mu* = 0) for these cases. Yet in this simulation, the null hypothesis is *false*, because you set *mu* = 1. What type of error have you made if you fail to reject the null hypothesis when it's false?

> Hold your cursor over the measures collection while the graphs are highlighted. The number of cases selected is shown in the lower-left corner of the window.

Q14 What percentage of cases are highlighted in the histogram of *tValue* that would have led to this type of error? Is that about what you would expect? Explain.

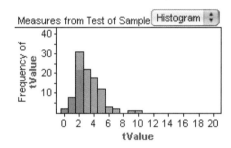

Q15 What is one way you can decrease this type of error? Try it with your collection and see if your idea truly decreases this error.

EXPLORE MORE

> To zoom in, press Ctrl (Win) Option (Mac) and click on the axis. To zoom out, press Shift as well.

1. Reset *mu* to 0 and collect more measures. Then, change the histogram of *tValue* to a dot plot. Be sure the bottom bin is selected on the histogram of *pValue*. Zoom in on the dot plot and decide on a value for *t* that separates the selected dots from the unselected dots. What is the value of *t* that you found? What does it represent? Add more cases if you like.

> You will need to change the scale on the histogram for *pValue*.

2. Repeat Explore More 1 for an alpha level of 0.01 instead of 0.05.

3. Plot *pValue* as a function of *tValue*. Sketch the scatter plot and explain it.

4. Set *mu* to 0 (or some other value) and set *SD* to 2. Predict what the distribution of *P*-values will look like. Graph the distribution and compare it to your prediction.

Significance Test for a Mean: Types of Errors — Activity Notes

Objectives

- Performing tests of significance for deciding whether it is reasonable to reject a claim that a sample was drawn from a population with a specified mean
- Exploring possible sample results when the null hypothesis is true and when it is false
- Generating a collection of *t*-values and *P*-values for hypothesis tests for comparison
- Deepening an understanding of Type I and Type II errors and how to reduce their probability
- Understanding that the significance level of a test is the probability that you will reject a null hypothesis when it is true—a Type I error—and that Type I errors are inevitable
- Understanding that a Type II error is not rejecting a null hypothesis that is false

Activity Time: 30–40 minutes

Setting: Paired/Individual Activity or Whole-Class Presentation (use **TypesOfErrors.ftm** for either)

Statistics Prerequisites

- Familiarity with the language for testing a hypothesis
- Familiarity with the structure and procedure for testing a hypothesis
- The *t*-test statistic and *P*-values
- Some familiarity with Type I and Type II errors

Statistics Skills

- Performing a hypothesis test for the sample mean with unknown SD
- Finding *P*-values and relating them to the plot of the distribution of the test statistic
- Determining critical values and significance levels
- Relating *P*-values to their corresponding *t*-values on plots
- Working in depth with Type I and Type II errors

AP Course Topic Outline: Part III D (2, 6, 7); Part IV B (1, 4)

Fathom Prerequisites: Students should be able to make graphs, use an inspector, and drag sliders.

Fathom Skills: Students use Fathom's hypothesis test tool, collect measures from a test, and use a slider to change the mean.

General Notes: This activity explores essential concepts in hypothesis testing. Students explore *P*-values, *t*-values, and types of errors by a simulation that repeatedly tests a null hypothesis, namely, $mu = 0$. In this simulation, students get a chance to see, for example, that they get the wrong answer 5% of the time when they do a test at the 5% significance level. This activity is much easier with some Fathom experience, although it makes few assumptions about the user's skill.

EXAMINE DATA

Q1 The dots in the plot are basically shifted horizontally with the slider. If the slider goes in the positive direction, the dots (mean) shift to the right. The *P*-values, *t*-statistic, mean, and so on, all change based on the new sample mean and SD.

Q2 The dots in the plot are basically stretched or squeezed (on average). If $SD > 1$, the dots tend to show a larger spread. If $0 < SD < 1$, the dots tend to show a smaller spread. The *P*-values, *t*-statistic, mean, and so on, all change based on the new sample mean and SD.

Q3 The student activity (after Q6) shows a histogram of 100 possible *P*-values for $mu = 0$ and $SD = 1$. The possible *P*-values range from 0 to 1.

Q4 The dot plot below is of 100 possible *P*-values for $mu = 2$ and $SD = 1$. Here, the possible *P*-values range from 0 to 0.0035.

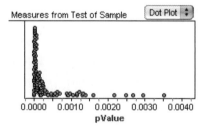

Q5 When $mu = 2$, we are still testing the null hypothesis that the mean is 0. The sample, however, is a random sample from a normal population with mean 2 and SD 1, so it is quite likely that the sample mean will be

Significance Test for a Mean: Types of Errors
continued

close to 2 and not close to 0, hence the large *t*-statistic and small *P*-value.

INVESTIGATE

Q6 The *tValue* distribution should be roughly mound-shaped with center 0 and spread of about 1. The *pValue* distribution should be roughly uniform with range 0 to 1.

Q7 The upper *pValues* are selected, typically *pValues* greater than 0.5. This is because the *tValues* are close to zero (or the sample mean is close to the standard).

Q8 The tails or outer bins are selected on the histogram of *tValue*. This is because when the *tValues* are far away from zero (or the sample mean is far away from the standard), the corresponding *pValue* is small.

Q9 Type I error

Q10 The percentage should be close to the significance level, which in this case is 5%. For the sample histogram shown in the student activity, the number of cases selected in the histogram was 6. So the percentage was 6%. This is what we'd expect because the significance level of a test, α, is the probability of rejecting the null hypothesis, if it is true. That is, α is the probability of making a Type I error.

Q11 You can decrease this type of error by reducing the significance level. The smaller the significance level (the larger the critical value) you choose, the stronger you are requiring the evidence to be in order to reject H_0. The stronger the evidence you require, the less likely you are to make a Type I error, that is, to reject H_0 when it is true.

Q12 See the answer for Q6 for descriptions of the histograms for $mu = 0$. For $mu = 1$, the distributions for *tValue* and *pValue* are both skewed right. The *pValue* distribution will have the biggest bins at the beginning and then a few bins way out in the tails. The *tValue* distribution will have the biggest bins in the interval from about 2 to 5.

Q13 Type II error. The probability of a Type II error is sometimes designated beta, or β.

Q14 The percentage should be close to about 18% or 19%. For the sample histogram shown in the activity, the number of cases selected in the histogram was 32. So the percentage was 32%, which is a little high. More than likely, students will not expect the percentage to be as high as they get. (How to find this percentage: we will reject if the *t*-value is less than 2.26, which translates to a sample mean being less than 0.7153. Given that the mean is 1 with SD 1, the probability of getting a sample mean less than 0.7153 is 18.4% using the normal distribution (because the SD is known and it is normal) or 19.5% using the *t*-distribution.)

Q15 To decrease the probability of making a Type II error, you can use a larger sample or increase the significance level.

DISCUSSION QUESTIONS

- The sample collection is a sample. But of what population?
- Explain why the distribution of *pValue* looks uniform for $mu = 0$.
- Roughly 5% of the samples will have $P < 0.05$ when $mu = 0$. How will that percentage change as *mu* increases? [It will increase, asymptotically approaching 1. This leads to the power function.]

EXPLORE MORE

1. *tValue* ≈ 2.262, the critical value.

2. *tValue* ≈ 3.240, the critical value for a significance level of 1%.

3. At *tValue* = 0, we get a *pValue* of 1. As the *tValues* increase (in absolute values), the *pValues* decrease as shown.

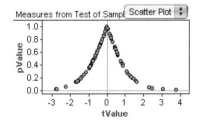

4. For $mu = 0$ and $SD = 2$, the distribution of *pValues* is roughly uniform. The graph of *tValues* vs. *pValues* appears periodic.

Confidence Interval for a Mean: σ Unknown—SAT Scores

You will need
- SATScores.ftm

SAT math scores are approximately normally distributed, with mean 518 and standard deviation 114. Suppose you don't know that and you're going to estimate the population mean score from a random sample of five

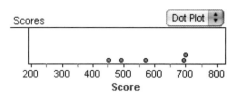

scores. You aren't supposed to know σ either, so you'll explore whether you can substitute *s* for σ and still calculate a "95%" confidence interval. (*Note:* In a real situation, you will never know σ.)

GENERATE DATA

1. Open the Fathom document **SATScores.ftm**. The collections in this document have been set up for you to get a nice display, so don't delete them or their attributes.

Hint: Learn about Fathom's randomNormal function.

2. The Scores collection has five cases and one attribute, *Score*. Define *Score* with a formula that generates random numbers from a normal distribution with mean 518 and standard deviation 114.

Q1 You have learned to construct a 95% confidence interval using the formula $\bar{x} \pm 1.96 \cdot \sigma/\sqrt{n}$. What percentage of the time should this confidence interval capture the true value of μ? Does the capture rate depend on the sample size?

3. There are five measures for the collection, none of which have formulas. For now, you want to define just two of the measures—the sample mean and standard deviation. Go to the **Measures** panel of the collection's inspector to define each with an appropriate formula. Your sample mean will appear on the dot plot. Write down your sample mean and sample deviation.

Q2 This time you aren't supposed to know σ. So, substitute your sample standard deviation *s* for σ (because that's all you have and all you'll ever have in a real situation) and calculate a "95%" confidence interval. Did your confidence interval capture the true value of μ?

Consider an "and" statement for Capture.

Q3 Now you want to define the last of the five measures for the collection: *Lower*, *Upper*, and *Capture*. *Lower* and *Upper* will use *s* just like you did in Q2. The table on next page describes the measures. Use the **Measures** panel of

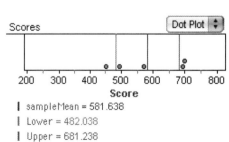

Confidence Interval for a Mean: σ Unknown—SAT Scores
continued

the collection's inspector to define these three measures with an appropriate formula, and record your formulas.

Measure	Description
sampleMean	Sample mean of the 5 scores
sampleSD	Sample standard deviation of the 5 scores
Lower	Lower bound of the "95%" confidence interval for the mean
Upper	Upper bound of the "95%" confidence interval for the mean
Capture	True if 518 is included in the confidence interval; otherwise, false

INVESTIGATE

The measures collection is set up to collect measures from the Scores collection and display the results graphically.

4. Click the **Collect More Measures** button on the measures collection. The Scores collection rerandomizes 100 times, and the results appear as gray or red bars in the measures collection. Gray means that the confidence interval captured the true mean of 518, and red means that it did not.

Q4 How many of the 100 bars are red? What percentage of the confidence intervals captured the true mean? Is this what you expected? Explain.

5. Re-collect measures two more times and write down your capture rates. Compare your results with those of others in your class.

Q5 What is your conclusion so far about the effect of replacing σ with *s*?

To see the range of the capture rates, you can make a new measure to record your capture rates.

6. Show the inspector for the measures collection and go to the **Measures** panel. Create a new measure named *Proportion_Captured* and define it with the formula proportion(Capture=true). On the **Collect Measures** panel, uncheck Animation on.

7. With the measures collection selected, choose **Collection | Collect Measures.** Fathom generates 5 new charts. For each new chart, Fathom calculates and records the capture rate like you did in Q4.

8. Make a dot plot for *Proportion_Captured* from the Measures from Measures from Scores collection.

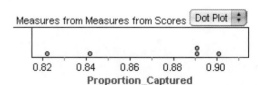

Confidence Interval for a Mean: σ Unknown—SAT Scores
continued

Make sure the inspector is not covering your dot plot or your chart.

9. Show the inspector for the new collection and go to the **Collect Measures** panel. Uncheck Animation on and change 5 measures to 20 measures. Click **Collect More Measures.**

10. On your dot plot, plot the mean proportion captured.

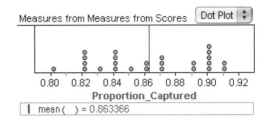

Q6 What is your mean capture rate for your 25 samples of 100 confidence intervals using s? What is your conclusion about the effect of replacing σ with s?

Now let's look at the sample standard deviation.

11. Make a histogram or box plot of *sampleSD* from the measures collection, and plot the mean of these 100 sample standard deviations. Plot the value of the population standard deviation as well.

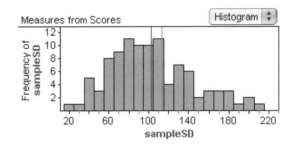

Q7 Describe the distribution of the sample standard deviation in terms of shape, center, and spread. How close is the mean value of s to σ? How many times is s smaller than σ? Larger?

Q8 Explain why the fact that the sampling distribution of s is skewed right means that the capture rate is less than advertised if you simply substitute s for σ.

Q9 When you use s to estimate σ, the capture rate is too small unless you make further adjustment. If an interval's true capture rate is lower than what you want it to be, do you need to use a wider or narrower interval to get the capture rate you want?

EXPLORE MORE

Hint: See your answer to Q9.

1. Perhaps using a multiplier other than 1.96 would improve the capture rate. Make a slider named t, and in your formulas replace 1.96 with t. Then use your simulation to estimate what multiplier would result in a 95% confidence interval.

2. Use your results from Explore More 1 to determine whether the same multiplier works for samples of size 3. How about samples of size 10? What do the results indicate?

Confidence Interval for a Mean: σ Unknown—SAT Scores

Activity Notes

Objectives

- Suspecting that the formula for the confidence interval for an unknown mean when the population standard deviation is unknown requires a value other than z (in this case, other than 1.96)
- Noticing that confidence intervals for the mean have widely varying widths because they are based on sample standard deviations, unlike previous confidence intervals, which were very close to the same width
- Understanding that the capture rate for a confidence interval when the population standard deviation is replaced by the sample standard deviation decreases unless an adjustment is made
- Developing a better understanding of confidence intervals and their graphical representation

Activity Time: 30 minutes

Setting: Paired/Individual Activity (use **SATScores.ftm**) or Whole-Class Presentation (use **SATScoresPresent.ftm**)

Statistics Prerequisites

- Definition and formula for the confidence interval for a mean
- Describing distributions graphically
- Some familiarity with confidence intervals that are shown graphically

Statistics Skills

- Constructing confidence intervals for the sample mean with unknown standard deviation
- Working with the definition of reasonably likely and rare events
- Sampling distribution of the sample mean
- Finding capture rates when using the sample standard deviation
- Creating and describing the distribution of the sample standard deviation

AP Course Topic Outline: Part III C, D (2, 6, 7); Part IV A (1–3, 6)

Fathom Prerequisites: Students should be able to make graphs and plot values, create measures and attributes, and use the formula editor.

Fathom Skills: Students work with true/false formulas (Boolean expressions), work with "and" statements, update a collection when the population changes, collect measures of measures, and use the randomNormal function.

General Notes: This activity deals with the fundamental idea that in estimating a parameter of a population, we construct a confidence interval for that parameter using a method that "captures" the true value of the parameter a certain percentage of the time. In the Fathom simulation, students will be able to see this capturing happen. Up to this point, they have seen that the capture rate is about 95% for a 95% confidence interval. In this simulation, they are not sure what capture rate they will get. The question is "Does substituting the sample standard deviation s for the population standard deviation σ in the formula $\bar{x} \pm z \cdot \sigma/\sqrt{n}$ produce the desired capture rate?"

Procedure: You can use this activity for a whole-class discussion with the document **SATScoresPresent.ftm**. Here, everything is set to collect measures. Simply click **Collect More Measures** on the measures collection. To see the histogram of the sample standard deviations, scroll the document window to the right. Also to the right is the dot plot of 100 capture rates for 100 confidence intervals. If you would like to generate more capture rates, turn animation off in the measures collection, then click **Collect More Measures** on the Measures from Measures from Scores collection.

GENERATE DATA

2. The formula should be randomNormal(518,114).

Q1 A 95% confidence interval, using the population standard deviation, will capture the true mean about 95% of the time. The capture rate does not depend on the sample size.

Confidence Interval for a Mean: σ Unknown—SAT Scores

continued

Activity Notes

Q2 Below are possible measures formulas. Other, equivalent formulas could be used. For this sample of 5 scores, the "95%" confidence interval would be 477.763 ± 1.96 · 41.4678/$\sqrt{5}$ ≈ 477.763 ± 36.348, or 441.415 to 514.111. So, in this instance, the confidence interval did not capture the true mean, 518.

Measure	Value	Formula
sampleMean	477.763	mean(Score)
sampleSD	41.4678	stdDev(Score)

Q3 The measures formulas for the mean and SD are above. Here are the rest of the measures formulas that students could use to complete the simulation. Other formulas could be used.

Measure	Value	Formula
Lower	441.415	sampleMean − 1.96 · sampleSD/$\sqrt{\text{count()}}$
Upper	514.111	sampleMean + 1.96 · sampleSD/$\sqrt{\text{count()}}$
Capture	false	(Lower < 518) and (518 < Upper)

INVESTIGATE

Q4–Q5 Typically, out of 100 runs of this simulation, about 9 to 16 intervals do not overlap 518. These should be about evenly divided between those that miss on the low side and those that miss on the high side. Because the capture rate should be about 95%, these intervals are too short. The proper adjustment for using s instead of σ will need to make the intervals a little longer than these intervals based on $z = 1.96$. In the example in the student activity, 19 intervals did not capture the true mean, so the capture rate for this example is 81%.

8. If your school's computers are rather slow, it would be wise to turn animation off before collecting the measures from the measures collection. If your computers are fairly quick, however, keep animation on at this point. That way, students can see the collecting process and the confidence intervals being made for each sample of 100 confidence intervals, which further reinforces what is going on.

Q6 The mean capture rate for the 25 samples of 100 confidence intervals shown in the student activity is 86.3%. Here is a plot of 100 samples of 100 confidence intervals. The mean capture rate is 87.6% and the range of capture rates goes from 79% to 93%. The conclusion is that these intervals are too short and something will need to be done to make the intervals a little longer.

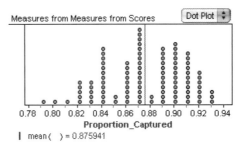

Q7 The histogram shows that the distribution of the sample standard deviations is slightly skewed right and the mean of s is smaller than σ. The mean of s is 102.108 and the standard deviation is 39.233. The sample standard deviation s is smaller than σ in 64 of the 100 samples and larger in 37 samples. (*Note:* To create the display of confidence intervals, the simulation actually needs to collect 101 measures. Hence, the graph contains 101 values, as shown by the count of values less than or greater than the true mean.)

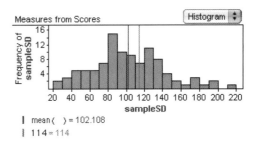

Confidence Interval for a Mean: σ Unknown—SAT Scores
continued

Activity Notes

Q8 Although the average value of s in repeated sampling is close to σ, the sampling distribution of s is skewed toward the larger values. Thus, the chance is greater than 0.5 that an observed sample standard deviation will be smaller than σ. This causes the confidence intervals to be too short more than half the time and the capture rate to be smaller than the advertised value.

Q9 If an interval's true capture rate is lower than what you want it to be, you can get a higher capture rate by using a wider interval.

DISCUSSION QUESTIONS

- What is meant by a *capture rate*? Does the capture rate depend on sample size? (The capture rate does not depend on sample size, but the interval that gives that capture rate does.)

- Are the segments in the display of confidence intervals the same length? Why or why not? Can you calculate the minimum and maximum lengths?
- Does substituting s for σ in the formula $\bar{x} \pm z \cdot \sigma/\sqrt{n}$ work? Why do you think it does or does not?
- In Explore More 1, what value did you find as a replacement for 1.96?

EXPLORE MORE

1. For sample size 5 (4 degrees of freedom), the value should be 2.776.

2. The multiplier does vary for different sample sizes (degrees of freedom), as described by a t-table. For $n = 3$, the value should be 4.303 and for $n = 10$, the value should be 2.262.

Confidence Interval for a Mean: Not Normal—Brain Weights

You will need
- BrainWeights.ftm

Your goal in this activity is to see what happens to the capture rate when you construct many confidence intervals based on random samples from a non-normal distribution. For example, the histogram shows the brain weights (in grams) of 68 species of animals under study by a certain zoo. Obviously, this distribution is not approximately normally distributed or even close to mound-shaped. What effect will that have?

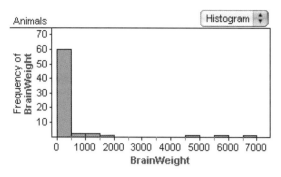

EXAMINE DATA

1. Open the Fathom document **BrainWeights.ftm.** The collections in this document have been set up for you to get a nice display, so don't delete them or their attributes.

2. The Animals collection contains the data shown in the histogram above—the brain weights of 68 species. Scroll through the case table to see the range of brain weights.

Q1 Describe the distribution in terms of shape, center, and spread.

3. The open collection Sample of Animals is set up to take samples of size 5, and the dot plot shows the brain weights of the 5 animals in the sample. Click **Sample More Cases** to verify that it takes a random sample of five species.

Q2 What is the range of the brain weights in your sample of size 5?

4. There are five measures for the Sample of Animals collection, none of which have formulas. For now, you want to define just two of the measures—the sample mean and standard deviation. Double-click the Sample of Animals collection to show the inspector. Go to the **Measures** panel and define both of these measures with an appropriate formula. Your sample mean will appear on the dot plot. Write down your sample mean and sample deviation.

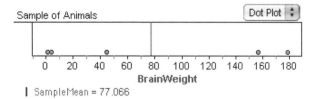

Confidence Interval for a Mean: Not Normal—Brain Weights
continued

Q3 Construct a 95% confidence interval estimate of the mean brain weight of all the species using the sample mean and the sample standard deviation that you wrote down in step 4, and the appropriate value for t^*. Does your confidence interval include the true mean weight of 394.49 grams? If not, does the interval lie below the true mean or above it? What value did you use for t^*?

5. Change the value of the t slider to the value you chose in Q3.

Consider an "and" statement for *Capture*.

Q4 Now you want to define the rest of the five measures for the collection: *Lower*, *Upper*, and *Capture*. The table below describes the measures. Use the **Measures** panel of the sample collection's inspector to define these three measures with an appropriate formula and record your formulas.

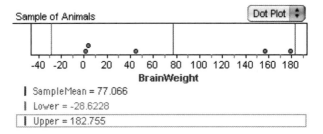

Measure	Description
SampleMean	Sample mean of the 5 brain weights
SampleSD	Sample standard deviation of the 5 brain weights
Lower	Lower bound of the 95% confidence interval for the mean brain weight
Upper	Upper bound of the 95% confidence interval for the mean brain weight
Capture	True if the true mean weight of 394.49 is included in the confidence interval; otherwise, false

INVESTIGATE

The measures collection is set up to collect measures from the Sample of Animals collection and display the results graphically.

6. Click the **Collect More Measures** button on the measures collection. The Sample of Animals collection rerandomizes 100 times, and the results appear as gray or red bars in the measures collection. Gray means that the confidence interval captured the true mean of 394.49, and red means that it did not.

Confidence Interval for a Mean: Not Normal—Brain Weights
continued

The true mean is represented by the black vertical line segment.

Q5 Approximately how many of the 100 bars are red? (Some of the bars may be too small for you to see.) For those that are red, does the interval lie above or below the true mean? Is the overall capture rate close to 95%? Is this what you expected? Explain.

7. Re-collect measures two more times and keep track of the capture rates. Compare your results with those of others in your class.

Q6 Looking at the capture rate from your simulation and the locations of the intervals that do not capture the population mean, write a brief statement on what happens to confidence intervals when techniques based on the normal distribution are used with distributions highly skewed toward large values.

You have learned that skewed distributions can almost always be made much more nearly symmetric by transforming to a new scale.

Make a plot of Transformed_BrainWt to see whether the distribution is symmetric or not.

8. Create a new attribute, *Transformed_BrainWt*, in the Animals collection. Try various transformations until you find a transformation that makes the distribution of *Transformed_BrainWt* approximately symmetric. Find the mean of *Transformed_BrainWt*.

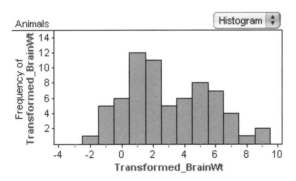

You will need to use the new mean that you found in step 8 to update your measure Capture.

9. Click **Sample More Cases** to update the sample collection. Show the sample collection's inspector. On the **Measures** panel, adjust the measures to reflect the change from *BrainWeight* to *Transformed_BrainWt*.

10. Drag the attribute *Transformed_BrainWt* from the sample collection to the horizontal axis of the dot plot for your sample.

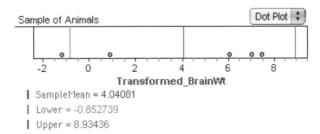

Confidence Interval for a Mean: Not Normal—Brain Weights
continued

Q7 Does your confidence interval (*Lower, Upper*) include the new true mean that you calculated in step 8?

11. Click **Sample More Cases** several times to see more confidence intervals for *Transformed_BrainWt*. Notice each time whether your new intervals capture the new true mean that you calculated in step 8.

Q8 Based on your observations in step 11, does your confidence interval include the new true mean that you calculated in step 8 more often or less often than before you did your transformation?

12. Being able to see the display of the confidence intervals in the measures collection depends on two sliders and the mean of the population. Scroll down in your document until you see two sliders, *scale* and *offset*. Set the sliders as shown.

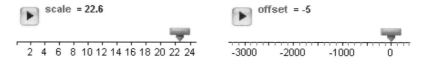

13. Now for the mean. Show the measures collection's inspector. On the **Display** panel, double-click the formula for *x* and change just the 394.49 to the new mean that you found in step 8.

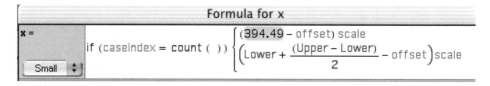

If your confidence intervals look a little squished in the window, try setting your *scale* slider to 50 and your *offset* slider to −2.2.

14. Now you're ready to see if your 95% confidence intervals capture your new mean 95% of the time. Click **Collect More Measures** on the measures collection.

Q9 Approximately how many of the 100 bars are red? For those that are red, does the interval lie above or below the true mean? Is the overall capture rate close to 95%? Is this what you expected? Explain.

Q10 Re-collect measures two more times and keep track of the capture rates. Compare your results with those of others in your class. Does the transformation improve the capture rate?

Confidence Interval for a Mean: Not Normal— Brain Weights

Activity Notes

Objectives

- Seeing that if a distribution is highly skewed, the capture rate for a confidence interval of the form $\bar{x} \pm t^* \cdot s/\sqrt{n}$ may be substantially lower than the advertised capture rate
- Noticing that confidence intervals for the mean have widely varying widths because they are based on sample standard deviations
- Finding out that although *t*-procedures are fairly robust—that is, they work well for modest departures from normality—they don't work well when the population is highly skewed and the sample size is small
- Learning to consider a transformation of the data *before* constructing a confidence interval
- Seeing that if a distribution is highly skewed to the larger values, the confidence intervals that fail to capture the true mean fall to the left of the mean

Activity Time: 40–50 minutes

Setting: Paired/Individual Activity (use **BrainWeights.ftm**) or Whole-Class Presentation (use **BrainWeights Present.ftm**)

Statistics Prerequisites

- Definition and formula for the confidence interval for a mean with unknown standard deviation
- Transforming data with natural and common logs
- Describing distributions graphically
- Some familiarity with confidence intervals that are shown graphically

Statistics Skills

- Learning why the approximately normal condition is necessary in inference
- Constructing confidence intervals for the sample mean with unknown standard deviation
- Sampling distribution of the sample mean
- Finding capture rates when using non-normal distributions and with transformed data
- Working with appropriate transformations

AP Course Topic Outline: Part III C, D (2, 6, 7); Part IV A (1–3, 6)

Fathom Prerequisites: Students should be able to make graphs and plot values, create measures and attributes, and use the formula editor.

Fathom Skills: Students work with formulas that change the display of a collection, work with "and" statements, and update a collection when the population changes.

General Notes: This activity deals with the fundamental idea that in estimating a parameter of a population, we construct a confidence interval for that parameter using a method that "captures" the true value of the parameter a certain percentage of the time—but only if the population meets certain conditions. In this activity the population is highly skewed toward the larger values. Skewed distributions can have a dramatic effect on the capture rate of a confidence interval or the error rate of a significance test. That is the essential message of this activity. You think you are using a procedure for constructing a confidence interval that has a 95% chance of capturing the true mean, but actually that rate may be only 80%—or even 50% as your students will see!

Procedure: You can use the activity as a whole-class discussion with the document **BrainWeightsPresent.ftm**. Here, everything is set to collect measures. Simply click **Collect More Measures** on the Measures from Sample of Animals collection. To work with the log transformation, click **Collect More Measures** on the Measures from Sample of Transformed Animals collection.

Students start with a partially constructed Fathom document, **BrainWeights.ftm**. Note that this simulation has three layers of abstraction—the population, a sample, and a collection of confidence intervals. Students don't need to make any changes to the population collection, Animals, until step 8.

If you'd like your students to work on the Extensions, they'll need to save the first part of their work after Q6, and then save their work again after Q10 as a separate file.

Confidence Interval for a Mean: Not Normal—Brain Weights

continued

Activity Notes

EXAMINE DATA

Q1 This distribution is highly skewed toward the larger values, with mean 394.49 and SD 1206.99. The median is 11.45 with an *IQR* of 177 − 2.845, or 174.155.

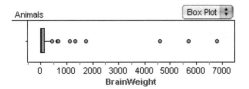

Q2 For the sample in the student activity, the range is from 0.14 to 177.

Q3 For the sample in the student activity, using $t = 2.776$, mean = 77.066, and SD = 84.1325, the confidence interval is $77.066 \pm 2.776 \cdot 84.1325/\sqrt{5} \approx 77.066 \pm 105.689$, or −28.623 to 182.755.

4., Q4 Here are the measures that need to be defined in the Sample of Animals collection. Other, equivalent formulas could be used. Note that the correct value of t^* is 2.776.

Measure	Value	Formula
SampleMean	111.1	mean(BrainWeight)
SampleSD	176.112	stdDev(BrainWeight)
Lower	−107.537	SampleMean − t · SampleSD/√count( )
Upper	329.737	SampleMean + t · SampleSD/√count( )
Capture	false	(Lower < 394.49) and (394.49 < Upper)

INVESTIGATE

Q5–Q6 After a simulation of 100 samples, students will definitely find that they get fewer than 95 captures. The capture rate may be as low as 35%! (A summary table may help count the number of false values for *Capture*.) Further, all of the intervals that miss will be on the left, below the true mean. Students should recognize that this is a result of the distribution being skewed right. This comes about because the sample mean and the sample standard deviation are correlated for distributions that are not normal. The correlation is positive for distributions skewed toward the larger values and negative for distributions skewed the other way. When taking a small sample from a positively skewed distribution, then, the sample mean is likely to be smaller than the population mean (as most of the data points lie below the population mean). The sample that produced the small mean is also likely to produce a small standard deviation, giving rise to a confidence interval that is too short and lies too far to the left of the population mean.

Here are the capture rates for 100 trials of 100 confidence intervals. The mean capture rate for the 100 trials is 51.4%, and the range of values goes from 37% to 62.5%.

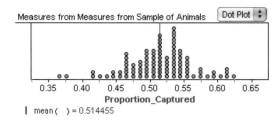

8.–9. The most appropriate transformation is a log transformation. Using natural logs, here are the measures.

Measure	Value	Formula
SampleMean	4.50912	mean(Transformed_BrainWt)
SampleSD	1.91872	stdDev(Transformed_BrainWt)
Lower	2.1271	SampleMean − t · SampleSD/√count( )
Upper	6.89115	SampleMean + t · SampleSD/√count( )
Capture	true	(Lower < 2.9772596) and (2.9772596 < Upper)

Q7–Q8 More than likely, their new confidence intervals will contain the true mean. With the natural log, the true mean is about 2.98 and for common logs, the new mean is 1.293. Hopefully, students will avoid the false assumption that the mean of the transformed brain weights will be the natural log of the mean of the brain weights. For the example in the activity (the dot plot), 2.98 is indeed between −0.853 and 8.93.

Confidence Interval for a Mean: Not Normal—Brain Weights

Activity Notes
continued

12. If students use common logs, they will want to use the values for the sliders given in the margin note for step 14.

Q9–Q10 The confidence intervals look like those shown in the student activity. Here the population mean of the natural logs of the brain weights is 2.98 and the observed capture rate from the simulation is 93/100, or 93%, much closer to the advertised 95%. The transformation helps indeed! Below are the capture rates for 100 trials of 100 confidence intervals. The mean capture rate for the 100 trials is 94.1%, and the range of values goes from 88% to 99%.

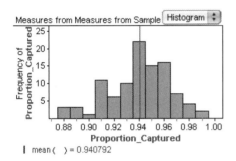

DISCUSSION QUESTIONS

- What are the dangers inherent in using a statistical method when its assumptions are not valid?
- What type of results would you expect for a distribution that was highly skewed left?

EXTENSIONS

1. Have students modify Sample of Animals and experiment with sample sizes other than 5 (in both of their documents). How does the capture rate for the confidence intervals depend on sample size?

2. A skewed distribution is just one example of deviation from normality. Have students experiment with some other non-normal distribution in their first saved document, and determine whether the non-normality has an effect on the capture rate of the confidence interval. Then, in the second saved document, they can try to find a transformation that makes the distribution nearly symmetric and look at the new confidence intervals. (They will need to again adjust the sliders *offset* and *scale,* and change the mean in the **Display** panel of the measures collection.)

Significance Test for a Proportion—Spinning and Flipping Pennies

You will need
- one penny
- **Pennies.ftm**

People tend to believe that pennies are balanced. They generally have no qualms about flipping a penny to make a fair decision. Is it really the case that penny flipping is fair? What about spinning pennies? In this activity you will record the results of 40 flips of a penny and 40 spins of a penny. You will then use the data to perform tests of significance.

COLLECT DATA

1. Open the Fathom document **Pennies.ftm.**

2. Begin spinning pennies. To spin, hold the penny upright on a table or the floor with the forefinger of one hand and flick the side edge with a finger of the other hand. The penny should spin freely, without bumping into anything before it falls. After each spin, record your results—heads or tails—in the attribute *Spin_Face*. Spin pennies until you have a total of 40 spins.

You can count the number of heads by making a bar chart and holding your cursor over the "H" bar.

Q1 Count the number of heads and compute $\hat{p}$. Do you believe heads and tails are equally likely when spinning pennies, or can you reject this standard? Explain.

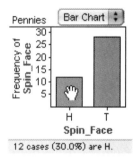

3. Begin flipping pennies. To flip, place the penny on the top of your thumb and flick it upward so that it spins so many times in the air, you couldn't begin to count the number. Let the penny fall and record whether it was heads or tails in the attribute *Flip_Face*. Flip the pennies until you have a total of 40 flips.

Q2 Count the number of heads and compute $\hat{p}$. Do you believe heads and tails are equally likely when flipping pennies, or can you reject this standard? Explain.

INVESTIGATE

Now you'll use Fathom to find out whether the "equally likely" probability model could generate results similar to the ones you observed.

Choose **Table | Show Formulas**, then double-click the shaded formula cell.

4. Define *Face* with the formula shown, which randomly selects heads or tails for each trial.

Pennies	Spin_Face	Face
=		randomPick("H","T")
1	H	T
2	H	T
3	H	H

Significance Test for a Proportion—Spinning and Flipping Pennies
continued

5. Show the collection's inspector. On the **Measures** panel, define a measure as shown.

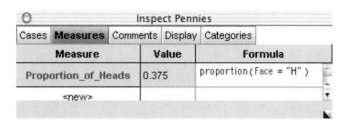

You can think of each measure as the result of spinning (or flipping) a penny 40 times.

6. Select the Pennies collection and choose **Collection | Collect Measures.** By default, Fathom collects five measures in a collection called Measures from Pennies.

7. Make a dot plot of *Proportion_of_Heads* from the Measures from Pennies collection.

To speed things up, uncheck Animation on.

8. Show the Measures from Pennies collection's inspector. On the **Collect Measures** panel, change the number of measures collected to 100 measures and check Replace existing cases. Click **Collect More Measures.**

Q3 How often did the sample proportion you got from your spins appear in the measures? How many values were greater than your value? Do you believe heads and tails are equally likely when spinning pennies, or can you reject this standard? Explain.

Q4 How often did the sample proportion you got from your flips appear in the measures? How many values were greater than your value? Do you believe heads and tails are equally likely when flipping pennies, or can you reject this standard? Explain.

So far, you have done an informal type of testing. Now you will use Fathom to formally test the hypothesis that heads and tails are equally likely when spinning pennies.

Q5 Check the conditions required to use a significance test for a proportion. Are the conditions satisfied? Discuss any concerns you might have about these data.

Now that you have checked the conditions, the next step is to state your hypotheses.

*You can choose **Edit | Show Text Palette** to format text and create mathematical expressions.*

Q6 State the null and alternative hypotheses for this situation. You can write out these two hypotheses in Fathom in a text object to be stored with your document. From the shelf, drag a text object into the document and type in your hypotheses.

Now it is time to compute the test statistic.

Significance Test for a Proportion—Spinning and Flipping Pennies
continued

9. Drag a test object from the shelf. From the pop-up menu, choose **Test Proportion.**

10. Drag *Spin_Face* from the case table to the top pane of the test where it says "Attribute (categorical): unassigned."

Q7 The last paragraph of the test above the "Note" describes the results and shows that the *P*-value for this example is 0.011. What are your *P*-value and test statistic?

> Test of Pennies — Test Proportion
> Attribute (categorical): Spin_Face
> Attribute: **Spin_Face**
> **12** out of **40**, or **0.3**, are **H**
> Alternative hypothesis: The population proportion for **H** is not equal to **0.5**.

Q8 Do your data support the standard that heads and tails are equally likely when spinning a penny? Write your conclusion in context, linked to your computations.

It is helpful to be able to visualize the *P*-value as an area under a distribution.

> If your *P*-value is small, you may need to adjust the axes to see the shading.

11. With the test selected, choose **Test | Show p_Hat Distribution.** The curve shows the binomial probability distribution curve for $n = 40$, $p = 0.5$. The shaded area corresponds to the *P*-value.

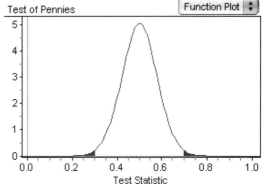

12. Drag *Flip_Face* from the case table to the top pane of the test and replace *Spin_Face*. Your test and distribution will update to reflect the change.

Q9 Do your data support the standard that heads and tails are equally likely when flipping a penny? Write your conclusion in context, linked to your computations.

Next you'll explore how the *P*-value depends on the standard, which is currently $p_0 = 0.5$.

13. Drag *Spin_Face* from the case table back to the top pane of the test to replace *Flip_Face*.

14. You can edit the blue text in the test. In the Alternative hypothesis, highlight the **0.5** of **is not equal to 0.5** and choose **Edit | Edit Formula**. Enter *p_Standard* (the value of the slider). In the test, 0.5 should be brown instead of blue.

> You can drag the slider's thumb to change the value, or edit the number itself.

Q10 What happens to the *P*-value and test statistic if you change *p_Standard* to 0.45? What's the new value? Is the *P*-value larger or smaller than before?

Significance Test for a Proportion—Spinning and Flipping Pennies
continued

Q11 If the actual probability of getting heads when spinning a penny were 0.45 (instead of 0.5), what would you now conclude from your sample data?

Now let's find the range of values for which your data are plausible. You can use Fathom's interval estimate tool to find this range directly.

15. Drag a new estimate from the shelf. Choose **Estimate Proportion** from the pop-up menu and drag *Spin_Face* into the top of the box, as you did with the test. You'll see information about the confidence interval.

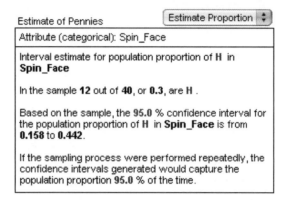

Q12 What is the range of population proportions for which your data are plausible? Interpret this confidence interval in terms of the context of the situation.

Q13 Drag the slider to change the values for *p_Standard* to the values you got in Q12. Is the *P*-value in the test 0.05 or larger? If so, explain why it is. If not, explain why there is a discrepancy.

EXPLORE MORE

1. Drag the slider *p_Standard* slowly and observe the changes that take place. For what value of the slider is all of the area under the curve shaded? Explain why it should be this particular value.

2. Repeat steps 13–15 for your flips. What is the range for which your data are plausible? Interpret this confidence interval in terms of the context of the situation.

Significance Test for a Proportion—Spinning and Flipping Pennies

Activity Notes

Objectives
- Using a test of significance to decide whether you should reject the claim that your sample has been drawn from a binomial population with a specified proportion of successes
- Understanding the concept of a confidence interval for estimating a population proportion, computing a confidence interval, and interpreting the results
- Discovering that critical z-values for hypothesis tests are special "edge" values of a statistic
- Exploring how hypothesis tests and confidence intervals for the population proportion are closely related
- Deepening the understanding of these terms in the context of inference for proportions: *confidence interval, null hypothesis* and *alternative hypothesis, test statistic, level of significance,* and *P*-values
- Computing and interpreting in context a *P*-value in testing a proportion
- Comparing actual results to some model to evaluate whether the observed results are consistent with the model
- Using simulation to estimate the probability of obtaining the observed results under the assumed model

Activity Time: 40–50 minutes

Setting: Paired/Individual Activity (use **Pennies.ftm**) or Whole-Class Presentation (use **PenniesPresent.ftm**)

Materials
- One penny per student (for American pennies, those from the 1960s are better than newer ones, if you can get them)

Statistics Prerequisites
- Familiarity with the language for testing a hypothesis
- Familiarity with the structure and procedure for testing a hypothesis
- The test statistic when testing a proportion
- Familiarity with the structure and procedure for constructing confidence intervals
- Familiarity with sampling distributions and reasonably likely outcomes
- Definition of probability and equally likely outcomes

Statistics Skills
- Finding confidence intervals for the sample proportion
- Checking conditions for testing a proportion
- Performing a hypothesis test for the sample proportion
- Finding *P*-values and relating them to the plot of the distribution of the binomial probability distribution
- Determining confidence levels, critical values, and significance levels
- Probability simulation
- Working with the definition of probability and equally likely outcomes
- Comparing actual data to a hypothesized model and detecting differences between models

AP Course Topic Outline: Part III A (1, 4, 5), C, D (1, 6); Part IV A (1–4), B (1, 2)

Fathom Prerequisites: Students should be able to make graphs and use case tables, create attributes and measures, use the formula editor, and use inspectors.

Fathom Skills: Students use random generators to create a collection that represents a random sample, use Fathom to do simulations, collect measures to compare models, use Fathom's hypothesis test and interval estimate tools, make a graph of the test statistic's distribution, and use a slider to change the null hypothesis.

General Notes: This activity explores essential concepts in hypothesis testing. You could use it as an introduction to traditional tests, but it may be more effective to wait until students have solved some straightforward problems using whatever technology is most common in your class. These activities do not, for example, explain what the test statistic is, so students who already know that—even shakily—may get more out of the experience.

This activity introduces Fathom's analyses—both hypothesis tests and confidence intervals (estimates). Students will explore *P*-values, critical z-values, and the correspondence between hypothesis testing and

Significance Test for a Proportion—Spinning and Flipping Pennies
continued

estimation. This activity is designed to be suitable as an introductory activity; that is, you don't have to know much about Fathom.

Your students may expect the probability of heads that results when a penny is spun to be 0.5. As they will learn, that isn't the case for U.S. pennies—the probability is closer to 0.4 for newer pennies and much less for pennies from the 1960s. For example, 1990 pennies have a probability of around 0.4 for heads, whereas 1961 pennies have a probability of only about 0.1 for heads.

Procedure: Your students will need the results from 40 spins of a penny and 40 flips of a penny. This can be done fairly quickly in class especially if they work in groups of two or three. If you don't want to use up class time, you can ask students to do the spinning and flipping of pennies at home and bring in their results, or you can use the document **PenniesPresent.ftm** for a whole-class discussion. This document contains spin data from one student who got 12 heads out of 40 spins. A random generator was used for the flips.

Try to get a classroom set of relatively old pennies of about the same date. Pennies from the 1960s work well.

COLLECT DATA

Q1 Heads and tails are not equally likely when a penny is spun. However, from a single sample of 40 spins, this may be difficult to conclude. One good strategy students may suggest is to construct a 95% confidence interval, based on their 40 spins, for the true proportion of heads when a penny is spun and see if $p = 0.5$ is in that confidence interval. Whether your students reject the model that spinning a penny is fair will largely depend on how far the proportion is from 0.5.

Q2 Heads and tails are roughly equally likely when a penny is flipped, and a 95% confidence interval has a 95% chance of capturing 0.5. This is true even for confidence intervals based on relatively small sample sizes.

INVESTIGATE

5.–8. Help students build a Fathom simulation to analyze the results, assuming that spinning a penny is fair. For step 8, students should arrange their Fathom screen so that they can see the case table and dot plot change. It is recommended that students leave animation on in step 8 while collecting measures so that they can see the dot plot grow.

Q3 If students are using newer pennies ($p = 0.4$), they probably won't reject the equally likely model. With only 40 spins, it is quite plausible to get a sample proportion of about 0.5. The plot below was made using the same type of simulation setup as the activity except with $p = 0.4$. In 15 cases out of 100, the proportion of heads was 0.5 or more when a penny was spun 40 times. The sample proportion in **PenniesPresent.ftm** is $12/14 = 0.3$. This outcome is unlikely to occur if heads and tails are equally likely, as can be seen in the dot plot in the activity. Only 2% of the trials in that document had a sample proportion of 0.3 or smaller.

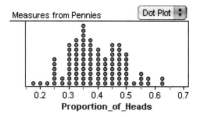

Q4 Heads and tails are roughly equally likely when a penny is flipped, and a 95% confidence interval has a 95% chance of capturing 0.5. This is true even for confidence intervals based on relatively small sample sizes. More than likely, the students' sample proportion will fall within the range of reasonably likely values on their dot plot.

Q5 The conditions are met for a significance test of a sample proportion. The sample is a simple random sample from a binomial population where we can consider the number of heads a binomial random variable. Both np_0 and $n(1 - p_0)$ are at least 10: $np_0 = 40(0.5) = 20$ and $n(1 - p_0) = 40(0.5) = 20$. The population of all possible spins is infinitely large so it is at least 10 times the sample size of 40.

Q6 H_0: The proportion of heads when a penny is spun is equal to 0.5, or $p = 0.5$.

H_a: The proportion of heads when a penny is spun is not equal to 0.5, or $p \neq 0.5$.

Significance Test for a Proportion—Spinning and Flipping Pennies
Activity Notes
continued

Q7–Q8 For the sample proportion in **PenniesPresent.ftm**, the P-value is 0.011 and the test statistic is −2.53. If spinning a penny results in heads 50% of the time, it is not reasonably likely to get only 12 heads out of 40 flips. Thus, we can reject the null hypothesis that spinning a penny is a fair process. The probability of getting a result as extreme as or more extreme than the one in the sample is 0.011. *If the probability of getting heads is 0.5,* there is a 0.011 chance of getting 12 heads or fewer or 28 heads or more in 40 spins.

Q9 For the sample proportion in **PenniesPresent.ftm**, the P-value is 0.34 and the test statistic is 0.9487. If flipping a penny results in heads 50% of the time, it is reasonably likely to get 23 heads out of 40 flips. Thus, we cannot reject the null hypothesis that flipping a penny is a fair process. The probability of getting a result as extreme as or more extreme than the one in the sample is 0.34. *If the probability of getting heads is 0.5,* there is a 34% chance of getting 23 heads or more or 17 heads or fewer in 40 spins.

Q10–Q11 For the sample proportion in **PenniesPresent.ftm**, the P-value jumps to 0.057 and $z = -1.907$. It is larger because the hypothesized proportion is now closer to the sample proportion for these data. The conclusion then becomes: With a P-value of 0.057, this sample result is reasonably likely to occur under the null hypothesis, so you cannot reject the null hypothesis that the probability of getting heads when spinning a penny is 0.45. The claim that the probability of getting heads when spinning a penny is 0.45 is plausible.

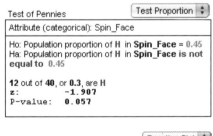

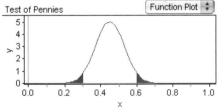

Q12–Q13 For the given sample proportion, the range is

$$0.3 \pm 1.96 \cdot \sqrt{\frac{0.3(1-0.3)}{40}} \approx 0.3 \pm 0.142$$

or 0.158 to 0.442. The plausible population proportions that could give a sample proportion of 0.3 are from 0.158 to 0.442. I am 95% confident that if I spin pennies indefinitely, the percentage that would be heads is somewhere in the interval from 15.8% to 44.2%.

If students got a fairly low sample proportion, one where $np < 10$, then the values in the confidence interval and the value from the slider will not match up because the conditions weren't met. If this is true, they need to use the binomial distribution formulas to calculate their confidence intervals. For example, with the **PenniesPresent.ftm** sample proportion of 0.3, the interval estimate gives 0.158. However, using the slider with $p\_Standard = 0.158$ results in a P-value of 0.019, not 0.025. This is because with $p = 0.158$ you can no longer use the normal approximation, so there is a discrepancy between the two values. You might want your students to use verbose mode on the test. The note on the bottom of the test will change when the normal approximation can no longer be used. The plot will also shift over to a discrete binomial probability plot.

EXPLORE MORE

1. The value of the slider that makes all of the area under the curve shaded is the sample proportion, in this instance, 0.3.

2. The values for the slider *p_Standard* and the interval estimate should be close, depending on the sample proportion. Any proportions below 0.25 or above 0.75 will give different estimates.

Inference Between Two Proportions—Plain and Peanut M&M's

You will need
- one bag of plain M&M's
- two bags of peanut M&M's

Like all statistics computed from samples, the difference of two sample proportions, $\hat{p}_1 - \hat{p}_2$, has a sampling distribution. In this activity you will construct a sampling distribution for the difference of two proportions. You will be sampling from the population of plain M&M's, which are 24% blue, and from the population of peanut M&M's, which are 15% green.

COLLECT DATA

Because bags of M&M's are filled from a huge vat with the colors already mixed in designated proportions, you may consider any sample that you pour from the bag of M&M's a random sample from the population of all M&M's of that type.

1. In your group, pour a sample of 40 plain M&M's and find the proportion that are blue.

2. In your group, pour a sample of 40 peanut M&M's and find the proportion that are green.

Q1 Subtract the proportion in step 2 from the proportion in step 1. What is your value for $\hat{p}_1 - \hat{p}_2$?

3. Plot your value for $\hat{p}_1 - \hat{p}_2$ on a dot plot with the other members of your class to construct your approximate sampling distribution of $\hat{p}_1 - \hat{p}_2$.

Q2 Estimate the mean and standard error for your approximate sampling distribution of $\hat{p}_1 - \hat{p}_2$. What is the shape?

Q3 What differences are you reasonably likely to get according to your plot?

Q4 Are the data consistent with the model that the difference between the percentage of blue plain M&M's and the percentage of green peanut M&M's is 9%, or can you safely reject that model?

INVESTIGATE

Building a Sampling Distribution

Now you'll use Fathom to find out whether the given model could generate results similar to the ones you observed.

To change the range of the slider, drag the axis or double-click it and change *Lower* and *Upper*.

4. Open a new Fathom document and drag two sliders from the shelf, one for each proportion. Make and set your sliders as shown.

Inference Between Two Proportions—Plain and Peanut M&M's
continued

> Consider using a random function.

5. Make a collection named M&Ms Sample with two attributes, *PlainBlue* and *PeanutGreen*. Add 40 cases to the collection.

6. Define the attributes with formulas that use the sliders to randomly assign the value true if the sampled candy is the right color or false if it is not.

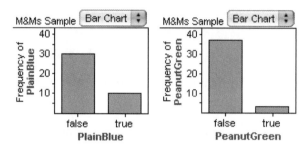

> To link the axes so that they have the same scale, select one of the graphs, then choose **Graph | Show Axis Links.** Drag the link icon from the axis you want linked onto the axis of the other graph.

7. Make two bar charts with the same scales.

> To rerandomize, select the collection and choose **Collection | Rerandomize.**

Q5 Rerandomize the collection a few times and observe the changes in the bar charts. Is it always true that the plain sample has more blues than the peanut sample has greens? Explain.

Now you'll define the sample statistic.

8. Show the collection inspector and go to the **Measures** panel. Create a measure named *Difference* that subtracts the proportion of peanut M&M's that are green from the proportion of plain M&M's that are blue.

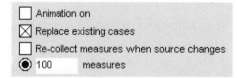

9. With the M&Ms Sample collection selected, choose **Collection | Collect Measures.** By default, Fathom collects five measures in a collection called Measures from M&Ms Sample.

10. Make a histogram or dot plot of *Difference* from the Measures from M&Ms Sample collection.

> To speed things up, uncheck Animation on.

11. Open the measures collection's inspector. On the **Collect Measures** panel, change the settings as shown. Click **Collect More Measures.**

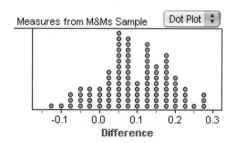

268 | 6: Inference for Means and Proportions

Inference Between Two Proportions—Plain and Peanut M&M's
continued

Q6 Describe the approximate sampling distribution of the difference of two proportions in terms of shape, center, and spread. How does this distribution compare to the sampling distribution you got in Q2?

Q7 According to this plot, what differences are you reasonably likely to get? Are these differences roughly the same as the ones you got in Q3?

Q8 How often did the sample difference from your sample of M&M's appear in the measures? How many values were greater than your value?

Q9 Are your data consistent with the model that the difference between the percentage of blue plain M&M's and the percentage of green peanut M&M's is 9%, or can you safely reject that model?

Constructing the Confidence Interval

Now it is time to construct a confidence interval for the difference of two proportions. The conditions that must be met are that the two samples are taken independently from two populations, each population is at least ten times as large as the sample size, and $n_1\hat{p}_1$, $n_1(1 - \hat{p}_1)$, $n_2\hat{p}_2$, and $n_2(1 - \hat{p}_2)$ are all 5 or more.

Q10 Are the conditions met for constructing a confidence interval for the difference of two proportions?

12. Drag a new estimate from the shelf. Choose **Difference of Proportions** from the pop-up menu. Highlight the appropriate blue text and replace it with the values you got from your sample in steps 1 and 2.

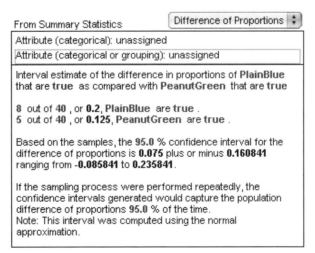

Q11 What is your 95% confidence interval for the difference of two proportions? Did your confidence interval capture the true difference? State a conclusion in the context of the situation.

Inference Between Two Proportions—Plain and Peanut M&M's
continued

This test is in non-verbose mode. Select the test, then choose **Test | Verbose**.

13. Now drag a new test from the shelf. From the pop-up menu, choose **Compare Proportions.** Enter the values from your sample as you did in step 12. Fathom will update the *P*-value at the bottom depending on those parameters. You can change the "is not equal to" to "is greater than."

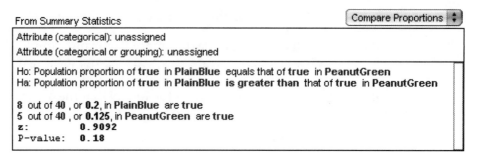

14. It is helpful to be able to visualize the *P*-value as an area under a distribution. With the test selected, choose **Test | Show Test Statistic Distribution.** The curve shows the normal distribution. The shaded area corresponds to the *P*-value.

Q12 Do your data support the claim that there are more blue plain M&M's than green peanut M&M's? Give appropriate statistical evidence to support your answer.

Inference Between Two Proportions—Plain and Peanut M&M's

Activity Notes

Objectives

- Being introduced to the concept of confidence intervals for a sampling distribution for the difference of two proportions
- Seeing that the properties of the sampling distribution of the difference of two proportions are analogous to the properties of the sampling distribution of the difference of two means
- Using simulation to construct an approximate sampling distribution for the difference of two proportions
- Finding the mean and standard error of the sampling distribution for the difference of two proportions
- Constructing and interpreting a confidence interval for the difference of two proportions
- Using a test of significance to decide whether to reject the claim that two samples were drawn from two binomial populations that have the same proportion of successes
- Comparing actual results to some model to evaluate whether the observed results are consistent with the model
- Using simulation to estimate the probability of obtaining the observed results under the assumed model

Activity Time: 40–50 minutes

Setting: Paired/Individual Activity (collect data and build simulation)

Optional Document: M&MsExt.ftm (Extensions 2–4 solutions)

Materials

- One bag of plain M&M's per group
- Two bags of peanut M&M's per group

Statistics Prerequisites

- Familiarity with sampling distributions and reasonably likely outcomes
- Probability concepts
- Familiarity with the language for testing a hypothesis
- Familiarity with the structure and procedure for testing a hypothesis
- Familiarity with the structure and procedure for constructing confidence intervals

Statistics Skills

- Working with the sampling distribution for the difference of two proportions
- Finding confidence intervals for the difference of two proportions
- Checking conditions for testing or constructing confidence intervals for the difference of two proportions
- Performing a hypothesis test for the sample proportion
- Finding *P*-values and relating them to the plot of the distribution of the normal distribution
- Probability simulation
- Comparing actual data to a hypothesized model and detecting differences between models

AP Course Topic Outline: Part III A (1, 4–6), B, C, D (4, 6); Part IV A (1–3, 5), B (1, 3)

Fathom Prerequisites: Students should be able to make graphs and sliders, use a case table, create attributes, use the formula editor and the random() function, and use inspectors.

Fathom Skills: Students use Fathom to do simulations of two different populations, collect measures to compare models, use Fathom's hypothesis test and interval estimate tools, create a test distribution plot to see the *P*-value, make a graph of the test statistic's distribution, and dynamically link axes to each other.

General Notes: In this activity, students construct an approximate sampling distribution for the difference of two proportions, $\hat{p}_1 - \hat{p}_2$. The first proportion comes from a random sample of size 40 taken from a population with 24% successes. The second proportion comes from a random sample of size 40 taken from a population with 15% successes.

Procedure: Students really enjoy this data-collecting activity. Ideally, each student or group of students has one bag of plain M&M's and two bags of peanut M&M's. You can instead, as a class, use a large bag of plain M&M's and

Inference Between Two Proportions—Plain and Peanut M&M's
continued

Activity Notes

two large bags of peanut M&M's. The number of bags and the number of students in your class will determine how many sample differences, $\hat{p}_1 - \hat{p}_2$, you have. Twice as many peanut bags is recommended because a small bag of peanut M&M's has, on average, about 24–26, whereas a small bag of plain M&M's has, on average, about 55.

You should check the web site http://us.mms.com/us/about/products/ to make sure the percentages given in the activity haven't changed. If they have changed, then pick one color from each type with the biggest possible difference. Given the low percentages, it is entirely possible that some of your students will get sample proportions that do not satisfy the conditions for constructing confidence intervals and doing a significance test. See the answer to Q10 for more on this.

COLLECT DATA

Q1 If, for example, the student gets 8 blue plain M&M's, the proportion would be 8/40 = 0.2. Then, if the student gets 5 green peanut M&M's, the proportion would be 5/40 = 0.125. The difference is 0.2 − 0.125 = 0.075.

Q2 Below is a set of typical results from a class of 25 students. There is a dot plot in step 11 of the activity of 100 simulated differences for comparison. Here, the distribution is beginning to look mound-shaped, except for the outliers; the center is 0.088 and the SD is about 0.11. The median is 0.075 and the IQR is 0.15 − 0.025, or 0.125.

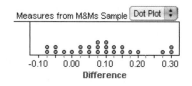

Q3 For this plot, the reasonably likely values are from about −0.06 to 0.3. (2.5% of 25 cases is about one dot.)

Q4 The sample data are consistent with the model that there is a 9% difference between the percentage of blue plain M&M's and the percentage of green peanut M&M's.

INVESTIGATE

6. Formulas:

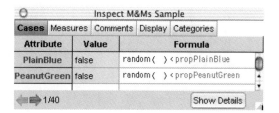

Q5 No; about 20% of the time, there will be more greens than blues.

Q6 The sampling distribution should look similar to the one in the activity. The mean of that particular distribution is 0.09225, and the standard deviation is approximately 0.088. The shape is approximately normal. This distribution is more normal than the one with only 25 cases. The mean and SD are closer to the theoretical values. The theoretical mean of the sampling distribution of $\hat{p}_1 - \hat{p}_2$ falls at $p_1 - p_2$, or $0.24 - 0.15 = 0.09$. The theoretical variance of the sampling distribution is the sum of the two variances, $\sigma^2_{\hat{p}_1-\hat{p}_2} = \sigma^2_{\hat{p}_1} + \sigma^2_{\hat{p}_2}$, so the standard error is

$$\sigma_{\hat{p}_1-\hat{p}_2} = \sqrt{\frac{p_1(1-p_1)}{n_1} + \frac{p_2(1-p_2)}{n_2}}$$
$$= \sqrt{\frac{0.24(1-0.24)}{40} + \frac{0.15(1-0.15)}{40}} = 0.088$$

Q7 Students should identify the values that make up the middle 95% of the points on their dot plots. On the theoretical distribution, the middle 95% falls between −0.0825 and 0.2625. Students could plot values on the graph or use a summary table to find mean()−1.96 s() and mean()+1.96 s(). On the sample dot plot, the reasonably likely values are from about −0.075 to 0.275.

Q8–Q9 For the sample difference of 0.075, there were 12 cases that were the same as and 49 cases larger than the sample difference. The model is consistent.

Q10 The conditions are met by the sample data. The samples were taken randomly and independently. There are certainly more than 400 M&M's in the world. For the sample data, $n_1\hat{p}_1 = 8$, $n_1(1-\hat{p}_1) = 32$, $n_2\hat{p}_2 = 5$, and $n_2(1-\hat{p}_2) = 35$ are all 5 or more.

Inference Between Two Proportions—Plain and Peanut M&M's
continued

Activity Notes

If a student gets fewer than 5 blues or 5 greens, then the conditions will not be met and he or she then might want to work with another student's proportion after trying to construct a confidence interval in Fathom. If the conditions aren't met, students will get #Evaluation error# for the estimate as shown. This gives you another opportunity to talk about the normal approximation to the binomial and explain why this condition is necessary.

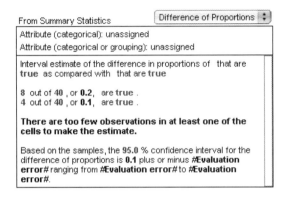

Q11 For the sample data, the confidence interval is from −0.086 to 0.236, which does capture the true percentage, 0.09. I am 95% confident that the difference between the percentage of all plain M&M's that are blue and the percentage of all peanut M&M's that are green is in the interval 7.5% ± 16%.

Q12 For the sample data, the P-value is 0.18, so we would fail to reject the null hypothesis that there is no difference. So, in this case, if there is no difference in the proportion of blue plain M&M's and the proportion of green peanut M&M's, then there is a 0.18 chance of getting a difference of 0.075 or larger with samples of these sizes. This difference is not statistically significant—it can reasonably be attributed to chance variation. We do not reject the null hypothesis and cannot conclude that there are more blue plain M&M's than green peanut M&M's.

DISCUSSION QUESTIONS

- Is it possible to get a negative difference of proportions? What are the maximum and minimum possible for the difference of proportions?
- How did your predictions for the sampling distribution compare with the actual results?
- What do you think is the theoretical mean of the sampling distribution for the difference of two proportions?
- Look at the distribution as a dot plot. Why is the distribution made of discrete values?
- What happens in Fathom if the conditions aren't met for constructing a confidence interval or for using a significance test?

EXTENSIONS

1. Have students experiment with Fathom's test and estimate tools by dragging *PlainBlue* and *PeanutGreen* from the M&Ms Sample collection into the test and estimate. Drag *PlainBlue* from the case table to the top pane of the test where it says "Attribute (categorical): unassigned." Drag *PeanutGreen* from the case table to the second pane from the top of the test where it says "Attribute (categorical or grouping): unassigned."

   Now, this time, they'll need to click on the blue text that says "PlainBlue where PeanutGreen is false" to bring up the pop-up menu. Change this to "PlainBlue." Then click on "false" and change it to "true."

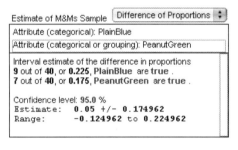

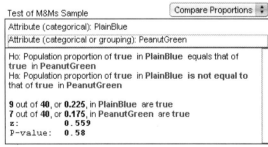

For Extensions 2–4, students may want to compare their original results with the modified simulation. One way to do this is to not use **Collect More Measures.** Instead, modify the sliders and/or collection for the new simulation, select the collection, then choose

Inference Between Two Proportions—Plain and Peanut M&M's
continued

Activity Notes

Collection | Collect Measures. This creates an entirely new collection of measures rather than emptying the original collection of measures. Now students can create and compare multiple histograms or dot plots. For sample results see the document **M&MsExt.ftm**.

2. What if the sample size were 80 instead of 40? Predict what you expect the sampling distribution for the difference of the two proportions to look like. Check your prediction with simulation. [Change the sample size by adding or deleting cases from the collection M&Ms Sample, then re-collect measures.]

3. Suppose that instead of finding the difference of the proportions, you find the ratio of the proportions. What should the sampling distribution look like? Check your prediction with simulation. [Change the formula for the measure to proportion(plainBlue)/ proportion(peanutGreen).]

4. In the population of peanut M&M's, 23% are blue and 23% are orange. Explain how you would take two *independent* random samples from the population of peanut M&M's to construct the approximate sampling distribution of the difference between these two proportions. If your sample size is 40 for each sample, describe the approximate sampling distribution of $\hat{p}_1 - \hat{p}_2$ in terms of shape, center, and spread. Then modify your Fathom document and do a simulation to check your prediction. [To sample independently, students would need to take each sample from separate bags of peanut M&M's. In Fathom, change the sliders so that they are both equal to 0.23.]

Inference for the Difference of Two Means: Unpaired—Orbital Express

You will need
- two kinds of paper
- stack of index cards
- tape measure
- marker for your center
- masking tape
- recording sheet
- pen or pencil
- OrbitalExpress.ftm

You work in the design and testing department for Orbital Express, a big delivery company. The company has decided to start delivering packages by dropping them from orbit onto the customers' houses. It is testing two competing designs (or two treatments) for the new re-entry vehicle. Your job here is to test the two designs (treatments) and report which is better. The two designs cost the same to build, so your objective is to find out which vehicle gets the package closer to its target.

COLLECT DATA

The two designs (treatments) you will test are made of two different types of paper, wadded up into balls. You'll test the designs by dropping them from "orbit" and measuring how far they land from a target. The target is the marker. You will measure the distance from the marker to where the wad of paper comes to rest, not where it first hits. (People aren't going to *catch* packages falling from orbit; they'll wait until the packages land, then walk out and pick them up.) *Note:* Before you begin collecting data, you may want to practice a few drops to make sure that you are being consistent.

1. After you have practiced a few times, test each design seven times using a random process. Keep track of your results (design type and distance) on your recording sheet.

Q1 Explain how you used randomization in your testing process to protect against confounding.

2. Record each distance with its design type on a piece of index card. Stretch your tape measure on the floor and tape it down. Set the cards opposite their recorded measurements—one design on one side of the tape, the other design on the other side.

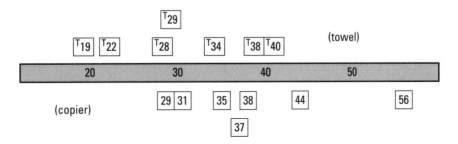

Q2 Compare the two distributions in terms of shape, center, and spread. Which is the better design? Explain why.

Exploring Statistics with Fathom
© 2007 Key Curriculum Press

Inference for the Difference of Two Means: Unpaired—Orbital Express
continued

3. You will need to justify your choice statistically. Develop a formula for a statistic—one single number that you think describes how much better one design is than the other. The statistic should be *small* if the two papers are *alike* and *larger* the more *different* they are.

Q3 Find the value of your statistic for your data—this value will be your test statistic. Does your statistic fit the rule given in step 3? Revise your formula as needed.

Q4 Does your statistic show that there really is a difference between the two designs? If it does, how small would it have to be before you were no longer convinced? If it does not show a real difference, how big would it have to be before you were convinced?

INVESTIGATE

Shuffling by Hand

Now you will investigate whether any difference you found is "real" or due to chance. Even if there were no relationship between the design and the distance—if the designs were really the same—there would still be a difference when you did the experiment, because there's some random variation. You will calculate how much difference to expect in your statistic.

4. Shuffle your index cards and deal them out, placing them by their measurement on the tape. The first seven cards should go on one side of the tape, the second seven on the other side. Compute (and record) your "shuffled" statistic.

5. Repeat step 4 four times, recording the "shuffled" statistic each time.

Q5 Make a dot plot of your five values. Revise your answer to Q4 based on your results. Are all of the values of the test statistic from your shuffles greater than the value from the original experiment? How did the shuffling experiment influence your opinion about the difference between the two designs, if at all?

Q6 What variables affect how close to the target the vehicle lands? How did you try to control most of them? Did you use any blocking?

Q7 How did your group decide to measure the distance? What did you do about "interference"?

Q8 Did anything surprise you in the shuffled data? What does the shuffling accomplish? Do you think shuffling five times is enough?

Q9 What is the null hypothesis in this situation?

Inference for the Difference of Two Means: Unpaired—Orbital Express
continued

Shuffling (Scrambling) with Fathom

You shuffled your cards and computed your statistic five times by hand. But to really analyze the situation, you should do it many more times. To be practical, this requires a computer.

6. Open the document **OrbitalExpress.ftm.** It has an empty setup, so you don't have to build everything yourself (though you could).

7. Enter the data into the Drop Data case table. You should have 14 cases—7 with each kind of paper.

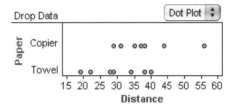

8. Make a graph that compares the two types of paper, as above. Sketch your graph.

Q10 Just based on your graph, does one design look better? Explain.

> The value of your measure should agree with the value you got in Q3.

9. Show the inspector for Drop Data. Go to the **Measures** panel and make a new measure, called *My_Measure,* and give it the formula you developed for your statistic.

Now you will set up the Scrambled Drop Data collection.

10. Show the inspector for the scrambled collection and go to the **Scramble** panel. Choose **Paper** from the pop-up menu. Click the **Scramble Attribute Values Again** button. Watch the case table to make sure that the paper types are shuffled.

Q11 Go to the **Measures** panel of the inspector. What's the value of *My_Measure* for this collection?

You need a place to collect the values of *My_Measure* for the scrambled data. This will be a *measures collection.*

11. Select the Scrambled Drop Data collection and choose **Collection | Collect Measures.** A new collection appears, called Measures from Scrambled Drop Data. Make a dot plot for this new collection.

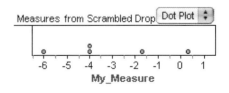

Q12 Are these five values roughly in the same range as your five by-hand shuffles, or are they in a completely different range?

Inference for the Difference of Two Means: Unpaired—Orbital Express
continued

Building the Distribution

Now you are ready to collect many measures.

12. Show the measures collection's inspector and go to the **Collect Measures** panel. Uncheck all the boxes and collect 95 measures. Click **Collect More Measures.** You'll see 95 new values of *My_Measure* appear—one from each scramble.

13. Make a histogram of *My_Measure*. Sketch the histogram.

The distribution in your new graph shows how your statistic, *My_Measure*, is distributed for situations in which the kind of design (paper) has no influence on the distance. It shows what the situation is like when the null hypothesis is true. You need to find out where your original test data are in this distribution.

Q13 Is the value you got in Q3 (and step 9) a reasonably likely value for the 100 measures of *My_Measure*, or is your value a rare event?

Select the histogram and choose **Graph | Plot Value.**

14. Plot the value of the statistic for your test data and add that line to your sketch.

Now you want to find out how many of the 100 shuffles are at least as extreme as your test statistic.

Hold down the Shift key as you click each bin.

15. Drag the edge of a histogram bin so that one edge of a bin lines up with your statistic. Select all the bins outside the line, that is, farther than the line from the center of the distribution. Hold the mouse over the measures collection (*not* Drop Data). Look in the status bar at the bottom of the Fathom window to see how many shuffles are selected.

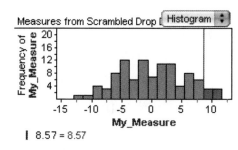

Q14 How many values in the histogram are as extreme as your value or more extreme than your value? What percentage?

16. Collect measures five more times and record what percentage of measures are as extreme as or more extreme than your test statistic.

Q15 Overall, about what percentage of the scrambled measures are as extreme as or more extreme than your original test statistic?

Q16 In your opinion, is one design of a re-entry vehicle for Orbital Express better than the other? Justify your answer.

Inference for the Difference of Two Means: Unpaired—Orbital Express

Activity Notes

Objectives

- Creating or developing a statistic that measures how much better one design is than the other
- Recognizing that a measured difference between two designs could result from chance variation and not necessarily from a true difference in designs
- Creating a sampling distribution of a statistic based on a null hypothesis
- Assessing how likely it is that any observed difference is a real difference or due to chance behavior
- *Optional:* Perform a test of significance and construct a confidence interval for the difference between means for an experiment (Extensions 1–3)

Activity Time: 60–120 minutes

Setting: Small Group Activity (collect data, use **Orbital Express.ftm**)

Materials

- Two kinds of paper with different weights, such as copier paper and paper towels. One sheet of each kind per group.
- 3 × 5 cards, cut into small, identically sized pieces (e.g., quarters); 20 pieces per group. You may want to use two colors to distinguish between the designs.
- One tape measure, at least 5 feet long, per group
- Small marker, one per group (a coin will work)
- Roll of masking tape, one per group
- Paper and pencil for recording data

Statistics Prerequisites

- Familiarity with summary statistics and comparing distributions
- Familiarity with reasonably likely outcomes and rare events
- *Optional:* Familiarity with the sampling distribution for the difference of two means (Extensions 1–3)
- *Optional:* Familiarity with the structure and procedure for testing a hypothesis (Extensions 1–3)
- *Optional:* Familiarity with the structure and procedure for constructing confidence intervals (Extensions 2 and 3)

Statistics Skills

- Working with experimental design—mainly using randomization to assign treatments
- Comparing actual data to a hypothesized model and detecting differences between models
- Working with the sampling distributions for two random variables (or the difference of two random variables)
- Creating a sampling distribution for a test statistic
- Scrambling—a randomized permutation test
- *Optional:* Performing a hypothesis test for the difference of two means (Extension 1)
- *Optional:* Finding confidence intervals for the difference of two means (Extension 2)

AP Course Topic Outline: Part I A–C; Part II A (3), C; Part III A (5, 6), B, C, D (5, 6). *Optional:* Part IV A (1, 6, 7), B (1, 4 or 5)

Fathom Prerequisites: Students should be able to enter data and make graphs, write formulas, and use the inspector.

Fathom Skills: Students use scrambling, collect measures, and compare their test statistic to the distribution. *Optional:* Students use test and estimate tools to compare means (Extensions 1 and 2).

General Notes: This activity is about hypothesis testing. Students test two competing designs for orbital re-entry vehicles by dropping wads of paper. The null hypothesis is that there is no difference between the designs; the analysis will see if we can reject that hypothesis. Students first test their design and do a preliminary analysis by hand. You can use the student worksheets or just give verbal instructions. Students then create a distribution of their statistic given the null hypothesis is true. That is, in order to see whether their statistic would be a reasonably likely outcome just by chance, they artificially make the two variables—*distance* and *paper*—independent. Their association is broken through randomization (by scrambling the values of one of the attributes).

In the Extensions section, your students can do a formal significance test and find confidence intervals for the difference of two means.

Inference for the Difference of Two Means: Unpaired—Orbital Express

continued

Activity Notes

Procedure: Paper towels and copier paper are quite different, so you may want to try materials that are more similar. Cheap, flimsy binder paper or scratch newsprint works well when contrasted with crisp, strong copier paper. Ideally, some groups will get significant differences, but not so gross that there is no point in doing statistics.

Find a place (indoors or sheltered from wind) where students can drop wads of paper. Standing and dropping is okay, but not as much fun as using stairwells or balconies. If you use outdoor bleachers, you'll need different re-entry vehicles to account for the wind.

Overview of Scrambling: Scrambling (or shuffling) was probably not part of your statistics education as an instructor, but it is a recognized nonparametric technique. Here students do it with concrete materials—cards on a tape measure—as well as with Fathom. Either way, they make the null hypothesis real by forcing the two attributes to be independent—by shuffling (or *scrambling*) one of them. When an attribute is randomly shuffled, there is no relationship between its values and those of any other attribute. There will still be chance variation, so by computing their measure (e.g., the difference of means) for a slew of null hypothesis situations, students can see how much the measure is likely to vary by chance.

Scrambling in Fathom requires three collections (see below). The original data are on the left. In the middle are the same data with the first attribute, *paper*, scrambled. At the right is the measures collection. (Students collect measures in step 12.) It has 100 values of *My_Measure*, which is the difference of means.

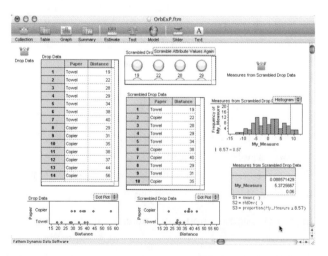

COLLECT DATA

Q1 To make this a true experiment, students need to practice first (to reduce variation) and to randomize in some way. Here are some ideas.

- Flip a coin once: if heads, all 7 towel designs are tested first; if tails, all 7 copier designs are tested first.

- Flip a coin between each trial, although students will then need to discuss what they do when they get to 7 tests for one design.

- Put 14 pieces of paper in a bag, 7 labeled towel and 7 labeled copier. Pick out a piece of paper, one piece at a time (do not replace), and this will be the test sequence.

Q2 The distributions in the activity could both come from normal populations, although the copier distribution is skewed toward the larger values because of the outlier at 56. The mean and median are higher in the copier design and the towel design has the smaller range, while the copier design has the smaller IQR. Depending on how different the two types of paper are, students may have a lot or a little evidence that one design is better than the other.

3. Students may need help writing formulas. However, at this point in your class, they might jump right to the difference of two means—especially if you have already covered combining random variables. They may need more help when they enter it into Fathom (see step 9), or they may need help writing down their procedure. Listen carefully to their ideas and see if you can explain things in terms of the ideas they already have. For example, when you ask what number they will use to describe how much better one design is than the other, some students answer, "The mean." If you ask, "The mean of what?" they answer, "The mean of *distance*." Let them try their formulas and see what happens.

Q3 There could be many different formulas in the class. Here are some ideas: difference of means, difference of medians, ratio of medians, ratio of means, the mean of the copier paper only, and the number of copier measurements that are larger than the overall mean. For the data in the student activity, the difference of means is 8.57 inches.

Inference for the Difference of Two Means: Unpaired—Orbital Express

Activity Notes
continued

Q4 For most groups, it will look as if the paper towel is better than the copier paper. Nevertheless, when you do the test, you will probably get a *P*-value above the orthodox 5%; that is, it will look somewhat unlikely—but still plausible—that the difference could have arisen by chance. One solution is to get more samples. One of the points of this activity is that random variation is more varied than we expect.

INVESTIGATE

Q5 Some groups may find that the values for their statistic are always larger from the shuffles. Have groups compare their results. For the dot plot in the student activity, the test statistic (8.57) was larger than all the shuffles and the closest simulated test statistic was about 5. So this supports the idea that there may be a difference in designs.

Q6 Wind, height dropped from, irregularities on the landing surface, and other factors can affect how close to the target the vehicle lands.

Q7 You may want to discuss measurement accuracy and significant figures with students.

Q8 Make sure that students understand the reason for shuffling. They are now creating a random division of the measurements and observing whether or not there seems to be a measurable difference. They may be surprised by the distribution of their shuffled cases.

Q9 The null hypothesis is that there is no difference between the two designs. Using means, the null hypothesis is that the mean distance from the target for the copier design is equal to the mean distance from the target for the towel design.

7. Each case will be *only one measurement*. Students should not put the two kinds of paper in the same row. Each collection will have 14 cases.

Q10 The split dot plot will look exactly like what they laid out with their tape measure in step 2, so their answer to this question will be the same as their answer to Q2.

9. Here, students may really need some help. For the example, if they chose the "mean of distance," their formula would be mean(distance). With this formula, every scramble gets the same statistic! What they really want is the *difference* of means:

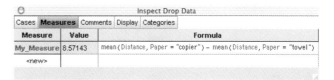

Because they haven't scrambled quite yet, they may feel that their value is fine. That is okay; they will then have to adjust their measure in Q11, because they will get the same value as before.

Q11 Make sure that students are checking the value of *My_Measure* in the scrambled collection. For the split dot plot in the activity, the value of *My_Measure* was 4 cm. Now that students have scrambled, if their value for *My_Measure* doesn't change, they'll need to fix their formula.

Q12 More than likely, the ranges will be very different. For example, the dot plot below shows the second scramble. It is quite different from the dot plot in step 11. The range there was from -6 to about 0.4, and here it is from about -6.5 to 8.

Q13–Q16 Answers will vary considerably depending on the students' data and test statistic. For the example in the activity, the value 8.57 was reasonably likely and not a rare event. There were 6 cases as extreme as or more extreme than the original difference of means, or 6% (using a two-tailed test would then increase this to a *P*-value of 0.12). So, although 8.57 seemed large, it wasn't large enough to reject the null hypothesis that there was not a difference between designs.

DISCUSSION QUESTIONS

- Why do you suppose the directions always say "at least as extreme" instead of "more extreme"? That is, why do ties count for the null hypothesis?

Inference for the Difference of Two Means: Unpaired—Orbital Express

continued

Activity Notes

- What makes a formula easy or hard to describe in Fathom? How might you express a hard formula more easily?
- Besides scrambling, how could you test whether the two papers were different? How is that method similar to the one we used here? What are the important differences between the two techniques?

EXTENSIONS

1. Have students test the claim that there is no difference between the two designs using a significance test in Fathom. They should check conditions, state the hypotheses, and write a conclusion in context. Here are the steps to use a test in Fathom:

   Drag a test object from the shelf. From the pop-up menu, choose **Compare Means.** Drag *Distance* from the case table to the top pane of the test where it says "First Attribute (numeric): unassigned." Drag *Paper* from the case table to the second pane from the top of the test where it says "Second Attribute (numeric or categorical): unassigned."

   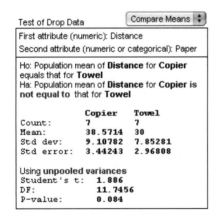

2. Have students construct a 95% confidence interval in Fathom to find an estimate for the true mean difference between the two designs. They should check conditions and write a conclusion in context. Here are the steps to create a confidence interval in Fathom:

   Drag a new estimate from the shelf. Choose **Difference of Means** from the pop-up menu and drag *Distance* and *Paper* into the top of the box, as you did with the test.

   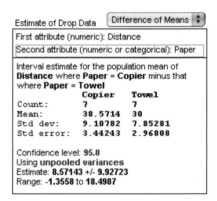

3. Ask, "Based on your work in Extensions 1 and 2, does your conclusion in Q16 change?"

4. Have students try to design a re-entry vehicle that's better than the existing designs. They should make a prototype and test it against the best "wad" design. They might also consider having a different researcher conduct the test.

Inference for the Difference of Two Means: Paired—Hand Spans

You will need
- centimeter ruler

The fictional detective Sherlock Holmes once amazed a man by relating "obvious facts" about him, such as that he had at some time done manual labor: "'How did you know, for example, that I did manual labour? It is true as gospel, for I began as a ship's carpenter.' Sherlock replied, 'Your hands, my dear sir. Your right hand is quite a size larger than you left. You have worked with it, and the muscles are more developed'" (from Sir Arthur Conan Doyle, *The Adventures of Sherlock Holmes*, ed. Richard Lancelyn Green (Oxford: Oxford World Classics, 1988)).

In fact, people's right hands tend to be bigger than their left, even if they are left-handed and even if they haven't done manual labor. But the difference is small, so to detect it you will have to design your study carefully.

COLLECT DATA

1. Measure your left and right hand spans in centimeters, to the nearest tenth. (An easy way to do this is to spread your hand as wide as possible, place it directly on a ruler, and get the distance between the tip of your little finger and the tip of your thumb.)

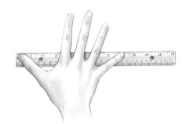

2. Record the data for each student in your class in a Fathom case table. Each case should represent one student with two attributes, *Right* and *Left*. Name the collection Hands.

3. Define a third attribute, *Individual_Diff*, with the formula Right−Left.

Q1 Make a graph of *Individual_Diff*. Describe the distribution in terms of shape, center, and spread.

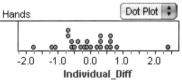

To find the standard error, use stdError() or s()/√count().

Q2 Use a summary table to find the mean, the standard deviation, and the standard error of *Individual_Diff*.

Q3 Does it matter that the right and left hand spans are paired? What do you think will happen to the standard error of the differences if the attributes are scrambled so that people's left hand spans are no longer paired with their right?

INVESTIGATE

Scrambling Paired Data

Next you will scramble the attributes so that a person's right hand span is no longer paired with their left hand span or with their individual difference.

Inference for the Difference of Two Means: Paired—Hand Spans
continued

4. Select the Hands collection and choose **Collection | Scramble Attribute Values**. This makes a new collection in which just the values for *Right* are scrambled.

5. Make a case table for the Scrambled Hands collection to see what happens when you scramble. Note that the formula for *Individual_Diff* is not carried over to the scrambled collection, so its values are the same as in the original case table.

6. Define a new attribute in the Scrambled Hands collection, *Difference*, with the formula Right–Left.

Q4 Make a graph of *Difference*. Describe the distribution in terms of shape, center, and spread. How does the distribution of *Difference* compare with the distribution of *Individual_Diff*?

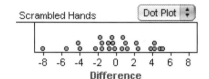

7. Drag the lower-right corner of the Scrambled Hands collection until you see the **Scramble Attribute Values Again** button. Click this button a few times and observe what happens in the case table. Check to see that the *Difference* updates but each person's individual difference does not.

Q5 Use a summary table to find the mean, the standard deviation, and the standard error for *Difference* in the scrambled collection.

> Make a new summary table, but don't drag an attribute to it. Instead, drag the name of the collection.

Q6 Go back to the unscrambled collection (Hands) and treat the left hand spans and right hand spans as two independent samples. Calculate the difference between the two sample means and the standard error of that difference using the formula

$$s_{\bar{x}_R - \bar{x}_L} = \sqrt{\frac{s_R^2}{n_R} + \frac{s_L^2}{n_L}}$$

Q7 Compare the standard errors from Q2, Q5, and Q6. Which is the smallest? Which two are closest to the same size? What bearing does the size of the standard error have on deciding whether there really is a difference in hand span?

Q8 Imagine making scatter plots of the data in your two collections, with *Left* on the horizontal axis and *Right* on the vertical axis. Sketch how you think they will look. Which would you expect to have the higher correlation? Why?

Assuming Independent Samples

Next you'll see the difference in the analyses if you assume that these data come from two independent samples rather than one sample of paired data.

8. Make the two scatter plots from Q8 and add a least-squares line to each.

Q9 Do the scatter plots look like what you predicted in Q8? How do they differ from your prediction? Which scatter plot has the higher correlation (in absolute value)?

Inference for the Difference of Two Means: Paired—Hand Spans
continued

Q10 Examine just the scatter plot for the Scrambled Hands collection. To use a significance test or construct a confidence interval to compare means, you must check whether the samples were independently selected or arbitrarily paired. Does your scatter plot show this? Explain.

The plus sign appears once you drag Left to the graph.

9. To check that the populations are approximately normally distributed, make a new graph. From the Scrambled Hands collection, plot *Right* on the horizontal axis. Then drag *Left* to the plus sign on the horizontal axis.

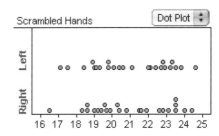

Q11 Are the conditions satisfied for a significance test for comparing two means?

*Choose **Difference of Means** from the pop-up menu.*

Q12 Drag a new estimate from the shelf and construct the confidence interval for the difference between the two means. Interpret the confidence interval in the context of the situation.

Q13 Examine just the scatter plot for the Hands collection. To use a significance test or construct a confidence interval for a paired comparison, you must check whether the samples were indeed paired. How does your scatter plot verify this? Explain.

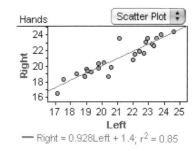

Q14 Are the differences in the Hands collection approximately normally distributed?

*Choose **Estimate Mean** from the pop-up menu.*

Q15 Drag a new estimate from the shelf and construct the confidence interval for the mean difference for the paired attributes. Interpret the confidence interval in the context of the situation. Is your conclusion here any different from your conclusion in Q12?

EXPLORE MORE

1. Carry out both the paired and two-sample *t*-tests for the measurements. Do the results here mirror the results you got with the confidence intervals? Is there a significant difference between a person's left and right hand spans?

2. Collect measures to study the distribution of the standard error of *Difference* in the scrambled collection. How likely is it that you would get a standard error as small as the original by scrambling?

3. Collect measures to study the distribution of the correlation between *Left* and *Right* in the scrambled collection. Interpret the meaning of this distribution.

Inference for the Difference of Two Means: Paired—Hand Spans

Activity Notes

Objectives

- Recognizing that paired data can greatly reduce variation over independent samples and produce a much more sensitive test (or estimate) of the true mean difference
- Recognizing whether the data production process results in two independent samples or one sample of paired observations
- Performing a test of significance and constructing a confidence interval for the difference between means from paired comparisons
- Deepening the understanding of these terms in the context of comparing means: paired data, dependence and independence of data values, and random sampling

Activity Time: 50–90 minutes (depending on data collection)

Setting: Paired/Individual Activity (collect data and use **HandSpanTemp.ftm**, or use **HandSpan.ftm**) or Whole-Class Presentation (use **HandSpanPresent.ftm**)

Materials

- One centimeter ruler per student

Statistics Prerequisites

- Familiarity with the language for testing a hypothesis
- Familiarity with the structure and procedure for testing a hypothesis
- Familiarity with the structure and procedure for constructing confidence intervals
- Independent versus dependent samples (correlation)
- Familiarity with the sampling distribution for the difference of two means

Statistics Skills

- Finding confidence intervals for the difference of two means (paired and unpaired)
- Checking conditions for testing the difference of two means (paired and unpaired)
- Comparing actual data to a hypothesized model and detecting differences between models
- Working with the sampling distribution for the difference of two means and the standard error
- Seeing that paired data can greatly reduce variation over independent samples and produce a much more sensitive test (or estimate) of the true mean difference
- *Optional:* Performing a hypothesis test for the difference of two means (paired and unpaired) (Explore More 1)

AP Course Topic Outline: Part I A–C, D (1–3); Part III B, C, D (5, 6); Part IV A (1, 6, 7), B (1, 4, 5)

Fathom Prerequisites: Students should be able to make graphs and use case tables, create attributes, and use summary tables, estimates, and tests. *Optional:* Students should be able to create measures (Explore More 2 and 3).

Fathom Skills: Students create a scrambled collection, use a summary table without attributes, and use hypothesis tests and interval estimates.

General Notes: This activity helps students see how pairing changes the standard errors of the differences between means when using paired differences, unpaired differences, and independent samples. It illuminates how important it is to pay attention to data production and, in the context of inference for means, helps students answer the key question "Paired data or independent samples?"

Procedure: If you have time, collecting their own data helps students see what paired data really means. Have the document **HandSpanTemp.ftm** open on one computer and be prepared to distribute the completed document for the activity. If you don't have time to collect data for your own class, you can use the sample data in the Fathom document **HandSpan.ftm** or you can use the document **HandSpanPresent.ftm** as a whole-class presentation. If you have your students use **HandSpan.ftm**, then they can proceed through the whole activity starting with step 3.

In **HandSpanPresent.ftm**, all the work is done, so students or you can click the **Scramble Attribute Values Again** button and see how the distribution of *Difference* changes as you rerandomize the attribute *Right*. Scroll down the document window to see the significance test and confidence intervals. Scroll right to see the sampling distribution of the standard error and the sampling distribution of the correlation for the scrambled collection (see Explore More 1 and 2).

Inference for the Difference of Two Means: Paired—Hand Spans
continued

Activity Notes

COLLECT DATA

Q1 The distribution is basically mound-shaped, with center about 0 and standard deviation about 0.8.

Q2 Summary tables are the natural tools for computing the statistics in this activity. The result below shows two equivalent methods for calculating standard error.

	Hands
Individual_Diff	-0.070833333
	0.83586933
	0.17062111
	0.17062111

S1 = mean( )
S2 = stdDev( )
S3 = stdError( )
$S4 = \frac{s(\ )}{\sqrt{count(\ )}}$

Q3 Yes, although students may not realize it at this point.

INVESTIGATE

Q4 The distribution is basically mound-shaped, with center about 0 and standard deviation about 3. The distributions are very similar except that the spread for *Difference* is much larger than the spread for *Individual_Diff*.

Q5 Scrambled Hands

Difference	-0.070833333
	3.0576318
	0.62413648

S1 = mean(Difference)
S2 = stdDev(Difference)
S3 = stdError(Difference)

Q6 Hands

	-0.070833333
	0.61666605

S1 = mean(Right) − mean(Left)
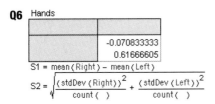

Q7 Students will probably notice a much lower standard error (reduced variation) for the differences from paired data in Q2 than for the differences from scrambled pairs or independent samples. Note that the differences after randomly ordering one set of the pairs (so that the data are no longer paired) give about the same standard error as would two independent samples, one from right hands and one from left. Paired data can greatly reduce variation over independent samples and produce a much more sensitive test (or estimate) of the true mean difference. The reduction in variation will be greatest when there is considerable variation between individuals but little variation in the differences from pair to pair.

Based on these sample results for paired data, it is not true that the right hand is bigger than the left, at least in terms of hand spans.

Q8–Q9 Students should expect that the left and right hand spans for the same person are highly correlated but that the scrambled pairs have no correlation. The scatter plots shown for Q10 (below) and Q13 (in the activity) support this hypothesis.

Q10 The scatter plot below shows the relationship between the left hand span and the right hand span for the arbitrary pairing of values from the scrambled Hands collection. The correlation is near 0, as it should be because the attribute *Right* has been randomly ordered. You would expect that your left hand tends to be about the same size as your right hand, but you wouldn't expect your left hand to be about the same size as the right hand of a randomly selected person.

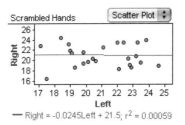

Q11 The dot plots of *Left* and *Right* show no reason to suspect that these data do not come from a population that is approximately normally distributed. The conditions appear to have been met except that this is not a random sample. So any conclusions will have to be about this population only.

Q12 A 95% confidence interval ($df = 23$) for the difference between the mean hand spans for *Left* and *Right* is $-1.31 < \mu(Right) - \mu(Left) < 1.17$. This interval

Inference for the Difference of Two Means: Paired—Hand Spans

Activity Notes — continued

overlaps 0, so there is no reason to suggest that one hand span is larger than another in this population.

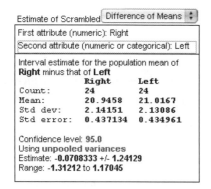

Q13–Q15 The scatter plot in Q13 shows the relationship between the left and right hand spans. This plot shows that the left and right hand spans for the same person are highly correlated—as it should, because these are dependent measurements. The two-sample *t*-procedure is not a valid option. The dot plot of the individual differences does look like these data could have come from an approximately normal distribution. Here are the summary statistics for the observed differences.

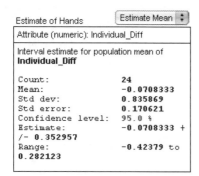

With 23 degrees of freedom, the confidence interval estimate of the true mean difference is −0.424 to 0.282. Any value of the true mean difference between left and right hand spans in this interval could have produced the observed mean difference as a reasonably likely outcome. This interval includes 0, so there is not sufficient evidence to say that the hand spans are different in this population. Although this interval overlaps 0, note that it is shorter than the confidence interval in Q12.

The conclusions are not different, but we get a better estimate for the true mean difference in hand spans for this population from the paired data.

DISCUSSION QUESTIONS

- What is the relationship between the scrambled pairs and the independent samples?
- What does the high correlation between *Left* and *Right* in the original data have to do with the problem of determining whether there really is a difference in hand spans?

EXPLORE MORE

1. The results here will mirror the confidence intervals. For the two-sample *t*-test, results will vary depending on the randomization of *Right*. The *P*-value, however, will be high in all cases.

2. Here is a plot of 100 standard errors for the scrambled collection. The mean is 0.612 and the standard deviation is 0.062. Note that none of these 100 measures were even close to 0.171.

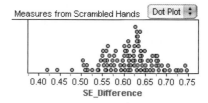

3. Here is a plot of 100 correlations for the scrambled collection. The mean is about 0 and the standard deviation is 0.22. Note that none of these 100 measures were even close to 0.92, the correlation of the sample data.

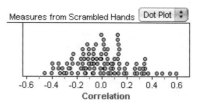

7
Chi-Square Tests

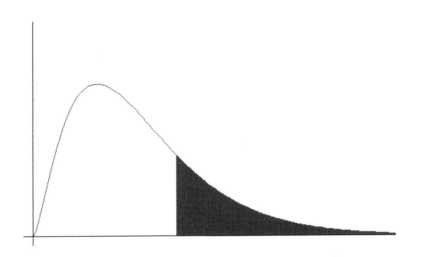

Measuring Fairness—Constructivist Dice

You will need
- die
- marker
- materials for making your own die
- **Constructivist Dice.ftm**

In this activity you will make your own die. There's no reason to believe your die will be fair, especially because it won't be a perfect cube. But is it really *that* unfair compared to real dice? What does "unfair" mean?

COLLECT DATA

1. Make your die. Here are some options:

   Clay: Mold your die by hand and then make the dots with a toothpick or pencil.

   Wood: Cut not-quite-cubes off the ends of strips (whose cross section is not quite square) with hand saws and then make the dots with permanent markers.

Fair die: When the die is rolled, each face has an equally likely chance of occurring.

Q1 Do you think your die is fair? Explain your answer. How could you figure out if your die is fair or unfair?

2. Roll your die 60 times and record the number of ones, twos, and so on that occur (the *observed frequencies*). Sketch a histogram of your results.

Q2 Do you think your die is fair now? Explain your answer. What in your data convinces you that your die is fair or unfair?

Q3 If you rolled a fair die 60 times, what would you expect to see or have happen? (These results are called the *expected frequencies*.)

3. Develop formulas for two test statistics that measure how unfair the die is. The statistics should be small if the die is close to fair, and larger the more unfair it is.

Q4 As a class, choose one test statistic. Then, calculate that statistic for the data you collected in step 2. Does the statistic fit the rule given in step 3?

Q5 Does the statistic show that your die is fair? If it does, how small would it have to be before you were no longer convinced? If it does not, how large would it have to be before you were convinced?

INVESTIGATE

Rolling a Fair Die

Now you'll look at what happens when you roll a fair die.

4. Roll the fair die 60 times and record the results.

Use the test statistic your class chose in Q4.

5. Calculate the test statistic for your rolls of a fair die and plot your value for the test statistic on a dot plot with the other members in your class.

Measuring Fairness—Constructivist Dice
continued

Q6 Compare the test statistic that you got in Q4 for the 60 rolls of your homemade die to your class's distribution. Is it far out on the tail, or is it in the pack?

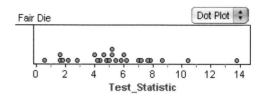

Q7 Sketch your class dot plot. Mark the position of the test statistic for the homemade die on the plot. How often did the value of your test statistic appear in the plot? How many values were greater than your value? How many values were less? Is this evidence that the die is unfair?

Q8 Based on the plot, what is the probability that a fair die would have a test statistic as extreme as or more extreme than your test statistic? Is this probability low enough to convince you that the die is not fair?

Simulating a Fair Die

Each person in your class computed the chosen test statistic for 60 rolls of a fair die, so depending on your class size, you might have 20 to 30 statistics. But to really analyze the situation, you should collect data many more times.

6. Open the Fathom document **ConstructivistDice.ftm.** You'll see a collection of 60 fair dice.

7. Make a histogram of *Face* and compare it to the histogram you made in step 2 for your homemade die.

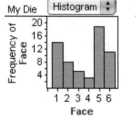

8. Click **Rerandomize** to roll the 60 dice. You can see the values of the fair dice change after each roll. Observe the histogram and try rerandomizing several times to get a histogram that looks like the histogram for your homemade die.

Q9 Were you able to find a histogram that looked like the histogram for your homemade die? If so, about how many times did you have to rerandomize?

You may want to use the other measures: count_1, count_2, and so on, in your formula. Be sure to check the formula for one set of fair-dice data.

Q10 Show the inspector and go to the **Measures** panel. Currently, the measure *HowUnfair* is defined as 0. Define the measure using the test statistic that you chose in Q4. What is your formula for *HowUnfair*?

Measuring Fairness—Constructivist Dice
continued

Q11 Click **Rerandomize** to get 10 values for your measure from different sets of random dice. Write them down. Compare those 10 values from fair dice to the value of your measure for your homemade die. What's your first impression: Is your die fair or not? Why?

Collecting Measures

Now you'll collect many samples of 60 dice.

9. Select the Fair Dice collection and choose **Collection | Collect Measures.** A new collection, Measures from Fair Dice, appears.

You'll see five values. By default, Fathom collects five measures.

10. Make a dot plot for *HowUnfair* from the Measures from Fair Dice collection.

11. You want more sets of 60. Show the measures collection's inspector and go to the **Collect Measures** panel. Uncheck Animation on, change to 95 measures, then click **Collect More Measures.** You should now have 100 measures.

*Select the graph and choose **Graph | Plot Value.***

12. You'd like to see where your test statistic for your homemade die appears in the distribution. Change the *HowUnfair* plot to a histogram and plot the value you got in Q4.

Q12 Sketch your histogram, including the plotted line.

Hold down the Shift key as you click each bin.

13. Drag the edge of a histogram bin so that one edge of a bin lines up with your statistic. Select all the bins outside the line, that is, farther than the line from the center of the distribution. Hold the mouse over the measures collection. Look in the status bar at the bottom of the Fathom window to see how many measures are selected.

Q13 What percentage of the measures are as extreme as or more extreme than your test statistic? Based on the original 60 rolls of your homemade die, and simulating 100 runs of rolling 60 fair dice, do you think your die is substantially different from a fair die? Explain.

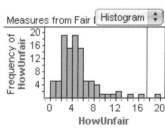

Measuring Fairness—Constructivist Dice

Activity Notes

Objectives

- Assessing a test situation by comparing it to a distribution of measures generated by a fair die. The fair die represents the null hypothesis—there's nothing unusual going on.
- Understanding that even with fair dice, there can be considerable variability. The expected frequencies, 10–10–10–10–10–10, are unusual. And even a fair die can look unfair in a set of 60 rolls.
- Realizing that the probability found (the *P*-value) is the chance that a fair die would give a result as extreme as or more extreme than the test statistic and that it is *not* the probability that the die is fair.

Activity Time: 80–100 minutes (40–50 minutes to make the dice and 40–50 minutes for the activity)

Setting: Paired/Individual Activity (make dice, collect data, combine data as a class, use **ConstructivistDice.ftm**)

Materials

- One fair die per student
- Marker
- Materials for making a die (clay or wood will work)

Statistics Prerequisites

- Familiarity with the idea of testing a hypothesis
- Familiarity with a test statistic when testing a hypothesis
- Familiarity with sampling distributions and reasonably likely outcomes
- Definitions of probability and equally likely outcomes

Statistics Skills

- Finding *P*-values on a plot to test a claim
- Comparing a test statistic to a distribution based on a fair die
- Creating a test statistic and then using it to make a decision
- Probability simulation
- Working with the definitions of probability and equally likely outcomes
- Comparing actual data to a hypothesized model and detecting differences between models

AP Course Topic Outline: Part I E (1, 4); Part III A (1, 4, 5), D (6, 8); Part IV A (1, 2), B (1, 6)

Fathom Prerequisites: Students should be able to make graphs, create measures, use the formula editor, and use inspectors.

Fathom Skills: Students write formulas, do simulations, collect measures to compare models, and make a graph of the test statistic's distribution.

General Notes: In this activity, students make their own die. There's no reason to believe these dice are fair, especially because they won't be perfect cubes. Students will answer the questions "Are the dice really *that* unfair—especially compared to real dice?" and "What does 'unfair' mean?"

Procedure: Each student should create her or his own die for this activity. This can be done days in advance. Here are a few ideas:

- From an art supply store, get clay that will harden when baked. Have students mold their dice, making the dots with a toothpick or pencil point. This clay comes in many terrific colors; perhaps you could bake it in a food lab at school or bring a toaster oven to class.
- Cut (or have someone cut) wood into strips whose cross section is not quite square. Then have students cut not-quite-cubes off the ends of these strips with hand saws. (Get help supervising students. Wood shop is one possibility.) Make the dots with permanent markers.

Always be careful about safety. And whatever scheme you choose, practice first!

After the dice are made, students can roll them in class or at home, depending on your time frame for the activity. The histogram they make is used in step 8, so have them keep it handy.

COLLECT DATA

Q1–Q2 Students will probably not believe that their dice are fair because they aren't perfect cubes.

Q3 Generally students expect to see each face come up about 10 times, though they should realize that this distribution will vary.

Measuring Fairness—Constructivist Dice
continued

Activity Notes

3. Encourage students to come up with at least two different statistics. Share the various statistics so that the class can choose one test statistic to work with. It is unlikely that anyone will choose the chi-square test statistic at this point. The most common statistic is the sum of the absolute differences from what we expect in a fair die.

Q4–Q5 One homemade die had this distribution, with a chi-square goodness of fit test statistic of 17.6.

My Die							
	Face						Row Summary
	1	2	3	4	5	6	
	14	8	5	3	19	11	60

S1 = count( )

INVESTIGATE

5. After this step, it is a good idea for the class to evaluate the test statistic. Does it still appear to be a reasonable test statistic? In other words, does it seem to be a measure or index of how far away the observed frequencies are from the expected frequencies? If not, have them choose another test statistic and redo Q4 and step 5 until they are satisfied with the class's test statistic.

Q6–Q8 For the sample test statistic and dot plot in the activity, 17.6 is way out of the range of the data, so we would feel more strongly that the homemade die is unfair. There were 25 dots in the plot, so the 95% dividing line would be somewhere between 10.25 and 14. So, anything below 10.25 would be reasonably likely and anything in this range or above would be considered suspect. For the 17.6, there are no values as extreme or more extreme, and thus, the probability according to the plot would be 0.

A more conservative ratio for probability—for more sophisticated students—is the number greater than or equal to the test statistic *plus 1* divided by the population *plus 1*. That is, it includes the test die as a candidate for fairness.

Quite uncubical dice can give results consistent with fairness. This should disturb students who are paying attention.

Q9 It is quite possible for students to find a histogram that looks like their histogram.

Q10 Students may need help writing the formula for the measure. One possibility is to add the absolute differences between the counts and 10 (the expected value). The formula would then be

$$|count\_1-10|+|count\_2-10|+|count\_3-10|+|count\_4-10|+$$
$$|count\_5-10|+|count\_6-10|$$

Q11 Each of the five measures in the sample dot plot in step 10 is well below the test statistic of 17.6, so the homemade die in the example still looks unfair.

Q12–Q13 A sample histogram appears in the activity. For that example, there was 1 measure out of 100 that was as extreme as or more extreme than 17.6. So the proportion (and *P*-value for this plot) would be 0.01, and we would reject the claim that the die is fair and conclude that there is evidence to support the claim that the die is different from a fair die.

DISCUSSION QUESTION

When we have a distribution of statistics, and we compare it to the test statistic, why do we calculate the proportion of that distribution which is greater than *or equal to* the test statistic and use that as our *P*-value, instead of simply greater than? [If we're making the case that our test statistic is unusual, it's only fair to include the "equal" ones in the proportion. That is, they "count" as being examples of naturally-occurring "that extreme" trials in the fair-dice situation.]

EXTENSIONS

1. One possible statistic is the sum of the absolute differences between the observed frequency and the expected frequency for each possible die roll. For example, if the die rolled 20–0–10–10–10–10, the value would be 20.0 because you compare those numbers to 10–10–10–10–10–10 and add the (absolute) differences. If you roll a die 60 times, the value for this measure is always an even integer. Why? [Both the expected frequencies and the observed frequencies sum to 60, and the expected counts are 10. If we roll, for example, 7 fours, then somewhere we have to pick up those 3 deficient rolls (because the sum of the observed frequencies is constant—60). So, in one instance we'll be down 3 and somewhere

else, we'll be up 3. Because we are using the absolute values, we'd get 3 + 3, or 6.]

2. If you didn't have a computer, you could still assess whether a set of rolls was reasonably fair if you knew what value of *HowUnfair* (for a particular measure) gave what *P*-value. To find this critical value, have students pick a test statistic (their own, for example) and pick a significance level that they find convincing. Use **ConstructivistDice.ftm** to make a distribution of the test statistic for samples of 60 die rolls. Figure out where the dividing line is. For example, if the significance level is 5%, find the line where only 5% of the measures of 60 rolls are that extreme or more extreme. That's the *critical value* for the test statistic and significance level. What is that value?

Take someone else's die and roll it 60 times. Calculate the same test statistic. Compare it to the critical value. Explain what the result means. [These will vary depending on the student's measure and significance level. For a 5% significance level, the chi-square critical value is 11.07.]

3. If students found the critical value in Extension 2, they can change their formula for *HowUnfair* so that it works for any number of rolls. For example, if you used the number 10 as an expected number of ones in 60 rolls, use count()/6 instead. What if there is a different number of rolls in the set? Repeat Extension 2, at least for 18, 30, and 120 rolls. Do you find that the critical value of *HowUnfair* that you got for 60 rolls gives the same significance level? What's the relationship? Have students compare their findings with those of others in the class, especially if they have studied different measures. [These will vary depending on the student's measure and significance level. However, no matter what they found in Extension 2, for most measures that students tend to pick, sample size will make a difference and there will be a different critical value for every sample size. The relationship will depend on their choice of measure as well. Some relationships may have formulas and others, like the chi-square, will be independent of the sample size and will depend only on the number of categories.]

CHI-SQUARE CONNECTION

The data in this activity are appropriate for a chi-square goodness-of-fit test. The expected value for each possible face is 10; if x_i is the number of times the face I came up, calculate chi-square by adding

$$\chi^2 = \sum \frac{(observed - expected)^2}{expected} = \sum_{i=1}^{6} \frac{(x_i - 10)^2}{10}$$

with 5 degrees of freedom. For example, if the data were 14–8–5–3–19–11, you'd get

$$\chi^2 = \frac{(14-10)^2}{10} + \frac{(8-10)^2}{10} + \frac{(5-10)^2}{10}$$
$$+ \frac{(3-10)^2}{10} + \frac{(19-10)^2}{10} + \frac{(11-10)^2}{10}$$
$$= \frac{16}{10} + \frac{4}{10} + \frac{25}{10} + \frac{49}{10} + \frac{81}{10} + \frac{1}{10} = \frac{176}{10} = 17.6$$

Is a chi-square of 17.6 unusual for 5 degrees of freedom? The *P*-value is 0.0035, so the set of rolls looks quite unusual. Here are the test and a plot of the distribution. You cannot see the shading for the *P*-value because 17.6 is too far out in the tail.

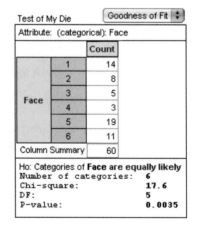

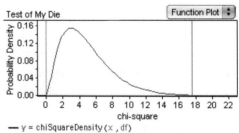

Chi-Square Goodness-of-Fit Test—Plain M&M's

You will need
- two small bags of plain M&M's
- PlainM&Ms.ftm

The way that M&M's are put into the bags makes each bag a random sample of all M&M's of that type. The M&M's, with colors already on them, are mixed in these proportions in a very big vat before they are put into the bags.

The percentages of almond M&M's are 20% blue, 10% brown, 20% green, 20% orange, 10% red, and 20% yellow. Do these same percentages apply to plain (milk chocolate) M&M's?

COLLECT DATA

1. Pour out a sample of 75 plain M&M's and count the number of each color.

2. Make a table of the observed number of plain M&M's of each color and calculate the expected number of plain M&M's of each color.

Q1 State how this scenario meets the criteria for a chi-square goodness-of-fit test.

Q2 State the null and alternative hypotheses.

Q3 Calculate the χ^2 test statistic. What is your value for χ^2? Based on your evidence so far, do you think the percentages that apply to almond M&M's also apply to plain M&M's? Explain.

INVESTIGATE

If the observed and expected counts were equal, χ^2 would be 0 and you would have no reason to doubt that the same percentages apply to almond and plain M&M's. If the observed and expected counts were very far apart, χ^2 would be large and you would have evidence to reject the claim that the same percentages apply to almond and plain M&M's. How can you assess whether a value of χ^2 is large enough to reject the null hypothesis? You need to see how much variation there is in the value of χ^2 when the distribution of colors in plain M&M's is the same as in almond M&M's.

3. Open the Fathom document **PlainM&Ms.ftm.** You'll see a collection containing 75 cases and a summary table.

4. Show the inspector and go to the **Cases** panel. Look at the formula for *Color*. It randomly assigns each M&M a color according to the given probability distribution, although you'll see the cumulative probabilities instead.

A switch statement acts like a sophisticated "if" statement. First, Fathom generates a random number between 0 and 1. It then substitutes that random number for the ? in each of the expressions, starting at the top, until it finds the *first one* that is true. Then it returns the value to the right of the colon for that expression. So, if

Chi-Square Goodness-of-Fit Test—Plain M&M's
continued

the random number is 0.45, the color green is chosen because "Green" is the *first* category where the statement is true.

> These are the expected percentages and counts for each color or category. They are not the expected percentages and counts for a particular case. You should see the same percentage and count next to every case that is the same color.

Q4 Look at the formula for *Exp_Percentage*. Explain how this switch statement assigns values. What different, simpler formula could have been used? Why do you think that formula was not used?

5. Double-click in the formula cell for *Exp_Count_pc* and write a formula that will calculate the expected frequency for each color. Make a case table to check if these values agree with the expected values you calculated in step 2.

6. Make a bar chart that shows the frequency of each color.

> Double-click the formula to edit it.

7. Make a second bar chart, but change the formula from count() to a formula that shows the difference between the observed count and the expected count.

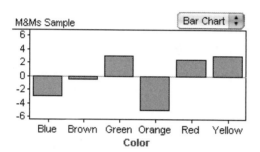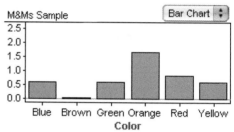

8. Finally, make a third bar chart with a formula that shows each bar's contribution to χ^2.

Q5 Click the **Rerandomize** button on the M&Ms Sample collection several times. Observe the relationship between the three bar charts. Keep rerandomizing until you get a sense of the relationship. Describe that relationship.

Now you'll collect some values for χ^2. You'll do this by using a summary table.

9. Look at the summary table. It is just like the one you created in steps 2 and 3, with all the calculations you need to find the chi-square value, only this time, it is set up for any random sample.

> You could use max(Exp_Count), min(Exp_Count), and other such functions too. Why?

Q6 In the summary table, S2=mean(Exp_Count_pc) is the expected count for each color. Why isn't the formula simply S2=Exp_Count_pc? (Try it.) Why does this happen?

You'd like to make the cells of this table a collection.

> When you collect the cells, the original collection will rerandomize again.

10. Select the summary table and choose **Summary | Create Collection From Cells**. Fathom creates a new collection named Cells from M&Ms Sample Table. Show the inspector and browse through the cases.

298 | 7: Chi-Square Tests

Chi-Square Goodness-of-Fit Test—Plain M&M's
continued

Each case in this new collection corresponds to a single cell (color) in the summary table. Each of the three formulas in the summary table corresponds to one attribute in this new collection. *S1* is the observed count, *S2* is the expected count, and *S3* is the term of the chi-square statistic for each color. You're interested in the sum of the values for *S3*.

11. Go to the **Measures** panel and create a new measure as shown. This number is the chi-square statistic.

    Now you are ready to collect measures of *Chi_Square*.

12. Select the Cells from M&Ms Sample Table collection and choose **Collection | Collect Measures.** Fathom will rerandomize the original collection 5 times, collecting the measures *Chi_Square* and putting them in a new collection, labeled Measures from Cells from M&Ms Sample Table.

13. Show the inspector for the new collection. On the **Collect Measures** panel, uncheck Animation on and collect 95 more measures.

Choose **Graph | Plot Value.**

14. Display the distribution of *Chi_Square* in a histogram and plot the χ^2 value you got in Q3.

Q7 Describe the distribution of *Chi_Square* in terms of shape, center, and spread.

Hold down the Shift key as you click each bin, or draw a selection rectangle around the bins that you want. Hold the mouse over the measures collection. Look in the status bar at the bottom of the Fathom window to see how many measures are selected.

15. Drag the edge of a histogram bin so that one edge of a bin lines up with your statistic. Select all the bins outside the line, that is, farther than the line from the center of the distribution.

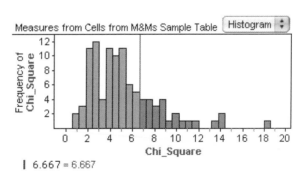

Q8 What percentage of the measures are as extreme as or more extreme than your test statistic—in other words, what is your *P*-value with this simulation?

Q9 Do you think that the same percentages apply to almond and plain M&M's? Explain.

Chi-Square Goodness-of-Fit Test—Plain M&M's
continued

EXPLORE MORE

Follow these steps to use Fathom to do a goodness-of-fit test with your observed counts and see how close your *P*-value from the simulation is to the *P*-value in the test.

a. Drag a new test from the shelf. From the pop-up menu, choose **Goodness of Fit.**

b. Edit the attribute name (in blue) to be *PlainM&Ms*. Enter the number of colors or categories (in this case, 6). The table will create six rows.

c. Change the row category names from their defaults to "Blue," "Brown," "Green," "Orange," "Red," and "Yellow." Enter your observed frequencies in the appropriate row in the column Count.

Notice that the test is, so far, assuming equal likelihood.

d. Click on the phrase "are equally likely" and choose **have probabilities given above** from the pop-up menu. Notice that the table gets a column for probabilities.

e. Enter the appropriate probabilities for each row. The chi-square statistic and associated *P*-value are displayed below the table.

f. With the test selected, choose **Test | Show Test Statistic Distribution** to display a plot of the chi-square distribution with shading to show the probability of getting a statistic as extreme as or more extreme than the one observed.

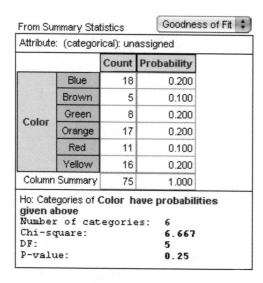
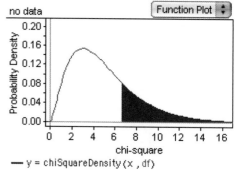

Chi-Square Goodness-of-Fit Test—Plain M&M's   Activity Notes

Objectives

- Becoming familiar with calculating the χ^2 statistic and its distribution
- Finding expected values and degrees of freedom
- Performing a chi-square goodness-of-fit test and understanding that it answers the question "Does this look like a random sample from a population in which the proportions that fall into these categories are the same as those hypothesized?"

Activity Time: 40–50 minutes

Setting: Paired/Individual Activity or Whole-Class Presentation (for either, collect data and use **Plain M&Ms.ftm** to build simulation)

Optional Document: ChiSquarePresent.ftm (see Extension)

Materials

- Two small bags of plain M&M's per student or 3–4 large bags of plain M&M's per class

Statistics Prerequisites

- Familiarity with the language for testing a hypothesis
- Familiarity with the structure and procedure for testing a hypothesis
- The test statistic for a chi-square goodness-of-fit test
- Familiarity with sampling distributions and reasonably likely outcomes
- Definitions of probability and equally likely outcomes

Statistics Skills

- Checking conditions for the chi-square goodness-of-fit test
- Performing a chi-square goodness-of-fit test
- Finding *P*-values and relating them to the plot of the simulated chi-square distribution
- Probability simulation
- Working with the definition of probability and with a probability distribution that does not have equally likely outcomes
- Comparing actual data to a hypothesized model and detecting differences between models

AP Course Topic Outline: Part I E (1, 4); Part III A (1, 5), D (8); Part IV B (1, 6)

Fathom Prerequisites: Students should be able to make graphs, create attributes and measures, use the formula editor, and use inspectors.

Fathom Skills: Students use random generators to create a collection that represents a random sample, work with switch statements, create a collection from the cells of a summary table and collect measures from that collection to compare models, and create a formula to calculate the chi-square test statistic. *Optional:* Students use the hypothesis test tool and make a graph of the test statistic's distribution (Explore More).

General Notes: This activity explores essential concepts in using a goodness-of-fit test and walks students through how to use Fathom to do a goodness-of-fit test when the categories are not equally likely—first with simulation and then using Fathom's built-in test. This activity is designed to be suitable as an introductory activity; that is, you don't have to know much about Fathom.

Procedure: It's important not to skip the hands-on Collect Data section because it gives students firsthand experience calculating and understanding χ^2. Students use two bags of plain M&M's to get a sample of size 75. They could do this either by using up the first bag (which has about 55 M&M's) and then getting the rest from the second bag or by choosing from both bags. You can also do this first section as a class with one large bag of plain M&M's.

Students start with a partially constructed Fathom document, largely because they are dealing with non–equally likely outcomes and that requires a few "fixes." With equally likely outcomes, the expected value is just count()/n, where *n* is the number of categories, but with non–equally likely outcomes, they need different probabilities for the different colors.

The Explore More section walks students through the procedure for entering their observed counts directly into a goodness-of-fit test. It is highly recommended.

COLLECT DATA

1. The expected number of M&M's for blue, green, orange, and yellow is 15; the expected number of M&M's for brown and red is 7.5.

Chi-Square Goodness-of-Fit Test—Plain M&M's
continued

Activity Notes

Q1 The conditions are met in this situation for a chi-square goodness-of-fit test. There were 75 outcomes in a random sample. Each outcome was only one color. You can compute the expected number of each color because you know the distribution of colors in a bag of almond M&M's. All of the expected counts were at least 5.

Q2 H_0: The distribution of colors in plain M&M's is the same as in almond M&M's. That is, there are 20% blue, 10% brown, 20% green, 20% orange, 10% red, and 20% yellow.

H_a: The distribution of colors in plain M&M's is not the same as in almond M&M's.

Q3 $\chi^2 = \sum \frac{(observed - expected)^2}{expected}$ with $6 - 1$, or 5 degrees of freedom.

For example, if the data were 18–5–8–17–11–16,

$$\chi^2 = \frac{(18-15)^2}{15} + \frac{(5-7.5)^2}{7.5} + \frac{(8-15)^2}{15}$$
$$+ \frac{(17-15)^2}{15} + \frac{(11-7.5)^2}{7.5} + \frac{(16-15)^2}{15}$$
$$= \frac{9}{15} + \frac{6.25}{7.5} + \frac{49}{15} + \frac{4}{15} + \frac{12.25}{7.5} + \frac{1}{15} = 6.67$$

Here, 6.67 doesn't seem very big, so students would probably conclude that they could not reject the claim that the distributions of colors were the same.

INVESTIGATE

Q4 The switch statement assigns each case the expected percentage of the color. Fathom compares the value of *Color* with each of the colors in quotes and returns the expression to the right of the first color that matches. If none match, it returns the value of the expression to the right of **else**. So, for example, if *Color* is "Green," then it returns the probability of "Green," which is 0.2.

5. The formula is Exp_Percentage·count().

7. The formulas count()−mean(Exp_Count_pc) and count()−Exp_Count_pc both work.

8. These formulas both work:
(count()−mean(Exp_Count_pc))²/mean(Exp_Count_pc)
   (count()−Exp_Count_pc)²/Exp_Count_pc

Q5 The first bar chart shows the actual observed counts for each color. The second shows how much each category is either over or under the expected value. The third bar chart shows the term of the chi-square statistic for that color. For example, for the graphs in the activity, "Blue" has 12 cases: This is 3 cases less than expected, so the second bar chart's "height" is at −3. The term for "Blue" in the chi-square statistic is 9/15, or 0.6, which is the height for "Blue" in the third bar chart.

Q6 Exp_Count_pc is the expected value for each color. Notice that each case gets this value, so the formulas for the summary table need a summary formula that gives just one of these values. Because all of these values are the same for a particular color, any summary measure will work because they all have the same value. For example, if you look in the case table at the cases that are "Blue," all of them have Exp_Count_pc = 15. So, the mean of this attribute for "Blue" is 15, as are the max, min, and so on. If students use just the attribute name in the formula for a summary table or a measure, they will get a message that says "Case attributes here must occur inside functions such as max or mean."

Q7 The histogram of 100 measures should be skewed right. Centers and spreads will vary. The histogram in the activity has median 4.6 and *IQR* 6.7 − 2.93, or 3.77. (*Note:* The mean is 5.25 with SD 3.13.)

Q8–Q9 For the histogram in the activity, the number selected was 25, so the *P*-value from the simulation is 0.25. Therefore, in this instance, we do not have enough evidence to reject the null hypothesis. The difference in the expected and observed counts can be attributed to variation in sampling alone. A *P*-value this small is reasonably likely to occur in random samples of this size if plain M&M's have the same distribution of colors as almond M&M's. So, we cannot conclude that the distribution of colors in plain M&M's is different from that in almond M&M's. (*Note:* It turns out that the distributions are different. The actual percentages are 24% blue, 13% brown, 16% green, 20% orange, 13% red, and 14% yellow.)

Chi-Square Goodness-of-Fit Test—Plain M&M's
continued

Activity Notes

EXTENSION

There is a presentation document, **ChiSquarePresent.ftm**, for the chi-square distribution and the goodness-of-fit test for equally likely outcomes (rolls of a die). It shows you the results for 100 measures for chi-square when rolling a fair six-sided die 60 times. The simulation uses the same process as the activity although, because the outcomes are equally likely, it is much easier to set up. You can change the number of sides of the die by changing the slider. The options are 4, 5, 6, 8, 10, 12, and 20 sides. After you change the slider, go to the **Cases** panel in the Dice collection, select the attribute *Face*, and at the bottom of the inspector change the category set to the appropriate set. For example, suppose you want to see how the chi-square distribution is different for a 12-sided die. Drag the slider to 12. Then, in the inspector, change **Six_Set** to **Twelve_Set**. Then click **Collect More Measures** to generate a distribution for chi-square with 11 degrees of freedom.

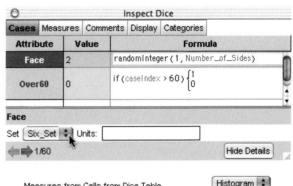

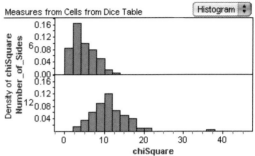

Chi-Square Test of Independence

You will need
- ChiSquareTest.ftm

The definition of independence of two variables is quite rigid. It requires $P(A \text{ and } B)$ to be exactly equal to $P(A) \cdot P(B)$ for *every* cell in a two-way table. This rarely happens with real data, even if there is no reason to believe the two variables are associated in any way. In this activity you'll work with the chi-square test of independence, which is much more useful when you are working with a random sample.

COLLECT DATA

1. Determine which eye is your dominant eye. Hold your hands together in front of you at arm's length. Make a space between your hands that you can see through. Through the space, look at an object at least 15 feet away. Now close your right eye. Can you still see the object? If so, your left eye is dominant. If not, your right eye is dominant. You will collect data for your class on their gender, whether the last digit of their phone number is even or odd, their dominant hand, and their dominant eye.

Q1 In step 2, you will count the number of females and males in your class. You will also ask if the last digit of their phone number is even or odd. Do you think that the categorical variables *Gender* and *Phone_Number* are independent? Explain.

Q2 Do you think the variables *Handedness* and *Eyedness* are independent? That is, if you know a person is right-handed, does that change the probability that the person is right-eyed? Explain.

To add many cases at once, choose **Collection | New Cases.**

2. Open the Fathom document **ChiSquareTest.ftm.** Collect the data described above in the empty Students collection. Make a two-way summary table and a ribbon chart for *Gender* and *Phone_Number*.

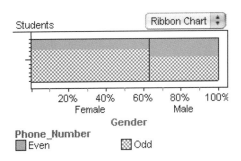

Exploring Statistics with Fathom
© 2007 Key Curriculum Press

Chi-Square Test of Independence
continued

> Two events, *A* and *B*, are independent if one of these statements is true:
> $P(A) = P(A|B)$,
> $P(B) = P(B|A)$,
> $P(A \text{ and } B) = P(A) \cdot P(B)$.

3. Using the definition of independence, determine if *Gender* and *Phone_Number* are independent events in the selection of a student at random from your class.

Q3 What can you conclude about the independence of the two variables in the population of all students based on your answer in step 3? Explain.

Q4 What can you conclude about the independence of the two variables in the population of all students based on your ribbon chart? Explain.

Q5 Even if *Gender* and *Phone_Number* are independent events in the population of all students, why are you likely to find that this is not the case in your class?

4. Make another two-way summary table and a ribbon chart for *Eyedness* and *Handedness*.

5. Using the definition of independence, determine if *Eyedness* and *Handedness* are independent events in the selection of a student at random from your class.

Q6 What can you conclude about the independence of the two variables in the population of all students based on your answer in step 5? Explain.

Q7 What can you conclude about the independence of the two variables in the population of all students based on your ribbon chart? Explain.

> Select the summary table and choose **Summary | Add Formula**.

6. For a chi-square test of independence, besides needing a random sample (which this survey is not), the expected values must be at least 5. Add the formula Expected to your Hand-Eye summary table.

Q8 Can you proceed with a chi-square test of independence? Explain.

INVESTIGATE

Testing for Independence

More than likely, in a class of 30 or fewer, there will be few left-handed people and your expected value in at least one cell will be less than 5. So you'll continue your investigation by looking at a random sample of 100 students.

7. Scroll down to the Random Students collection.

Q9 What can you conclude about the independence of the two variables in the population of all students based on the ribbon chart? Explain.

Q10 State how this scenario meets the criteria for a chi-square test of independence. Then state the null and alternative hypotheses.

Chi-Square Test of Independence
continued

*Choose **Test | Verbose** to see a compact version of the test.*

8. Drag a test object from the shelf and choose **Test for Independence** from the pop-up menu. From the Random Students collection, drag *Handedness* and then *Eyedness* to the top portion of the test.

If the test statistic is large and the P-value small, you may need to rescale your graph.

9. With the test selected, choose **Test | Show Test Statistic Distribution.** The shaded area under the right portion of the curve corresponds to the *P*-value for the observed chi-square statistic.

Q11 What can you conclude about the independence of *Handedness* and *Eyedness* in the population of all students based on this hypothesis test? State your conclusion in the context of the situation.

Forcing Independence by Scrambling

The null hypothesis states that there is no relationship between the two attributes *Handedness* and *Eyedness*. What if you were to take all the values for the attribute *Eyedness* and scramble them so that "Left" and "Right" got reassigned randomly to each case? Any relationship that might exist between the two attributes would be wiped out by the scrambling. Any remaining relationship would have to be due to chance alone.

10. Select the Random Students collection. Choose **Collection | Scramble Attribute Values.** Make a case table for the new Scrambled Random Students collection.

11. By default, Fathom scrambles the first attribute in the collection, which is *Gender*. You need to scramble a different attribute. Show the scrambled collection's inspector. Go to the **Scramble** panel and choose **Eyedness** from the pop-up menu.

12. Make a ribbon chart for the scrambled collection. Then click the **Scramble Attribute Values Again** button a few times. You should see the values in the *Eyedness* column of the case table change each time. As you scramble, you can see the variation in the relative proportions. This variation is due solely to chance.

13. Make a new test for independence. This time, drop *Handedness* and *Eyedness* from the scrambled collection into the test. Each time you scramble again, the chi-square statistic and the *P*-value are recomputed. Because you're simulating the conditions of the null hypothesis, the chi-square values will not be very large, and the *P*-values will not be very small.

Now you want to collect many chi-square values from the scrambled collection. You will build up a distribution of these values and see what shape it has.

Chi-Square Test of Independence
continued

14. Select the scrambled test and choose **Test | Collect Results As Measures.** Fathom scrambles the scrambled collection five times, each time collecting values computed by the test for independence and putting them in a new collection, called Measures from Test of Scrambled Random Students.

> Collecting 95 measures may take a while. A status bar will give you an idea of the progress.

15. Show the inspector for this new measures collection. The two important attributes are *chiSquareValue* and *pValue*. Go to the **Collect Measures** panel. Uncheck Animation on and specify that you want 95 measures instead of 5. Click **Collect More Measures.**

> The shape of the chi-square histogram resembles the plotted chi-square distribution you made in step 9.

16. You should have 100 *P*-values and 100 chi-square statistics. Make a histogram of each of these attributes. Plot the value of the chi-square statistic you got for the original collection of students.

Q12 What percentage of the measures are as extreme as or more extreme than your test statistic—in other words, what is your *P*-value with this simulation?

Q13 What is your conclusion about these two variables?

Q14 Why do *chiSquareValue* and *pValue* seem limited to certain intervals? If you collected more measures, would that change? Try it. Speculate on what you find.

Q15 What is the range for the distribution of *P*-values? Select the lowest bar in the *pValue* histogram. What happens? Explain why this happens.

Q16 Use your histogram of the distribution of *P*-values to find the *critical value* for chi-square in this simulated distribution of chi-square. What is that value for your histogram? Compare your value with those of others in your class.

EXPLORE MORE

1. Scroll down the document until you see the Scottish Children collection. This is a random sample of 500 children from Scotland. Test to see if *Eye_Color* and *Hair_Color* are independent.

2. Repeat the scrambling process that you did in steps 10–16 for the Scottish Children. Compare your chi-square value to the distribution of chi-square values that you get by collecting measures. What percentage of the measures are as extreme as or more extreme than your test statistic?

3. The distributions of *chiSquareValue* and of *pValue* for the Scottish Children will be quite different from the ones you made in step 16. Why?

Chi-Square Test of Independence

Activity Notes

Objectives

- Reviewing the theoretical definition of independence, yet realizing that with a real sample it is almost impossible to meet the criteria exactly
- Understanding the need for a different test for independence
- Performing a chi-square test of independence and understanding that it answers the question "Does this sample look like it came from a population in which these two categorical variables are independent?"

Activity Time: 30–50 minutes

Setting: Paired/Individual Activity (collect data, analyze, use **ChiSquareTest.ftm** to build simulation)

Statistics Prerequisites

- Familiarity with the language for testing a hypothesis
- Familiarity with the structure and procedure for testing a hypothesis
- The test statistic for a chi-square test
- Familiarity with sampling distributions and reasonably likely outcomes
- Definitions of probability and independence

Statistics Skills

- Checking conditions for the chi-square test of independence
- Performing a chi-square test of independence
- Finding *P*-values and relating them to the plot of the simulated chi-square distribution
- Probability simulation
- Practice using the multiplication rule
- Comparing actual data to a hypothesized model and detecting differences between models

AP Course Topic Outline: Part I E; Part III A (1, 5), D (8); Part IV B (1, 6)

Fathom Prerequisites: Students should be able to make graphs (ribbon charts) and summary tables, use case tables to record data, and use inspectors.

Fathom Skills: Students use ribbon charts to check for independence, create a scrambled collection to force independence, collect measures from a test to compare models, use the hypothesis test tool, and make a graph of the test statistic's distribution.

General Notes: This activity starts with a review of the definition of independence from probability and then walks students through inference with a chi-square test of independence and through scrambling and collecting measures from a test.

With Fathom, students can simulate conditions under which the null hypothesis is true and repeatedly perform the sampling and computation of a chi-square statistic. Although this does not tell them anything more about the particular experiment, it does shed light on the process of statistical inference. The null hypothesis here states that there is no relationship between the two attributes *Handedness* and *Eyedness*. If you take all the values for the attribute *Eyedness* and scramble them so that "Left" and "Right" get reassigned randomly to each case, any relationship that might exist between the two attributes would be wiped out by the scrambling. Any remaining relationship would have to be due to chance alone.

Procedure: It's important not to skip the hands-on Collect Data section because it gives students firsthand experience with the reason a test of independence is necessary. If your class does not contain at least 30 students, collect or have students collect additional data from other students to bring the total up to 30 or more, with about half males and half females. If there are only males or only females in your school, you can substitute another categorical variable that divides the class, such as hair color, whether or not they have a dog, or whether or not they're wearing sneakers.

For the Investigate section, if your expected values were all at least 5, then your class can continue to use your document and your data. If not, they'll need to use the document provided. Also, the sample of students in the document is indeed a random sample and not a survey, so it satisfies that condition for the test as well.

COLLECT DATA

Q1 These two variables should be independent. There is no reason to suspect that males are more or less likely than females to have even phone numbers.

Chi-Square Test of Independence
continued

Q2 Most students will think that these variables are not independent. Most will guess that if someone is right-handed, then that person is right-eyed.

Q3 For each cell in the table, students should determine whether the probability that a randomly selected student in that cell is equal to the probability of being in that row times the probability of being in that column. For example, in the sample two-way table in the student activity, $P(Male$ and $Even) = 5/30$, or 0.17, but $P(Male) \cdot P(Even) = 11/30 \cdot 10/30 = 0.12$.

It is unlikely that the variables will be independent. However, because this is a relatively small sample, we wouldn't be willing to say that the two variables aren't independent in the population.

Q4 In the sample ribbon chart, the proportions are fairly close to being equal across the categories, so this suggests that the variables could be independent.

Q5 This is only a sample from the population, so we don't expect things to work out exactly. In fact, sometimes it is not even possible for the numbers to work out exactly.

Q6 For each cell in the table, students should determine whether the probability that a randomly selected student in that cell is equal to the probability of being in that row times the probability of being in that column. Again, it is unlikely that the variables will be independent according to the definition.

Q7 For the sample ribbon chart, it looks like these two variables are not independent. The proportions across the categories are different, which suggests an association between the two variables.

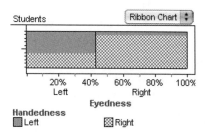

Q8 More than likely, in a class of 30 or fewer, there will be few left-handed people and your expected value in at least one cell will be less than 5. So you will not be able to proceed with a chi-square test of independence.

INVESTIGATE

Q9 The proportions across the categories are quite different, suggesting that the two variables are not independent.

Q10 This situation satisfies the conditions for a chi-square test of independence: The students come from a simple random sample taken from one large population. Each student in the sample falls into one eye-dominance category and one hand-dominance category. Here are the expected values under the assumption of independence. The expected number in each cell is 5 or more.

Random Students		Handedness		Row Summary
		Left	Right	
Eyedness	Left	13 / 6.97	28 / 34.03	41 / 41
	Right	4 / 10.03	55 / 48.97	59 / 59
Column Summary		17 / 17	83 / 83	100 / 100

S1 = count( )
S2 = Expected

H_0: Eyedness and handedness are independent.

H_a: Eyedness and handedness are not independent.

Q11 Reject the null hypothesis. These are not the results you would expect for a sample from a population where there is no association between eyedness and handedness. A value of χ^2 this large is rather unlikely to occur in a sample of this size if eyedness and handedness are independent. Examining the table and ribbon chart, it appears that right-eye dominance tends to go with right-handedness and left-eye dominance tends to go with left-handedness.

Q12–Q13 For the sample histogram below, none of the chi-square values are as large as 10.65. So the percentage would be 0. We would again conclude that the variables are not independent.

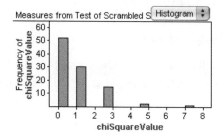

Chi-Square Test of Independence
continued

Q14 Collecting more measures will not change what happens. Because there is a limited number of left-handed students, there is a limited number of possible intervals that the chi-square value will fall into; thus there is a limited number of possible intervals that their associated P-values will fall into.

Q15–Q16 The distribution of P-values is spread over the interval from 0 to 1. When you select the lowest bar in the *pValue* histogram, the highest bars in the *chiSquare Value* histogram are selected. By selecting only those P-values less than or equal to 0.05, you can read off an approximation for the so-called *critical value* for chi-square in the chi-square graph. The critical value for the given histogram is 4.5. (The theoretical value is 3.84.)

EXPLORE MORE

1. This situation satisfies the conditions for a chi-square test of independence: The students come from a simple random sample taken from one large population. Each student in the sample falls into one eye-color category and one hair-color category. The test below shows the expected values under the assumption of independence. The expected number in each cell is 5 or more. (There is a warning that one is too small. That value, 4.99, is close enough to 5 to proceed.)

   H_0: Eye color and hair color are independent.

   H_a: Eye color and hair color are not independent.

   Test of Scottish Children — Test for Independence
   First attribute (categorical): Eye_Color
   Second attribute (categorical): Hair_Color

		Eye_Color				Row Summary
		Blue	Dark	Light	Medium	
Hair_Color	Black	1 (5.0)	26 (10.1)	0 (11.0)	11 (11.9)	38
	Dark	11 (14.7)	52 (29.9)	14 (32.3)	35 (35.1)	112
	Fair	27 (15.0)	11 (30.4)	52 (32.9)	24 (35.8)	114
	Medium	24 (26.8)	40 (54.4)	64 (58.8)	76 (64.0)	204
	Red	4 (5.5)	7 (11.2)	17 (12.1)	14 (13.2)	42
Column Summary		67	136	147	160	510

   First attribute: Eye_Color
     Number of categories: 4
   Second attribute: Hair_Color
     Number of categories: 5

   Warning: 1 out of 20 cells have expected values less than 5.

   Ho: **Eye_Color** is independent of **Hair_Color**
   Chi-square: 114.6
   DF: 12
   P-value: < 0.0001

2. Reject the null hypothesis. These are not the results you would expect for a sample from a population where there is no association between eye color and hair color. A value of χ^2 this large is extremely unlikely to occur in a sample of this size if eye color and hair color are independent. Examining the table and ribbon chart, it appears that lighter eyes tend to go with lighter hair and darker eyes tend to go with darker hair.

   Scrambling the children and then collecting the measures from the test of the Scrambled Children collection gives histograms similar to these:

   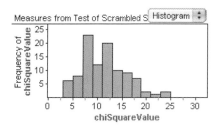

   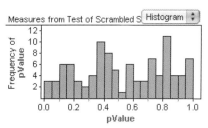

   None of these chi-square values come close to 114.6, so again, we have sufficient evidence to reject the null hypothesis that the two variables are independent.

3. These distributions have no obvious gaps like the ones for the Students collection. This is because each category has quite a few choices and there are more categories to scramble—which gives many different possible numeric combinations for expected and observed values. This will lead to more diverse values for chi-square.

Inference for Regression

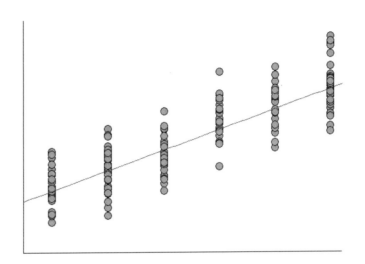

Sampling Distribution of the Slope—How Fast Do Kids Grow?

On average, U.S. children between the ages of 8 and 13 grow at the rate of 2 inches per year. Heights of 8-year-olds average about 51 inches. At each age, the heights are approximately normal, with standard deviation roughly 2 inches. (Source: National Health Statistics and Nutrition Examination Survey, May 30, 2000, www.cdc.gov/nchs/about/major/nhanes/growthcharts.htm.)

GENERATE DATA

To add cases, choose **Collection | New Cases.**

1. In a new Fathom document, create a case table with the attributes below. Add cases for ages 8 through 13.

   Children

	Age	Height	Deviation	Observed_Height
1	8			
2	9			

2. Use the information about average heights to fill in *Height*. You can write a formula for the relationship or enter numerical values.

To add a least-squares line, select the graph and choose **Graph | Least-Squares Line.**

3. Make a scatter plot of *Height* versus *Age*. Add a least-squares line. This true regression line is in the form $\mu_y = \beta_0 + \beta_1 x$, which relates average height, μ_y, to age, x.

Q1 Interpret β_0 and β_1 in the context of the situation.

4. Now suppose you have a randomly selected child of each age. To find how much the height of your child of each age deviates from the average height, define *Deviation* with a formula that randomly selects a deviation from a normal distribution with mean 0 and standard deviation 2.

5. To find the heights of your children, define *Observed_Height* with a formula that sums *Height* and *Deviation*.

6. Make a scatter plot of *Observed_Height* versus *Age*. Add a least-squares line. This regression line is in the form $\hat{y} = b_0 + b_1 x$, which estimates the true population parameters based on your children. Record your estimated slope, b_1.

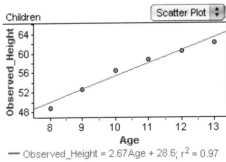

Q2 How close is your estimated slope, b_1, to β_1? Compare your estimated slope with those of others in your class.

Sampling Distribution of the Slope—How Fast Do Kids Grow?
continued

INVESTIGATE

Distribution of Estimated Slopes

Because *Observed_Height* is generated randomly, there will be variation in your estimated slopes. You'll investigate the distribution of the estimated slopes.

> To rerandomize, select the collection and choose **Collection | Rerandomize**.

7. Rerandomize the collection several times and observe the range of estimated slopes that appear for the regression line.

Q3 Predict the shape, center, and spread for the distribution of estimated slopes.

> The command linRegrSlope finds the slope of the least-squares line. It's under Functions: Statistical: Two Attributes.

8. Show the inspector and go to the **Measures** panel. Define a new measure as shown.

Measure	Value	Formula
Slope	1.01706	linRegrSlope(Age, Observed_Height)
<new>		

9. With the Children collection selected, choose **Collection | Collect Measures**. Fathom rerandomizes the Children collection five times and collects the slopes for each new randomization.

10. Make a dot plot of *Slope* from the measures collection.

> Make sure you can see your dot plot before you click **Collect More Measures**.

11. Show the inspector for the Measures from Children collection. Go to the **Collect Measures** panel. Uncheck Animation on and change the number of measures to 195. Click **Collect More Measures**.

Q4 Change your dot plot to a histogram. Describe the shape, center, and spread of your plot of estimated slopes.

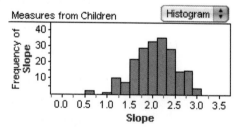

Q5 Suppose that instead of creating your data by simulation, you used actual data by choosing one child from each age group at random and measuring their heights. In what ways is the model in step 3 a reasonable model for this situation? In what ways is it not reasonable?

Conditional Distributions

Now you'll explore the conditional distribution of y given x—all of the values of y for a fixed value of x. What does that mean? Good question!

Sampling Distribution of the Slope—How Fast Do Kids Grow?
continued

*Select the collection and choose **Edit | Copy Collection**. Click in an empty space and choose **Edit | Paste Collection**.*

12. Make a copy of the Children collection. Go to the **Cases** panel in the new collection and define *Age* as shown. Add cases so that there is a total of 200 cases in this collection.

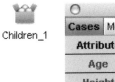

13. Make a new scatter plot of *Observed_Height* versus *Age*. The vertical column of points for all of the 8-year-old children is the conditional distribution of the height, *y*, given that the age, *x*, is 8.

Q6 Each conditional distribution of height for a given age has a mean, called μ_y for a population and $\bar{y}$ for a sample. For your scatter plot, estimate the mean of the conditional distribution given that the age is 8. Estimate the mean for age 10.

This measure of variability is roughly the spread of the conditional distribution given an age.

Q7 Each conditional distribution of height for a given age has a measure of variability, called σ for a population and *s* for a sample. For your scatter plot, estimate the measure of variability of the conditional distribution given that the age is 8. Estimate the measure of variability for age 10.

A linear model is appropriate for a set of data if the conditional means fall near a line (the true regression line) and the variability is about the same for each conditional distribution.

The standard deviation of the y's for a given x is not quite the same as the measure of variability. But it is pretty close and gives a general impression of spread.

14. Make a new summary table. Drag *Age* to the summary table while holding down the Shift key, then drag *Observed_Height* to the summary table. Replace the formula count() with the formula for the standard deviation. Then choose **Summary | Add Formula** and enter the formula for the mean.

Children_1

	Age						Row Summary
	8	9	10	11	12	13	
Observed_Height	1.9878859	2.078024	2.7146537	1.9539885	2.0208527	1.9375555	3.840601
	51.069993	53.090181	55.171262	57.075578	58.492948	60.958706	56.041375

S1 = stdDev( )
S2 = mean( )

One way to tell if the conditional means fall near a line (the true regression line) is to find the slope and *y*-intercept for the least-squares line through the means.

Sampling Distribution of the Slope—How Fast Do Kids Grow?
continued

15. Select the summary table and choose **Summary | Create Collection From Cells.** This will create a new collection with 6 cases (one for each age) with the attributes, *Age, S1,* and *S2,* where *S1* and *S2* are as you defined them in step 14.

> If *S1* is your mean and *S2* is your standard deviation, then replace *S2* with *S1* in the formulas.

16. As you did in step 8, define the measure *Slope.* Also define a new measure named *Intercept* as shown. These values are the slope and intercept of the least-squares line for the means of the conditional distributions for this sample of 200 children.

Measure	Value	Formula
Slope	2.02523	linRegrSlope(Age, S2)
Intercept	34.6264	linRegrIntercept(Age, S2)

    Inspect Cells from Children_1 Table

Q8 Do the means look like they fall near the true line that you found in step 3? Does it look like the variability is about the same for each conditional distribution? What is that variability?

Q9 Drag *Deviation* to the summary table for Children_1 and drop it on the arrow with *Observed_Height.* What is the approximate mean *Deviation* for each conditional distribution?

EXPLORE MORE

1. Increase the range of values in the formula for *Deviation* in the Children collection. Collect more measures (be sure to replace existing cases). What effect does this have on the distribution of estimated slopes?

2. Collect 200 measures from the Cells from Children_1 Table collection. Make a histogram of *Slope.* Compare it to the histogram you described in Q4, in terms of shape, center, and spread. Make a histogram of *Intercept* and describe its shape, center, and spread.

3. Make a new graph. Drag *Observed_Height* to the horizontal axis, and then drop *Age* onto the vertical axis while holding down the Shift key. Is each conditional distribution of *y* given *x* approximately normal?

Sampling Distribution of the Slope— How Fast Do Kids Grow?

Activity Notes

Objectives
- Understanding that the regression line, $\hat{y} = b_0 + b_1 x$, must sometimes be thought of as an estimate of a true, underlying linear model, $y = \beta_0 + \beta_1 x + \varepsilon$
- Recognizing that the slope of a regression line fitted from sample data varies from sample to sample
- Learning about the conditional distribution of y given x and realizing that the variability in the values of y associated with a given x is equal for all values of x

Activity Time: 30–45 minutes

Setting: Paired/Individual Activity or Whole-Class Presentation (build simulation for either)

Statistics Prerequisites
- The equation of a line: slope and y-intercept
- Working with the least-squares regression line
- Describing distributions in terms of shape, center, and spread
- Mean and standard deviation

Statistics Skills
- Slope and intercept in context
- Linear regression models
- Distribution and variation of the slope
- Variation in the slope
- Normal distributions
- Relationship between the regression model and the true linear model
- Conditional distribution of y given x; means, measure of variability, and deviations from the mean for each conditional distribution

AP Course Topic Outline: Part I D (1–3); Part III C; Part IV A (1), (getting ready for A (8) and B (7))

Fathom Prerequisites: Students should be able to make case tables and enter data, make graphs and summary tables, fit data using a least-squares line, and use the randomNormal function.

Fathom Skills: Students use the commands linRegrSlope and linRegrIntercept to define measures, copy and paste a collection, create a collection from the cells of a summary table, collect measures, use the randomInteger function and add cases, and use a numeric variable as a categorical variable to see the conditional distribution of y given various x's.

General Notes: This activity is essential for understanding the assumptions of the linear regression model. Using Fathom makes it quick and easy to calculate least-squares lines and collect a sampling distribution of estimated slopes. This activity also introduces students to the conditional distribution of y given x and walks them through its various properties and how it relates to the linear model.

GENERATE DATA

Q1 The scatter plot will be perfectly linear, and the true regression line is $y = 51 + 2(x - 8)$, or $y = 35 + 2x$. The intercept, β_0, means that the height of a child aged 0 tends to be 35 inches, which is nonsense, so the line does not model heights well outside a certain age range. The slope, β_1, means that every year the average child tends to grow about 2 inches.

5. Here is the completed case table.

Children

	Age	Height	Deviation	Observed_Height
=		51 + 2 ⟨Age – 8⟩	randomNormal ⟨0, 2⟩	Height + Deviation
1	8	51	3.51051	54.5105
2	9	53	0.436208	53.4362
3	10	55	2.26809	57.2681
4	11	57	-0.0061059	56.9939
5	12	59	-0.352446	58.6476
6	13	61	3.53624	64.5362

Q2 The scatter plot of *Observed_Height* versus *Age* will also show a linear pattern. The estimated slope, b_1, will be very close to the theoretical slope of 2. For the example in the activity, the slope is 2.67.

INVESTIGATE

Q3 Some students will realize that the center should be around 2, but most won't have any ideas about the spread or shape.

Q4 The plot should be mound-shaped and centered at 2 (the theoretical value of the slope) with standard

Sampling Distribution of the Slope—How Fast Do Kids Grow?
continued

Activity Notes

deviation about 0.48 (the theoretical standard error of the slope). The histogram in the activity has mean 2.06 and standard deviation 0.46.

The theoretical SE of the slope is

$$\sigma_{b_1} = \frac{\sigma}{\sqrt{\Sigma(x_i - \bar{x})^2}}$$

$$= \frac{2}{\sqrt{(8-10.5)^2 + (9-10.5)^2 + \cdots + (13-10.5)^2}}$$

$$\approx 0.47809$$

Q5 The model is reasonable because children's heights at each of these ages are known to be approximately normal with the given means. It's not reasonable because the standard deviations are not constant but increase with age.

Q6 Each conditional distribution has a mean, μ_y, which lies on the theoretical line and so changes with x according to the equation $\mu_y = \beta_0 + \beta_1 x$. For Age = 8, the mean should be close to 51 and for Age = 10, the mean should be close to 55. For the scatter plot shown in the activity, the mean for age 8 is 50.6 and the mean for age 10 is 55.08.

Q7 Each conditional distribution has a measure of variability, σ. It is an assumption of inference for regression that the variability, σ, of the conditional distribution is equal for each value of x. For this situation that value should be 2. Although the standard deviation is not exactly the measure of variability, it's a suitable measure of spread that gets the general idea across and the values are close enough to 2 to be rather convincing. The SD's for the scatter plot in the activity are shown below.

Children_1

	Age						Row Summary
	8	9	10	11	12	13	
Observed_Height	1.6268216	1.5069672	2.4647387	1.8777492	1.9564266	2.0142302	4.2075794

S1 = stdDev ( )

Q8 The slope should be around 2 and the intercept should be close to 35. The summary table in the student activity shows that the slope of the line of best fit through the six means is 2.03 and the intercept is 34.63.

Q9 The mean of the deviations for each conditional distribution should be about 0. The average deviation from any mean is 0, so the average deviation of the conditional distribution at each age is 0.

DISCUSSION QUESTIONS

- Did you use a formula or numerical values for *Height*? If you used a formula, how did you arrive at the formula? How does the formula relate to the true regression equation?
- How well did you predict the shape, center, and spread in Q3? Did you get the results you expected?

EXPLORE MORE

1. Increasing the standard deviation will make the measure of variability larger for each conditional distribution of *y* given *x*, and the distribution of estimated slopes will have more variability.

2. For *Slope*, the mean is 1.999 and the standard deviation is 0.087, which is very close to the SE of the slope. The shape is fairly normal. For *Intercept*, the mean is 35.009 and the standard deviation is 0.93. The shape is somewhat skewed, however.

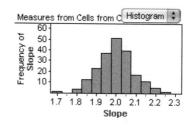

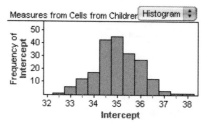

3. The conditional distributions at each age are approximately normal.

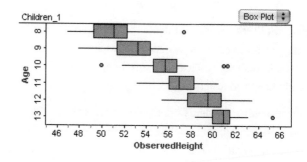

Variation in the Slope

In this activity you will create a scatter plot with only two values for the predictor, *x*, and you will generate four values of the response, *y*, for each *x*. The response values will come from normal distributions with specified means and standard deviations. Your goal is to discover what causes there to be more or less variation in the slope of the least-squares regression line through the eight points.

GENERATE DATA

1. In a new Fathom document, make three sliders named *x1*, *y1*, and *SD*.

2. Make a collection with eight cases. Define two attributes as shown.

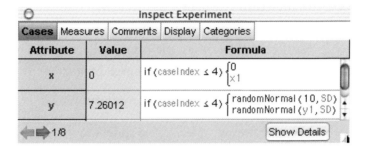

3. Make a scatter plot of *y* versus *x*. Drag each slider so that you understand how it affects the eight points in the scatter plot.

Q1 Explain the effect on the eight points when you drag the slider for *x1*; the slider for *y1*; the slider for *SD*.

INVESTIGATE

Now you'll look at how changing the sliders affects the least-squares regression line and its slope.

4. Select the graph and choose **Graph | Least-Squares Line**.

To rerandomize, drag the collection's lower-right corner until you can see the **Rerandomize** button. Then click on the button.

Q2 Experiment with dragging slider *SD*. Pick a value for *SD*. Then rerandomize the collection several times. Then drag the slider *SD* to another value and rerandomize several more times. Observe the variation in the estimated slopes. How are the values of *SD* and the amount of variation in the estimated slopes related?

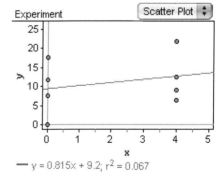

Exploring Statistics with Fathom
© 2007 Key Curriculum Press

8: Inference for Regression   321

Variation in the Slope
continued

Q3 Repeat Q2 with *x1*. How are the values of *x1* and the amount of variation in the estimated slopes related?

Now you're going to gather 50 estimated slopes for each of four settings of the sliders.

5. Show the collection inspector and go to the **Measures** panel. Define these three measures for the Experiment collection.

6. With the collection selected, choose **Collection | Collect Measures.** By default, Fathom rerandomizes the collection five times, each time collecting the measures you defined and storing them in a new collection named Measures from Experiment.

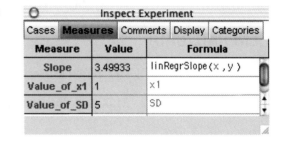

Now you will take the four different samples.

7. For the first sample, set the sliders to $x1 = 1$, $y1 = 12$, and $SD = 3$.

8. Show the measures collection's inspector and go to the **Collect Measures** panel. Uncheck Animation on and check Replace existing cases. Change the number of measures collected to 50. Click **Collect More Measures.**

9. Now that you have collected the measures for the first sample, uncheck Replace existing cases so that you'll have all your measures in the one collection.

Note: You have already done sample 1.

10. For each sample (row) in this table, change the values of the sliders, then click **Collect More Measures** to collect 50 measures at these values.

Sample	x1	y1	SD
2	1	12	5
3	4	18	3
4	4	18	5

11. Make a histogram of the attribute *Slope* from the measures collection.

You want to see just the values for $x1 = 1$. To do that, you need to add a filter.

12. Select the graph and choose **Object | Add Filter.** Enter the formula Value_of_x1=1.

Now you want a histogram with $x1 = 4$.

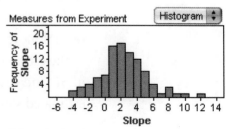

Variation in the Slope
continued

Double-click the filter expression to show the formula editor.

13. Select the histogram and choose **Object | Duplicate Graph.** Change the filter on the new graph to Value_of_x1=4.

14. Split each histogram by dropping *Value_of_SD* onto its vertical axis while holding down the Shift key. That way, *Value_of_SD* will be treated as a categorical variable.

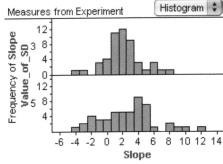

To adjust the bin widths, double-click the graph to show the inspector.

15. If you adjust the axes and bin widths of the two histograms so that they have the same scale, you can see the differences in the four distributions better. To do this without dragging bins, select one of the graphs and choose **Graph | Show Axis Links.** Drag the link on one vertical axis to the vertical axis on your other histogram. Then drag the link on the horizontal axis to the other horizontal axis. Now, if you adjust one axis, it will change the corresponding axis in your other graph.

16. Make two more plots of the measure *Slope* from the measures collection. Filter the first with the formula Value_of_SD=3 and the second with Value_of_SD=5. Then split each histogram by dropping *Value_of_x1* onto its vertical axis while holding down the Shift key.

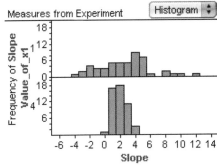

Q4 Based on the histograms, how does the variation in the estimated slopes change with an increase in the variation, *SD*, of the responses, *y*?

Q5 Based on the histograms, how does the variation in the estimated slopes change with an increase in the spread of the explanatory variable, *x1*?

EXPLORE MORE

1. Based on your histograms, does it look like each of the distributions of estimated slopes is approximately normal?

2. Describe the relationship between the standard errors of the estimated slopes and *x1* and/or *SD*. Can you think of a formula for the standard error of the estimated slope?

Variation in the Slope

Activity Notes

Objectives
- Recognizing that the slope of a regression line fitted from sample data will vary from sample to sample
- Understanding that the formula for the SE of the slope reflects the fact that having a wider spread in the values of x and a smaller spread in the distances of y from the regression line decreases the variability of the slope
- Working with the conditional distribution of y given x and using the fact that the variability in the values of y associated with a given x is equal for all values of x
- Seeing that the variability in the values of y from a sample is measured in terms of the difference of y and its predicted value, $\hat{y}$, and so is equivalent to the variability in the residuals

Activity Time: 30–45 minutes

Setting: Paired/Individual Activity (build simulation) or Whole-Class Presentation (use **VariationPresent.ftm**)

Statistics Prerequisites
- The equation of a line: slope and y-intercept
- Working with the least-squares regression line
- Describing distributions in terms of shape, center, and spread
- Mean and standard deviation

Statistics Skills
- Linear regression models
- Distribution and variation of the slope
- Comparing sampling distributions of the slope for different conditions
- Variation in the least-squares regression line
- Relationship between the variability of the distribution of the estimated slope and the variability of the response
- Inverse relationship between the variability of the distribution of the estimated slope and the variability of the predictor
- Conditional distribution of y given x; means, measure of variability, and deviations from the mean for each conditional distribution

AP Course Topic Outline: Part I D (1–3); Part III C; Part IV A (1), (getting ready for A (8) and B (7))

Fathom Prerequisites: Students should be able to make new collections, make and drag sliders, make scatter plots and other graphs and summary tables, fit data using a least-squares line, and use the formula editor.

Fathom Skills: Students use the command linRegrSlope to define a measure, duplicate a graph, collect measures, use filters, use a numerical variable as a categorical variable, use split graphs, and link axes and adjust bin widths.

General Notes: This activity looks at the slopes of least-squares regression lines and their distribution with repeated sampling. By controlling parameters of the population, students can study the relationship between the parameters and variation in the measured slope.

Procedure: In steps 1–4, students set up a collection based on the randomNormal function and sliders. They then explore in an unstructured fashion by dragging sliders and observing the effect. If you are using the presentation document **VariationPresent.ftm**, the simulation is set up already, so all you need to do is click **Rerandomize** to see the changes, then drag the sliders to another setting and rerandomize again.

In steps 5–16, students explore the same collection within a structured environment—choosing four settings for the population parameters and collecting measures for each of these samples.

If you are using **VariationPresent.ftm**, this, too, is set up, so all you need to do is look at the already made histograms. If you would like to start over and collect your own measures, there are two ways to accomplish this. The first is to select the measures collection, then choose **Edit | Select All Cases** and **Edit | Delete Cases.** The second way to do this is to set your sliders for the first sample ($x1 = 1$, $y1 = 12$, $SD = 3$), then show the measures collection's inspector. Check Replace existing cases and click **Collect More Measures.** Then proceed with step 9 and skip the instructions for making the histograms.

Variation in the Slope
continued

GENERATE DATA

3. Be sure students explore the effects of the sliders and fully understand how they change the scatter plot.

Q1 Dragging the slider for $x1$ moves one set of four points in the direction the slider moves. For example, if you drag the slider from 1 to 4, the four points shift from $x = 1$ to $x = 4$. The slider for $y1$ moves those same four points either up or down. For example, if you drag the slider from 5 to 8, the four points shift from a conditional distribution centered at $y = 5$ to a conditional distribution centered at $y = 8$. The slider for SD varies the spread in the conditional distributions of both sets of four points. For smaller values, the points cluster closely around the mean y; for larger values of SD, the points are farther from the mean y.

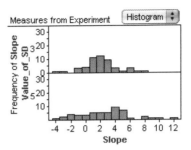

Value_of_x1 = 1

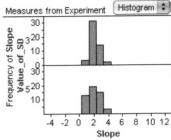

Value_of_x1 = 4

INVESTIGATE

5.–6. Students can make clearer observations if they set slider SD to 1, for example, and rerandomize ten times, then change SD to 5 and rerandomize ten times, then change SD to 10, and so on.

Q2 There is a direct relationship between the variability of the slopes and the variability (standard deviations) of y. As the SD gets larger, the slopes vary more and more between rerandomizations. As the SD gets smaller, the slopes vary less between rerandomizations.

Q3 There is an inverse relationship between the variability of the slopes and the variability of $x1$. As $x1$ gets closer to 0, the slopes vary considerably, whereas as $x1$ gets larger (in absolute value), the slopes vary less with rerandomization.

11.–15. Data gathered from a population will vary from sample to sample, and so will measures computed for these samples. Compare the histograms vertically to analyze the effects of slider SD, and compare them horizontally to analyze the effects of slider $x1$. Sample histograms:

16. Compare the histograms horizontally to analyze the effects of slider SD, and compare them vertically to analyze the effects of slider $x1$. Sample histograms:

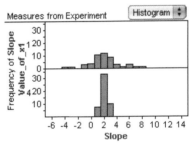

Value_of_SD = 3

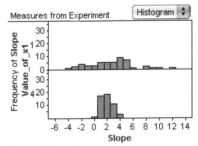

Value_of_SD = 5

Variation in the Slope
continued

Here are the summary statistics for this sample.

Sample	Mean	Median	SD	Theory	SE
1	50	1.96	1.95	2.18	2.12
2	50	2.69	2.77	3.36	3.53
3	50	2.10	2.01	0.55	0.53
4	50	1.95	1.86	0.86	0.88

The theoretical slope, β_1, is 2 in each case, and all four simulated sampling distributions have a mean very close to 2.

The theoretical standard errors for the slopes can be calculated using

$$\sigma_{b_1} = \sigma/\sqrt{\Sigma(x_i - \bar{x})^2}$$

because σ is known for each case; these are shown in the Theory column. Again, observe that the standard deviations calculated from the four simulated sampling distributions of the values of b_1 are very close to the theoretical standard errors.

As the variation in the conditional responses (the given standard deviation) increases, so will the variation in the slopes. As the explanatory variable, x, spreads out more, the variation in the possible slopes of the regression line will decrease, assuming the conditional variation in the y's does not change.

Q4 As the variation (the given standard deviation *SD*) in the conditional responses, y, increases, so will the variation in the slopes.

Q5 As the spread in the explanatory variable, *x1*, increases, the variation in the possible slopes of the regression line will decrease, assuming the conditional variation in *SD* does not change.

DISCUSSION QUESTIONS

- What relationships did you find between the population parameters and variability of the slope? How did you discover them? Were your findings in Q4 and Q5 consistent with your findings in Q2 and Q3?
- What is the theoretical slope, β_1, for all four settings in the table? How does the theoretical slope present itself in the histograms?
- Can you give a commonsense explanation as to why the variability in estimated slope increases as variability in the response increases?
- Similarly, can you give a commonsense explanation as to why the variability in slope decreases as spread in the predictor increases?

EXPLORE MORE

1. Each distribution of the slopes does appear approximately normal.

2. The variability of the slopes varies directly with the variability (standard deviations) of y. The variability of the slopes inversely varies with the variability of *x1*. The theoretical standard errors for the slopes can be calculated using

$$\sigma_{b_1} = \sigma/\sqrt{\Sigma(x_i - \bar{x})^2}$$

Inference for a Slope—How Tall Are You Kneeling?

You will need
- measuring tape or meterstick

Leonardo da Vinci (Italian, 1452–1519) was a scientist and an artist who combined these skills to draft extensive instructions for other artists on how to proportion the human body in painting and sculpture. One of his guidelines was that a person's kneeling height is three-fourths the person's standing height. In this activity you will investigate this rule.

COLLECT DATA

Q1 What is the theoretical regression line, $\mu_y = \beta_0 + \beta_0 x$, that Leonardo is proposing for this situation? Use height as the explanatory variable.

1. Work with a partner to measure your height and your kneeling height.

2. Record the data for each student in your class in a Fathom case table. Each case represents one student with two attributes, *Height* and *Kneeling_Ht*. Name the collection Students.

Select the graph and choose **Graph | Least-Squares Line.**

3. Check Leonardo's guidelines visually by making a scatter plot of *Kneeling_Ht* versus *Height*. Add the least-squares regression line to the graph.

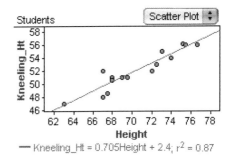

Q2 Interpret the slope and *y*-intercept in context. Compare the estimated slope and intercept to the theoretical slope and intercept (from Q1). Are they close?

4. Define a new attribute, *Residual*, with the formula

    linRegrResidual(Height,Kneeling_Ht)

5. In a new graph, make a residual plot for the line you fitted through the raw data in step 3.

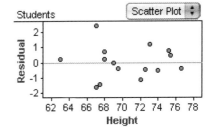

Q3 Is fitting a regression line appropriate for the pattern displayed in the scatter plot? Explain by inspecting the original plot as well as the residual plot.

Q4 What can you conclude about the relationship between the two variables in the population of all students based on your scatter plot and residual plot? Explain.

Inference for a Slope—How Tall Are You Kneeling?
continued

INVESTIGATE

Significance Test for a Slope

More than likely, you found a trend of some sort in your data. A significance test for a regression slope asks, "Is that trend real or could the numbers come out the way they did just by chance?" You will use a significance test for a slope to find out whether the positive linear association between the two variables that you have observed is "real" or due to chance variation.

Q5 Check the residual plot that you made in step 5. Do the residuals stay about the same size across all values of *x*?

6. Make a dot plot or box plot of *Residual* to see if it's reasonable to assume that the values came from a normal distribution.

Q6 If you assume for the moment that your sample is a simple random sample, can you proceed with a significance test for a slope? (Be sure to discuss all conditions!)

Q7 State the null and alternative hypotheses for this situation. You can write out these two hypotheses in Fathom in a text object. From the shelf, drag a text object into the document and type in your hypotheses.

To format text and create mathematical expressions, choose **Edit | Show Text Palette**.

Now it is time to compute the test statistic.

7. Drag a test object from the shelf. From the pop-up menu, choose **Test Slope**.

Choose **Test | Verbose** to see a more compact version of the test.

8. Drag *Kneeling_Height* to the top pane of the test where it says "Response attribute (numeric): unassigned." Then drag *Height* to where it says "Predictor attribute (numeric): unassigned."

> Test of Students — Test Slope
> Response attribute (numeric): Kneeling_Ht
> Predictor attribute (numeric): Height

Q8 What is your *P*-value, *df*, and the value of the test statistic? Do your data support the standard that the slope is zero and that there is no linear association between the two variables, or do they suggest that the trend is real? Write your conclusion in context, linked to your computations.

It is helpful to be able to visualize the *P*-value as an area under a distribution.

If your *P*-value is small, you might need to rescale your axes to see the shading.

9. With the test selected, choose **Test | Show Test Statistic Distribution**. The curve shows the probability density for the *t*-statistic with 13 degrees of freedom. The shaded area corresponds to the *P*-value.

Inference for a Slope—How Tall Are You Kneeling?
continued

Simulating the Null Hypothesis

The null hypothesis states that there is no "real" linear relationship between the two attributes *Height* and *Kneeling_Ht*. What if you took all the values for the attribute *Kneeling_Ht* and scrambled them so that each *Kneeling_Ht* got randomly reassigned to a *Height*? Any linear trend that might exist between the two attributes would be wiped out by the scrambling. Any remaining relationship would have to be due to chance alone.

10. Select the Students collection. Choose **Collection | Scramble Attribute Values.** Make a case table for the new collection, Scrambled Students.

11. By default, Fathom scrambles the first attribute in the collection—which is *Height*. You need to scramble a different attribute. Show the scrambled collection's inspector and go to the **Scramble** panel. Choose **Kneeling_Ht** from the pop-up menu.

12. Make a scatter plot for the scrambled collection and add a least-squares regression line. Then click the **Scramble Attribute Values Again** button a few times. As you scramble, you can see the variation in the slopes of the regression line. This variation is due solely to the random assignment.

13. Make a test for the scrambled slope. Select your original test and choose **Object | Duplicate Hypothesis Test.** This time, drop *Kneeling_Ht* from the scrambled collection into it. Fathom updates both *Height* and the plot of the test statistic distribution.

14. Click **Scramble Attribute Values Again** at least ten times. Observe the test statistic and the *P*-value.

Q9 What range of values did you get for your test statistic? For your *P*-value?

Now you want to collect many *t*-values from the scrambled collection. You will build up a distribution of these values and see what shape it has.

15. Select the scrambled test and choose **Test | Collect Results As Measures.** Fathom scrambles the collection five times, each time collecting values computed by the test for a slope and putting them in a new collection, labeled Measures from Test of Scrambled Students.

Collecting 95 measures may take a while. A status bar will give you an idea of the progress.

16. Show the inspector for this new measures collection. The attributes you want to look at are *slope*, *tValue*, and *pValue*. Go to the **Collect Measures** panel. Uncheck Animation on and specify that you want 95 measures. Click **Collect More Measures.**

Inference for a Slope—How Tall Are You Kneeling?
continued

17. You should have 100 slopes, 100 *P*-values, and 100 test statistics. Make a histogram or dot plot of each of these measures. Plot the value of the test statistic you got in Q8 for the original collection of your class's data.

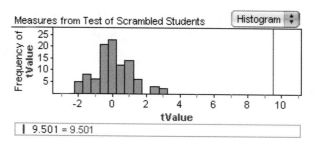

Q10 What percentage of the measures are as extreme as or more extreme than your test statistic—in other words, what is your *P*-value with this simulation?

Q11 Based on your plot of *slope*, what is the range of reasonably likely slopes when the null hypothesis is true? Is your slope (from Q2) even close to this range?

Confidence Interval

Now you'd like to find the range of plausible slopes for which your slope is reasonably likely.

Model

18. To construct a 95% confidence interval for a slope, drag a new model from the shelf and choose **Simple Regression** from the pop-up menu.

19. From the original Students collection, drag *Kneeling_Height* and *Height* to the top pane of the model.

Q12 What is the 95% confidence interval for your class's data? Interpret the confidence interval in the context of the situation.

Q13 Did your confidence interval capture the slope from Leonardo's guideline (Q1)?

EXPLORE MORE

Explore the relationships between different attributes in the Fathom document **WorldsWomen.ftm**. For each pair of attributes below, test whether there is a linear relationship between those two attributes. Then find the confidence interval for the slope.

1. Find one pair of attributes that satisfy the conditions for a significance test for a slope.

2. Find one pair of attributes that need a transformation to satisfy the conditions for a significance test for a slope and apply the transformation.

Inference for a Slope—How Tall Are You Kneeling? Activity Notes

Objectives
- Checking the conditions that are needed before doing inference for a slope
- Performing a test of significance for the slope
- Constructing and interpreting a confidence interval for the slope

Activity Time: 40–50 minutes

Setting: Paired/Individual Activity (collect data and use **KneelingTemp.ftm** or use **Kneeling.ftm,** build simulation) or Whole-Class Presentation (use **Kneeling.ftm**)

Optional Documents: WorldsWomen.ftm (Explore More 1 and 2), **WorldsWomenExp.ftm** (Explore More 1 and 2 solutions)

Materials
- One measuring tape, yardstick, or meterstick for each pair of students

Statistics Prerequisites
- Familiarity with the language for testing a hypothesis
- Familiarity with the structure and procedure for testing a hypothesis
- The equation of a line: slope and y-intercept
- Working with the least-squares regression line
- Describing distributions in terms of shape, center, and spread
- Familiarity with sampling distributions and reasonably likely outcomes
- Familiarity with the conditions that need to be checked for a significance test for a slope

Statistics Skills
- Slope and intercept in context
- Linear regression models and the relationship between the regression model and the true linear model
- Checking conditions for a significance test for a slope
- Performing a significance test for a slope
- Finding t-values and relating them to the plot of the simulated slopes
- Simulating a situation where the null hypothesis is true
- Constructing a confidence interval for the slope
- Comparing actual data to a hypothesized model and detecting differences between models

AP Course Topic Outline: Part I D; Part III D (7); Part IV A (1, 3, 8), B (1, 7).

Fathom Prerequisites: Students should be able to make case tables and enter data (if collecting data), make scatter plots and plot values, fit data using a least-squares line, and use inspectors and the formula editor.

Fathom Skills: Students use the command linRegrResidual to define an attribute, use the hypothesis test and linear model tools, create a scrambled collection to simulate the null hypothesis, collect measures from a test to compare models, and make a graph of the test statistic's distribution.

General Notes: This activity starts with students collecting their own data and then walks students through inference for a slope. They then simulate what would happen if the null hypothesis were true by scrambling and collecting measures from a test. Lastly, they construct a confidence interval for the slope of their data.

Procedure: As usual in activities that involve pooling data, the mechanics of pooling can be time-consuming. You may want to collect data on one day, enter the data on a single computer, then distribute the saved Fathom document. You could also use the document **KneelingTemp.ftm** as a template that students can enter their data into directly. Alternatively, if you don't have time for your students to collect data, they can use the sample data in **Kneeling.ftm.** This set of data is a random sample of 15 students and has additional attributes your students can explore.

If you do have time, students enjoy doing this activity, though be aware that some students are sensitive about their height. Have them work with a partner to measure their height and kneeling height. Make sure students realize that neither Leonardo's guidelines nor the regression line should be regarded as a human ideal.

COLLECT DATA

Q1 Because Leonardo says kneeling height is three-fourths a person's standing height, the theoretical model is $y = 0.75x + \varepsilon$, where x is the height and y is the kneeling height.

Inference for a Slope—How Tall Are You Kneeling?
continued

Activity Notes

3. When students plot the attributes, they should notice approximately linear trends in the scatter plot. In the sample data, there does appear to be a linear relationship between *Height* and *Kneeling_Ht*. The relationship is fairly strong with a positive trend and tends to be constant in strength.

Q2 For the sample data and plot shown in the activity, the least-squares regression equation for predicting the kneeling height from the height is *Kneeling_Ht* = 0.705*Height* + 2.36; $r = 0.87$. The slope is 0.705, which means that for every 1 inch increase in height, there tends to be a 0.705 inch increase in kneeling height. Leonardo predicted a 0.75 inch increase. About 87% of the variation in kneeling height between these students can be attributed to their height. The y-intercept is 2.36, which is the kneeling height of a person who is 0 inches tall. Obviously, this model isn't accurate for values well outside the range of data. Leonardo predicted 0 for the y-intercept.

Q3 For the sample data, the residual plot does show random scatter about the $y = 0$ line. There is no obvious curvature in either the scatter plot or the residual plot. A linear model appears to be appropriate.

Q4 Based on the scatter plot and residual plot, there appears to be a strong positive linear association between *Height* and *Kneeling_Ht*.

INVESTIGATE

Q5 For the sample data, the residuals stay about the same size across all values of x—in other words, the vertical spread is about the same as you go from left to right across the scatter plot.

Q6 Besides having a random sample, the conditions that must be checked before proceeding with a significance test for a slope are

- Linear relationship—addressed in Q3
- Residuals have equal standard deviations across all values of x—addressed in Q5
- Residuals are normal at each fixed x—answers will vary depending on the plot students made in step 6.

For the sample data, the residuals look like they could reasonably have come from a normal distribution. So, the conditions are met.

Q7 Sample text object with hypotheses:

> H_0: $\beta_1 = 0$
> where β_1 is the slope of the true linear relationship between height and kneeling height for students.
>
> H_a: $\beta_1 \neq 0$

7.–8. This is one test where you must have the raw data and can't enter the summary statistics. Also, you can't edit the null hypothesis, so this only tests the null hypothesis that the slope is 0.

Q8 For the sample data, the test statistic is 9.501 with a P-value less than 0.0001. There are 13 degrees of freedom. No, these data do not support the null hypothesis that there is no linear association between height and kneeling height. Thus, there is strong evidence to reject the null hypothesis and say that the slope of the regression line of *Kneeling_Ht* versus *Height* is different from zero. A linear model having zero for the true slope probably would not have produced these data, as students see in the next section. If the null hypothesis were true here, less than one sample in ten thousand would give such a large value of the test statistic. (*Note:* The actual P-value in this instance is 0.00000016.)

Q9 Here are sample dot plots for 100 *tValues* and 100 *pValues*. The measure *tValue* ranges from −2 to 3 and the measure *pValue* ranges from 0 to 1.

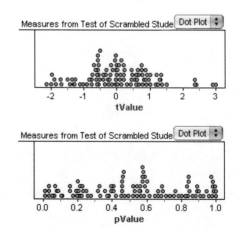

Inference for a Slope—How Tall Are You Kneeling?
continued

Q10 More than likely, there will be no measure as extreme as or more extreme than their test statistic. For the sample data, $t = 9.5$, which is well outside the range of 100 test statistics. So, in this instance, the P-value is 0.

Q11 Here is the result of 100 slopes. Taking off the bottom and top 2.5%, the reasonably likely slopes given the null hypothesis is true would be in the interval from about -0.363 to about 0.43. The sample slope of 0.705 is not within this interval.

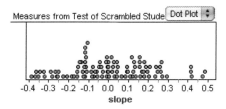

Q12 For the 95% confidence interval for the slope with 13 degrees of freedom, the critical value is $t = 2.160$. The 95% confidence interval is then 0.705 ± 0.160305, or 0.545 to 0.865. You are 95% confident that the slope of the true linear relationship between height and kneeling height is between 0.545 and 0.865, or the increase of 1 inch in height would tend to give an increase in kneeling height somewhere between 0.55 inch and 0.87 inch. In other words, if one person is 1 inch taller than another person, his kneeling height tends to be taller by between 0.55 inch and 0.87 inch. This result means that any true slope b_1 between 0.545 and 0.865 could have produced such data as a reasonably likely outcome. A value of b_1 outside the confidence interval could not have produced numbers like the actual data as a reasonably likely outcome.

Q13 More than likely, their confidence interval will capture Leonardo's slope of 0.75.

EXPLORE MORE

1. There are a few relationships that could be tested: *LifeExpW* versus *InfantMortG*, *LifeExpM* versus *InfantMortB*, and *SchoolEnrollment* versus *FertilityRate* are a few. For the first, assuming this is a random sample, the conditions are met for a significance test for a slope. The scatter plot and residual plot show a fairly strong negative association between the two attributes with no hint of curvature. The residuals are of constant strength across all values of *x*, and the residuals look like they could reasonably have come from a normal distribution. See the document **WorldsWomenExp.ftm** for the significance test and the 95% confidence interval for the slope.

2. The document **WorldsWomenExp.ftm** shows the regression of infant mortality for girls versus fertility rate. The association is positive and looks like it follows a linear trend, but the plot is heteroscedastic—countries with larger fertility rates tend to have larger variation in the mortality rate for infant girls. Further, the box plot of the residuals is skewed left and shows four outliers.

The transformation log(*InfantMortG*) versus *FertilityRate* eliminates the heteroscedasticity and the outliers in the residuals. The log transformation gives a symmetric box plot of the residuals.

Assuming this is a random sample, the conditions are met for a significance test for a slope. The scatter plot and residual plot for the transformed attributes show a fairly moderate positive association between the two attributes with no hint of curvature. The residuals are now of constant strength across all values of *x*, and the residuals look like they could reasonably have come from a normal distribution. The solution document shows the significance test and the 95% confidence interval for the slope.